Ecological Studies

Analysis and Synthesis

Edited by

J. Jacobs, München · O. L. Lange, Würzburg
J. S. Olson, Oak Ridge · W. Wieser, Innsbruck

Volume 1

Analysis of
Temperate Forest Ecosystems

Edited by

David E. Reichle

With 91 Figures

Springer-Verlag New York · Heidelberg · Berlin 1970

Series Foreword

A series of concise books, each by one or several authors, will provide prompt, world-wide information on approaches to analyzing ecological systems and their interacting parts. Syntheses of results in turn will illustrate the effectiveness, and the limitations, of current knowledge. This series aims to help overcome the fragmentation of our understanding about natural and managed landscapes and waters — about man and the many other organisms which depend on these environments.

We may sometimes seem complacent that our environment has supported many civilizations fairly well — better in some parts of the Earth than in others. Modern technology has mastered some difficulties but creates new ones faster than we anticipate. Pressures of human and other animal populations now highlight complex ecological problems of *practical importance* and *theoretical scientific interest*. In every climatic-biotic zone, changes in plants, soils, waters, air and other resources which support life are accelerating. Such changes engulf not only regions already crowded or exploited. They spill over into more natural areas where contrasting choices for future use should remain open to our descendents — where Nature's own balances and imbalances can be interpreted by imaginative research, and need to be.

Ecological Studies will bring together insights about the functioning and organization of ecosystems on the scale of the whole *biosphere, communities, populations* and *individual components* and *their interfaces*. Here methods are interpreted broadly. They include not only techniques of measurement, sampling and experimentation, but also the thoughtful strategy of investigation to choose between multiple working hypotheses or to suggest fresh ideas.

Analysis includes biological, physical and chemical analyses of the parts and processes of ecosystems. Mathematical analyses not only aid these studies, but provide means of expressing relations between environmental and biological variables in ways that help us think about what is happening in Nature.

Synthesis includes drawing scattered and new information to bear on answering or at least clarifying specific questions. Also it includes generalizing over classes of environmental conditions, organisms or ecosystem types. Word pictures, flow charts and mathematical models are just steps in making any general propositions explicit. They help predictions. These hasten tests, and changes if necessary, to make our concepts and decisions more appropriate when our models are still rough approximations of reality.

Thus individual volumes or chapters may bring quite different sciences to bear on one problem or system. Small problems or "subsystems" in turn may show better ways of dealing with larger ecological problems. These we must understand better if we are to make far-sighted policy and wise technical management of future environments.

J. Jacobs, O. L. Lange
J. S. Olson, W. Wieser

Preface

The expanding population and technology of civilized man steadily encroach upon the quality, and very existence, of his environment. Some believe that this assult upon the integrity of the environment is the price which society must pay for technological benefits. Others disagree. Man's crisis is his survival in a changing world. He has learned neither to respect, understand, nor affect the consequences of his technology on the living environment. Eutrophication of waters, pollution of the atmosphere, dispersal of exotic pest species, and broad-scale alteration of the landscape are but examples of the growing evidences of disruption of the ecological balance in the environment. At the same time, the challenge to meet the ever-increasing food needs of the world's population will require optimization of the productivity of managed natural and agricultural ecosystems with minimal degradation of environmental quality.

Political decisions on environmental policies in the absence of ecological knowledge are untenable. To remedy this situation, attention must be focused upon the urgent problem of understanding the functional properties of natural ecosystems. The basic ecological problem in natural resource development is how the characteristics of natural ecosystems — high productivity and long-term stability — can be transferred to managed ecosystems where yield is in a more usable form. Under what conditions and to what degree do such factors as geography, climate, topography and soils determine the ecosystem type; and how do ecosystem attributes such as food chain complexity, species diversity, and species composition contribute to productivity and stability? Modern science has created a dangerous illusion: that we have become increasingly independent of the balance of nature. Just the opposite is true. Technology, often brought into use before the ultimate hazards to the environment were known, has strained and many times disrupted this balance. Short-term economic gain often has shrouded the fact that man's capability as a destructive force in the environment has increased tremendously, while his knowledge, understanding, and ability to preserve his environment remain underdeveloped.

The International Biological Program (IBP) traces its recent origin to these very problems of human welfare. The IBP is a cooperative effort on behalf of the world's scientists to understand, through research and synthesis, the basic processes of environmental systems which support life on their planet. Emphasis is directed to the basic production relations which regulate natural ecosystems and those which control agricultural production to provide data on land-use efficiency in terms of solar energy conversion, biomass production, and mineral utilization. In late July 1968, the Terrestrial Production (PT) Section of the IBP and Oak Ridge National Laboratory on behalf of the U. S. IBP Analysis of Ecosystems Project co-hosted a workshop-seminar in Gatlinburg, Tennessee, USA. The objectives of this meeting were to summarize much of the existing data and to establish a conceptual framework for the analysis of an ecosystem — emphasizing the temperate deciduous forest. The collected papers form the basis for chapters in this first volume of *Ecological Studies*. The contributors are distinguished not only by their international reputations but also by their active involvement in IBP-related analysis of ecosystem research.

It would be presumptuous to think that this volume, or even many volumes, could consider the total complexity of an ecosystem or all aspects of its particular components. What the authors have striven for has been the conceptual development and practical application of an "attitude" — the interrelation of ecosystem parts which form a whole greater than the summation of its structures and processes, and that this whole, the ecosystem, is the meaningful organizational unit for ecological understanding of the environment.

The text consists of eighteen chapters. Following the introduction, three chapters consider the development of systems analysis models in ecosystem research, the effects of environmental factors on ecosystems, and phenology in productivity studies. The next section (three chapters) is devoted to primary production — harvest methods, gaseous exchange techniques, and productivity models. Secondary production is covered in three chapters on canopy insects, birds and mammals. Next follow the floral and faunal components of decomposer systems, and their effect on organic matter breakdown and nutrient recycling. Three chapters are then devoted to soil-plant relations and the cycling of nutrient elements in the forest ecosystem. The section on water flux in temperate forests concerns the inputs, outputs, and models of the hydrologic cycle. The chapters are not all-inclusive; principles developed in one section often apply equally to other sections, but are not repeated to avoid reiteration of more then a few major points. Thus, the methods of systems analysis introducted in Chapter 2 apply generally to all subject areas, although they are developed only in particular chapters where distinctive contributions of data or literature sources can be made. The concepts of food chain and population energetics, discussed in the chapter on the soil fauna, apply equally as well to canopy insect, mammal and bird populations.

From the very beginning, my labors with this volume have been eased by assistance from my family, the International Biological Program staff, and my colleagues in the Ecological Sciences Division. My biggest debt, however, is due the many contributors without whose efforts this text would have been impossible.

Oak Ridge, 1970   DAVID E. REICHLE

Contents

Contributors

BOTKIN, DANIEL B., Dr., School of Forestry, Yale University, New Haven, Connecticut, USA

BOURLIÈRE, FRANCOIS, Prof., Convener IBP (PT), Département de Physiologie, Faculté de Médecine, Paris, France

CROSSLEY, D. A. JR., Dr., Department of Entomology, University of Georgia, Athens, Georgia, USA

CURLIN, JAMES W., Dr., Ecological Sciences Division, Oak Ridge National Laboratory, Oak Ridge, Tennessee, USA

DENAEYER-DE SMET, S., Chef de Travaux, Université Libre de Bruxelles, Brussels, Belgium

DUVIGNEAUD, PAUL, Prof., Director, Laboratoire de Botanique Systematique et d'Ecologie, Université Libre de Bruxelles, Brussels, Belgium

EDWARDS, C. A., Dr., Entomology Department, Rothamsted Experimental Station, Harpenden, Herts, England

FORTESCUE, J. A. C., Dr., Soils Section, Petawawa Forest Experiments Station, Chalk River, Ontario, Canada

FRANKLIN, RUDOLPH T., Dr., School of Forest Resources and Department of Entomology, University of Georgia, Athens, Georgia, USA

GOODALL, DAVID W., Dr., Ecology Center, Utah State University, Logan, Utah, USA

HADLEY, MALCOLM, Dr., Département de Physiologie, Faculté de Médecine, Paris, France

LIETH, HELMUT, Prof., Department of Botany, University of North Carolina, Chapel Hill, North Carolina, USA

MADGWICK, HERBERT A. I., Dr., Division of Forestry and Wildlife Sciences, College of Agriculture, Virginia Polytechnic Institute, Blacksburg, Virginia, USA

MARTEN, G. G., Soils Section, Petawawa Forest Experiment Station, Chalk River, Ontario, Canada

McCULLOUGH, DALE R., Dr., Department of Wildlife and Fisheries, School of Natural Resources, University of Michigan, Ann Arbor, Michigan, USA

NELSON, DANIEL J., Dr., Ecological Sciences Division, Oak Ridge National Laboratory, Oak Ridge, Tennessee, USA

OLSON, JERRY S., Dr., Ecological Sciences Division, Oak Ridge National Laboratory, Oak Ridge, Tennessee, USA

RAFES, PAUL M., Dr., Laboratory of Forest Science, Academy of Sciences of the USSR, Uspenskoye, Moscow, USSR

REICHLE, DAVID E., Dr., Ecological Sciences Division, Oak Ridge National Laboratory, Oak Ridge, Tennessee, USA

SATOO, TAISITIROO, Prof., Department of Forestry, University of Tokyo, Tokyo, Japan

SMITH, FREDERICK E., Dr., School of Design, Harvard University, Cambridge, Massachusetts, USA

STANHILL, GERALD, Dr., Department of Agricultural Meteorology, Volcani Institute of Agricultural Research, Rehovot, Israel

STEUBING, L., Prof. Dr., Botanisches Institut, Justus-Liebig-Universität, Gießen, Germany

WOODWELL, GEORGE M., Dr., Department of Biology, Brookhaven National Laboratory, Upton, Long Island, New York, USA

Combination of Qualitative and Quantitative Approaches

FRANÇOIS BOURLIÈRE and MALCOLM HADLEY

A central issue in the study of the ecosystem's structure and functioning is the analysis of the processes governing the production of organic matter, the flow of energy and the cycling of nutrient resources. To achieve these goals for ecosystems in different climatic regions, and to describe and compare the efficiency of various plant and animal communities under different environmental conditions, is a formidable task. The intention of the following remarks is to present a number of factors which must be considered in ecosystem analysis if studies on energy flow and organic production are to be meaningful. The relationships between plant and animal populations, and between these populations and the abiotic environment, are so complex that much of our knowledge of these relationships is still of a purely qualitative nature. One of the goals of ecologists is to find ways of quantifying these relationships. The problems involved in this task are immense, but we must take encouragement from the words of MACFADYEN (1967) who writes, ". . . the theme of energy flow study has allowed us to quantify certain aspects of community functioning which only twenty-five years ago were subject to no more than descriptive generalization".

Factors Limiting to Production

If the actual and potential production of various components of ecosystems are expressed only in terms of energy budgets, the importance of certain limiting factors on production may be grossly misrepresented. The use of energy budgets, expressed in calories, permits comparison of the efficiencies of various ecosystems in fixing solar energy and the transfer of energy between organisms and populations within these ecosystems. However, the heterogeneity of microenvironments in these systems imposes certain limits on the pattern and rates of flow of materials through the different trophic levels which constitute living communities.

Nutrients are a major limiting factor on production which cannot be measured in terms of calories. As OLSON (1968) has written, "the flow of energy and the very existence and role of the living species in the ecosystem depend on the flow of nutrients". It is, therefore, essential that transfers of both energy and nutrients should be an integral part of productivity studies. The symptoms and ways of combating nutrient deficiencies in plants and animals are sufficiently well known not to require detailed illustration; suffice it to say that more information is needed on nutrient dynamics, particularly on the transfer of different nutrients from one population or trophic level to another.

Although the mechanisms of nutrient uptake by plants often are not known in detail, all nutrients are not taken up by plants in proportion to the element concentration in soil, nor in relation to the minimum nutrient requirements of plants

(OVINGTON, 1968). Particular combinations of nutrients are important in animal growth (DAVIS, 1968), and animal species vary in their specific requirements of nutrients and other dietary essentials (PROSSER and BROWN, 1961). The importance of particular dietary elements in the nutrition of heterotrophs can be a limiting factor in energy flow. All calories do not have the same nutritive value, and it is as important to measure the quality of organic material (e.g., in protein, fat, nitrogen and certain other elements) as to measure the production of dry weight per unit area. The requirements, and therefore food sources, of many consumers are known to differ according to particular life stages. Many dietary essentials are needed in larger amounts for growth than for adult maintenance. The nestling and fledgling stages of many granivorous birds, for example, need insect food during the period of rapid growth (MOREL et al., 1957; WARD, 1965); hence the necessity of synchronization between the production of dietary essentials and the breeding period of many consumers.

Temporal Aspects of Production

The relationship between breeding season and food supply illustrates the importance of the temporal aspect of production; it is essential to measure production both seasonally and from year to year. There is no ecosystem in which energy flows at a constant rate. Daily fluctuations in temperate habitats and seasonal changes in certain tropical ecosystems (e.g., lowland rain forest) are of small magnitude, and temporal differences may be thought of as "regular pulsations" in productivity and energy flow. In contrast to this relatively constant pattern of production, arctic and subarctic, mountain and desert ecosystems are characterized by short periods of surplus primary production. The effects of particular environmental factors in these regions (e.g., cold in tundras and lack of rainfall in deserts) result in drastic changes in the temporal pattern of production and rate of energy flow. These "bottlenecks" to production are of major significance to the functioning of these different types of ecosystems.

In habitats where there are large seasonal fluctuations in climate and production, "sinks" of organic and nutrient matter may occur, which cannot be fully utilized by other trophic levels. Accumulation of dead organic material in peatland areas may continue for many years before a balance is approached between income and loss. The storage of nutrients within animal biomass which is likely to be accentuated by overpopulation may, especially if followed by large scale emigration, seriously decrease continued productivity of the ecosystem.

Ecologists should be aware of the dangers involved in using oversimplified energy budgets to quantify bioenergetic relationships. In biomass studies of mammals, for example, it is important to distinguish between seasonal and cyclic changes in weight and average weight increases from one year to another. If regular changes in seasonal biomass occur over a number of years, then growth curves can be constructed by taking the means of measurements taken in each particular year, or other suitable unit of time. When such intensive studies are not possible, growth curves may be produced from only one set of annual measurements. Here it is important to take into account the animal's phenology, and each set of values should be taken at the same point in the animal's seasonal cycle.

Even if these considerations are acknowledged the result is still only a series of discontinuous measurements of a particular component of the ecosystem. Techniques must be developed whereby dynamic sequences of the functioning of the ecosystem can be obtained. Living communities are characterized by an everchanging pattern of flow of energy and cycling of materials. A simplified energy budget is not necessarily the most meaningful representation of the functional organization of an ecosystem.

Population Interactions

One of the most difficult problems in productivity studies is the estimation of the influence of interactions between different populations. How does a change in the abundance and species composition of consumers affect their food resources? How does manipulation of a specific group of primary consumers change the productivity of primary producers? Many studies have described such interactions in qualitative terms; more effort needs to be directed toward finding ways to quantify these phenomena.

It is often feasible to estimate the net primary production of a particular area by summing the calorific value of the plant standing crop and the calorific value of the primary production consumed by animals. This estimated figure for production may differ widely from production in the absence of primary consumers. VARLEY (1967) has discussed how grazing activities can have multiple and far-reaching effects on primary production. Holes in the leaves of woody plants will change the Leaf Area Index, affect the amount of light reaching lower leaves, and may also induce sap flux in growing leaves. The activities of sap-sucking insects may affect primary production by injection of saliva which results in growth abnormalities or even death, by being the vectors of plant diseases and by reducing the plant's resistance to attack by other heterotrophs. Sap-sucking spittlebugs have been shown to feed on amino-acids being transported in the xylem of plants (WIEGERT, 1964), and ANDRZEJEWSKA (1967) has written that the calorific value of plant losses resulting from only three-weeks feeding by the homopteran *Cicadella viridis* was from 28—36% higher than the calorific values of food which the insects had consumed.

VARLEY (1967) has drawn attention to another potentially important consequence of the feeding activities of sap-sucking insects, namely the removal of water from plants. Aphids which consume phloem excrete considerable quantities of water and excess amino acids and sugars. The calorific content of these solutes often will be assessed as a part of the energy budget, but no allowance will be made for the water which has no nutritive value to the insect. If this water is replaced by the plant, however, the cost of osmosis in the root hairs and energy needed to lift water up the plant should not be neglected.

Sucking insects such as aphids and spittlebugs may have a greater effect on plant growth per unit of energy consumed than chewing herbivores. Similarly, the consumption of nectar and other nonphotosynthesizing parts such as flowers or fruits would generally affect primary production less than removal of amino acids from xylem or consumption of photosynthetic leaf tissues. This is not to say, however, that feeding activities of this kind do not greatly influence the quantity and quality of primary production. In a study on the effect of the moose *(Alces alces)* on the production of a mixed coniferous broad-leaved forest near Moscow, DINESMAN (1967)

1*

found that these animals reduced the production of young plant growth by about 50%. Only 5% of the total reduction caused by moose was attributable to direct consumption of plant material; the greater part was due to the decrease in growth rates of damaged plants. Moreover, the moose produced a major influence on tree quality. Damage mainly affected the best specimens of young trees, causing inhibition of growth and resulting in less robust trees becoming dominant. Thus, the consumption of calories by moose would give an entirely erroneous picture of the influence of these animals on primary forest production. It is clear that knowledge of a consumer's productivity provides no necessarily accurate index of the effect of the consumer on its food source.

Many other interactions are likely to be important in influencing the rates of primary production and energy flow. The effect of the activities of burrowing mammals on soil fertility, the influence of different grazing pressures of large herbivores on species composition, and the effect of predation of herbivorous invertebrates by birds are three such examples. The great problem at present is to find ways of quantifying these activities. Manipulative treatments will help to further greatly our understanding of the influence of consumers on the continued production of the food source. The effect of grazing may be assessed by taking comparable areas and eliminating grazing in one of them by the physical exclusion of large herbivores and chemical treatments of herbivorous invertebrates. An alternative method is comparison of the effects on plant production of natural variations in grazing during successive years.

Energy-Saving Mechanisms

Implicit in the analysis of ecosystems is the study of relationships between the structure and functioning of natural communities. One facet of this relationship is the way in which animal populations are adapted, both behaviorally and physiologically, to different environmental conditions. The importance of behavior patterns and physiology of animals in the utilization of available food resources largely has been neglected in energy flow studies, but further knowledge of energy saving mechanisms and devices are needed to better understand how ecosystems operate.

Behavioral adaptations which reduce the amount of muscular work necessary for an animal to assimilate a given amount of energy will impart a selective advantage to that animal. Social structures are one important device by which animal populations regulate the intensity of their demands on the environment, and by which they may better utilize available resources. In many wild ungulates a common phenomenon is the formation of large herds during migration, their splitting up upon arrival at feeding pastures and extreme herd fragmentation observed under true desert conditions or during unfavorable seasonal conditions. The structure of many bird populations also exhibits striking adaptations to particular habitat types. In the thorn bush belt of Senegal, for example, plant production only occurs during the short rainy periods which are restricted to 2 or 3 months of the year (Bourlière, 1961; Morel and Bourlière, 1962). Populations of resident birds are therefore kept at a very low level by the scarcity of food during the long period of drought. These sparse populations are unable to utilize all the food produced during the wet season and leave an important seasonal surplus, which is exploited by large populations of nomadic species and long-range migrants. Similarly, resident consumer populations in tropical savannas and tundra regions appear unable to consume all the food during peak periods of production, since the low carrying capacity of the environment during critical stages of the yearly cycle also keeps their populations low. Migration and nomadism may therefore be considered as adaptations for the efficient utilization of seasonal food surpluses which sedentary species are unable to adequately exploit. These examples further stress the importance of considering the temporal aspects of production; they also reflect how man's pattern of food utilization in a particular habitat type often follows closely that adopted by wild consumer populations.

WALTER (1968) has recently provided a striking example of the subtle ways in which a species' breeding behavior and migration pattern may be closely adapted to its food supply. Eleonora's falcon *(Falco eleonorae)* normally feeds upon insects and small nonmigratory vertebrates, and spends the nonbreeding season in Madagascar. The species has a very limited and distinctive breeding area, being restricted to small islands and cliffs from the Canary Islands to Cyprus. This distribution acts like a giant "mist net" cast over the whole Mediterranean, and enables the falcon to exploit efficiently the seasonal flux (July to September) of summer migrants flying from Northern Europe to Africa. The breeding season of this falcon is delayed, and is adjusted to correspond to the presence of the large and readily available summer food supply. Moreover, the species has developed a highly specialized hunting technique, by which the birds direct themselves against the wind and maintain their position with normal wing beat frequencies. Large numbers of male falcons also have been observed to build a "vertical wall" above the sea, thus blocking the passage of migrating passerines, and increasing the chances of successful predation.

The importance of the many physiological adaptations which enable organisms to live in otherwise unfavorable habitats or allow them to survive during seasonal periods of stress must be recognized in bioenergetic comparisons. Examples of these adaptations are the dormancy of seeds, the anhydrobiosis of rotifers and tardigrades, the hibernation and aestivation of many vertebrates, and the water saving mechanisms of desert rodents. Behavioral adaptations which act as "energy saving devices" include the storage of seeds by many ants (SUDD, 1967); the storage of honey by bees (BUTLER, 1954); the cultivation of fungi by ants (SUDD, 1967) and termites (BRIAN, 1965; GRASSÉ and NOIROT, 1958, 1961); the mixed foraging parties of many tropical birds, sometimes associated with mammals and army ants (WILLIS, 1966); local and long-range migrations of birds, which enable many species to exploit seasonal food surpluses (BOURLIÈRE, 1961), and the open social structure of many forest primates, which is a seasonal adaptation to take advantage of the patchy production of fruits in the tropical forest canopy (REYNOLDS, 1965). These and other similar phenomena must be considered if the relationship between structure and functioning of ecosystems is to be understood.

The four topics briefly outlined demonstrate the pitfalls of a reductionist approach when studying complex natural ecosystems. Mathematical generalizations will be necessary to describe the major pathways of energy flow and to estimate the efficiency of various trophic levels. But, once this initial stage has been completed, a closer analysis of ecosystem metabolism will be required. It is only through such study that the structural and physiological adaptations of the various species to their different habitats can be analyzed. Such an understanding is indispensable for a sound management of natural biotic communities and establishment of man-made ecosystems in different climatic zones of the world which possess the optimum combinations of efficiency, useable yield and stability.

References

ANDRZEJEWSKA, L.: Estimation of the effects of feeding of the sucking insect *Cicadella viridis* L. (Homoptera-Auchenorrhyncha) on plants, pp. 791—805. In: Secondary productivity of terrestrial ecosystems. (PETRUSEWICZ, K. Ed.) Panstwowe Wydawnictwo Naukowe, Warsaw-Krakow 1967.

BOURLIÈRE, F.: Symposium sur les déplacements saisonniers des animaux. Introduction. Rev. suisse Zool. **68**, 139—143 (1961).

BRIAN, M. V.: Social insect populations. London-New York: Academic Press 1965.

BUTLER, C. G.: The world of the honeybee, London: Collins 1954.

DAVIS, G. K.: Mineral elements in the nutrition of larger mammals. Amer. Zool. **8**, 169—174 (1968).

DINESMAN, L. G.: Influence of vertebrates on primary production of terrestrial communities. pp. 261—266. In: Secondary productivity of terrestrial ecosystems. (PETRUSEWICZ, K. Ed.) Panstwowe Wydawnictwo Naukowe, Warsaw-Krakow 1967.

GRASSÉ, P. P., and C. NOIROT: La meule des termites champignonnistes et sa signification symbiotique. Ann. Sci. nat. Zool. **20**, 113—128 (1958).

— — Nouvelles recherches sur la systématique et l'éthologie des termites champignonnistes du genre *Bellicositermes*. Insect. Soc. **8**, 311—357 (1961).

MACFADYEN, A.: Methods of investigation of productivity of invertebrates in terrestrial ecosystems, pp. 383—412. In: Secondary productivity of terrestrial ecosystems. (PETRUSEWICZ, K. Ed.), Panstwowe Wydawnictwo Naukowe, Warsaw-Krakow 1967.

MOREL, G., and F. BOURLIÈRE: Relations écologiques des avifaunes sédentaire et migratrice dans une savane sahelienne du bas Sénégal. Terre et Vie. **16**, 371—393 (1962).

— M. Y. MOREL, and F. BOURLIÈRE: The black-faced weaver bird or dioch in West Africa, an ecological study. J. Bombay Nat. Hist. Soc. **54**, 811—825 (1957).

OLSON, J. S.: Use of tracer techniques for the study of biogeochemical cycles, pp. 271—288. In: Functioning of terrestrial ecosystems at the primary production level, (ECKARDT, F. E. Ed.). Paris: UNESCO 1968.

OVINGTON, J. D.: Some factors affecting nutrient distribution within ecosystems. In: (ECKARDT, F. E. Ed.): Functioning of terrestrial ecosystems at the primary production level, p. 95—105. Paris: UNESCO 1968.

PROSSER, L. G., and F. A. BROWN: Comparative animal physiology, Second Edition, p. 84. Philadelphia-London: Saunders 1961.

REYNOLDS, V.: Budongo, an African forest and its chimpanzees. New York: Natural History Press 1965.

SUDD, J. H.: An introduction to the behaviour of ants. London: Arnold 1967.

VARLEY, G. C.: The effects of grazing by animals on plant productivity, pp. 773—778. In: Secondary productivity of terrestrial ecosystems, (PETRUSEWICZ, K. Ed.) Panstwowe Wydawnictwo Naukowe, Warsaw-Krakow, 1967.

WALTER, H.: Zur Abhängigkeit des Eleonorenfalken *(Falco eleonorae)* vom mediterranen Vogelzug. J. Ornithol. **109**, 323—365 (1968).

WARD, P.: Feeding ecology of the Black-faced Dioch *Quelea quelea* in Nigeria. J. Ornithol. **107**, 173—214 (1965).

WIEGERT, R. G.: Population energetics of meadow spittlebugs (*Philaenus spumaris* L.) as affected by migration and habitat. Ecol. Monogr. **34**, 217—241 (1964).

WILLIS, E. O.: The role of migrant birds at swarms of army ants. Living Bird **5**, 187—231 (1966).

2

Analysis of Ecosystems

Frederick E. Smith

The immediate goal of ecosystem analysis studies is to understand ecosystems and ecosystem processes; many of these goals are directly related to such applied problems as organic matter production or the influence of management practices on various ecosystem parameters. A beginning point for such studies must be research design. We first ask what has to be done to accomplish our goals. Oftentimes, subject matter will dictate the strategy. In this case, scientists must work together in multidisciplinary teams, study whole ecosystems (parts cannot be left out for lack of interest, personnel, or technology), share data immediately and completely, and devote a considerable effort to the process of synthesis. Systems oriented techniques with high-speed computers offer the only means by which this synthesis can be accomplished.

Criteria for Developing a Systems Model

Conceptual definitions must be converted into operational definitions lacking assumptions that may prove to be false in the real world. Conceptually the ecosystem is viewed as a functional unit with recognizable boundaries and an internal homogeneity. Operationally, we first recognize that boundaries are arbitrary and that functional unity may exist like beauty only in the eye of the beholder. We have, therefore, defined the ecosystem as everything that exists and happens within a precisely bounded region. Two sets of criteria are used for the location of the boundary. For many purposes the region must be large enough to contain a full set of ecosystem processes and their interactions; secondly, the boundary should be placed where inputs and outputs across it are most easily measured.

Once an ecosystem is defined by the location of its boundary (perimeter, roof, and floor), the next stage is to identify all its significant components. The air, land, and water can be subdivided into a number of components, and the plants and animals can be broken down to their species or to major species and groups of minor species. By lumping or splitting in various ways, the total number of components in an idealized ecosystem can vary from 5 to 50,000 the only restriction being that they must always add up to the whole ecosystem. The subdivision or fusion of components is itself one of the activities that will continue throughout the research program, and is in fact guided by the research. The ecosystem boundary also may be altered as we learn more about the system. If categories can be lumped, I think we will prefer to lump them. If they cannot be lumped, because the distinction between them is meaningful to the operation of the system, we will have to keep them separate. Hopefully, the total number of components needed to account for the significant ecosystem processes will not be more than several hundred.

The four major groups of components are the producers, the consumers, the decomposers, and the abiotic environment. So far, in the development of our programs, these account for approximately equal portions of the total cost of field and laboratory research.

Program goals require that research plans be organized to a degree unprecedented in ecology. Component projects must be compatible and, collectively, complete. The bulk of the research is conventional, however, so that most of the new emphasis is on organization. One method of formalizing the system is to reduce it to a series of tables, as follows.

Let us suppose that our system has n components, each of which can be described quantitatively according to many criteria: e.g., caloric content, carbon, phosphorus, nitrogen, and water (many more exist and are important). For each one of these, the following set of tables can be considered. Let us consider only phosphorus for the moment, remembering that additional sets of tables can be constructed for the other parameters.

The amount of phosphorus in each component (x_i) yields the top row of estimates in Table 1. The sum of this row is the total amont of phosphorus in the ecosystem.

Table 1. *The amounts (x) of phosphorus in each of the n components of an ecosystem, and the rates at which phosphorus is entering (a) and leaving (z) the system via each component. In any real system many of the rates will be virtually zero*

Component	1	2	3	.	.	.	n
Amount	x_1	x_2	x_3	.	.	.	x_n
Inflow	a_1	a_2	a_3	.	.	.	a_n
Outflow	z_1	z_2	z_3	.	.	.	z_n

Phosphorus may be entering the system, being added to some or all of the components. These rates (a_i) are the inflows, and form the second row of estimates in Table 1. Their sum is the total rate of inflow of phosphorus into the system. Phosphorus also may be leaving the system, being subtraced from some or all of the components. These rates (z_i) are the third row of estimates in Table 1, and their sum is the total rate of outflow from the system.

The location of a boundary for an ecosystem study will influence greatly how many of these rates are significant, and also how difficult they may be to estimate. One approach is to locate the boundary where the number of zero rates is maximal (where the system is least open in its connections with adjacent systems), such as the boundary between field and forest or between lake and land. It has been the custom, in fact, to draw boundaries where discontinuities are the most clear. Here inflows and outflows may be minimized. Here they will also be the least alike. That is, inflows will differ in both kind and amount from outflows.

A second approach is to locate the boundary where inflows and outflows are most similar. An example is to delimit a region of forest within a much larger forest. Since inflows and outflows will tend to be equal and opposite, their difference may have little net effect. The result, however, may be to maximize the total flow — i.e., selection of the most open system available. Finally, the significance of inflow and outflow also decreases as the system is made larger, leading to the choice of the largest system which is feasible for study.

In addition to inflow and outflow, phosphorus may be transferred from one component to another within the ecosystem. Movement through the food web is a good example. These rates can be presented as a square table of n rows and

Table 2. *Transfer rates of phosphorus among n components of an ecosystem. In any real system many of the rates will be zero*

		Transfer to each component							
		1	2	3	4	.	.	.	n
Transfer from each component	1	—	y_{12}	y_{13}	y_{14}	.	.	.	y_{1n}
	2	y_{21}	—	y_{23}	y_{24}	.	.	.	y_{2n}
	3	y_{31}	y_{32}	—	y_{34}	.	.	.	y_{3n}
	4	y_{41}	y_{42}	y_{43}	—	.	.	.	y_{4n}
	.	.	.	.	.	.	.	.	.
	n	y_{n1}	y_{n2}	y_{n3}	y_{n4}	.	.	.	—

columns, as shown in Table 2. Each row represents losses from a particular component, and the row sum is the total rate of loss to other components. Each column shows gains to a particular component, and the column sum is the total rate of gain from all other components. Here again many entries will be zero or negligible. An advantage of this format is that it can handle any degree of complexity within the system. No assumption of discrete trophic levels is needed, and the confusion of a complicated flow diagram can be avoided. The portion of this table that comprises the food web is the "who-eats-whom" matrix.

With this information it is possible to write an equation for the rate of change of each parameter for each of the components. For example, for component 3:

$$dx_3/dt = a_3 - z_3 + (y_{13} + y_{23} + \cdots + y_{n3}) - (y_{31} + y_{32} + \cdots + y_{3n}).$$

If component 3 were leaf litter, this equation might represent the rate of change with time in the amount of phosphorus due to: litter moved inward across the ecosystem boundary, minus that moved outward, plus litter received from the various plant components (other y_{i3} terms may be zero), minus litter removed by consumers or decomposers, and minus litter transformed into humus (other y_{3j} terms may be zero).

Now assume that we have made a study based on this approach, and have estimated the average amounts of phosphorus in all of the components and also all of the average rates of movement in and out of the system and among components. Then, Tables 1 and 2 can be filled with simple numbers representing amounts and rates. What have we learned? First, we have discovered where activity does and does not occur, and for most ecosystems this alone may be a considerable increase in information. We also can examine the study for completeness, searching for unevaluated components and rates or those subject to doubt. Finally, we can set up this system in the computer, beginning with all the x_i's at their estimated levels, and letting them change through time according to the equations, dx_i/dt.

At this point we will discover that these data are not at all adequate for the goals of the program. Each component will change in the same direction at the same rate for as long as the computer simulation continues. In the real world the ecosystem *responds* to change with changes in the rates, and the nature of these responses is what we are after. The use of an average rate (simple number) rather than a variable rate (mathematical function) destroys the system properties that are one goal of the study.

It turns out in fact that the new problem is an order of magnitude more complicated than the estimation of averages. In particular, the rates must be expressed as *functions* of the system, and not as simple numbers, if we wish to learn anything about the system. Each transfer rate, y_{ij}, and each outflow, z_i, is a *set* of functions, an equation, which relates the rate to all those direct causal factors that govern it. It is only then that we can predict how each rate will change if the system is changed. Furthermore, if all of these are combined in the computer, we can predict how the system will respond to change. This is the only way that we can discover how ecosystems operate.

This complication invokes a vast amount of experimental research, much of it at the level of physiological and behavioral ecology. Again, the system model helps define where the research must be done. We cannot decide casually to study the behavior of this or the physiology of that. We must discover and study *that which is relevant* in the system. For example, the rate at which grasshoppers eat grass may depend upon both the amount and quality of the grass, the number and size of the grasshoppers, the temperature, rainfall, humidity, wind, harassment from predators, and many other factors. Success of the study depends very strongly on locating and analyzing the more important processes relating cause and effect in the system. For this reason it is necessary for investigators to work together in constant communication sharing creative ability and data, and to work as closely with synthesists as with field researchers. Each investigator must concentrate on his particular problem with aims for maximal information input to all related problems. The increased efficiency with which each can pursue his research is the individual benefit from teamwork; the coordination of work permits synthesis and the study of whole ecosystems — in effect the creation of a whole new level of ecological science.

The expressions of rates as sets of functions require the estimation of many parameters other than amounts and rates of phosphorus. In a simplified form these can be arranged as shown in Table 3. For each component, in addition to x_i, a_i, z_i, and y_{ij},

Table 3. *Additional descriptors needed for the specification of functional relations in an ecosystem of n components*

Component	Component attributes (open list)					
1	w_{11}	w_{12}	w_{13}	.	.	.
2	w_{21}	w_{22}	.	.	.	
3	w_{31}	w_{32}	w_{33}	w'_{43}	.	.
.						
.						
.						
n	w_{n1}	w_{n2}	w_{n3}	.	.	.

External attributes (viz., climate, weather; open list)
A_1, A_2, A_3, A_4, etc.

there may be a list of descriptors, additional attributes that are needed. These w_{ij}'s may include the average size and number of individuals in a component, age structure, distribution in space, etc. These, like the y_{ij}'s and the z_i's, may be sets of functions. In addition, there will be a set of inputs (A_i) to the system that are not included in Table 1. These are external variables such as those of climate and season that effect the system as a whole. These, together with the inflows, a_i, are the externally controlled variables that influence the system but are not in turn affected by the system. The remaining rates, y_{ij} and z_i, measure system processes that may be functions of any or all of the other amounts and rates that have been presented.

In general terms, the immediate goal of many of the research projects is to find appropriate mathematical functions for the *effect of external variables* and for *relationships among internal variables*. A complete set of these comprise the ecosystem model. The validity of the model can then be estimated by beginning with an initial distribution of amounts, x_i, and a program through time of the input variables (A_i and a_i) and observing how well the component amounts predicted from the computer follow observations in the field. The effect of a treatment such as fertilization or irrigation can also be followed both in the computer and in the field. Once a valid model has been constructed it becomes a powerful tool for prediction and management, as well as for obtaining insight into the governing properties of the ecosystem.

A System Model

So much for the grand scheme. An extremely simple hypothetical ecosystem will show how the methods of systems analysis can contribute to the design of research and the synthesis of results. Imagine a controlled ecosystem of three components: water, an aquatic plant population, and an herbivore population. Let it be internally homogeneous with the external environment held constant. Diagrammatically the system can be presented as a box (Fig. 1) with three compartments. This emphasizes that the ecosystem is a physical whole, a three-dimensional piece of the biosphere whose components must add up to that whole. The only inflow is water (with its phosphorus), while the only outflows are water and herbivores (with both outflows independent of each other). Assume that the herbivores do not take phosphorus from the water, and that the plants do not eat the animals. The ten remaining non-zero quantities are shown on the figure and are listed again in Table 4 in the format used for Tables 1 and 2.

Let us run this system with a constant inflow of 100 mg phosphorus per day, and wait until the outflow equals the inflow — i.e., until steady state has been achieved. This requires monitoring the outflow in water and in herbivores. We might find that, after equilibration, 19 mg/day flow out in the water while 81 mg/day are lost through herbivore emigration. Next, we might ask how the phosphorus is distributed within the system, and find that 9.5 mg are in the water, 1.4 in the plants, and 9.0 in the herbivores, giving a total "standing crop" of 19.9 mg of phosphorus in the ecosystem. Finally, with careful study using tracers we might find that plants are taking up phosphorus from the water at the rate of 133 mg/day, and losing it back to the water through excretion and similar processes at the rate of 7 mg/day. Similarly, we may find that herbivore grazing amounts of 126 mg/day, with 45 mg/day returned to

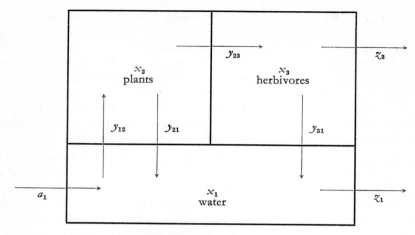

x_1 = amount of P in water
x_2 = amount of P in plants
x_3 = amount of P in herbivores
a_1 = rate of inflow of P in water
z_1 = rate of outflow of P in water
z_3 = rate of outflow of P in herbivores
y_{12} = rate of uptake of P from water by plants
y_{21} = rate of loss of P from plants to water
y_{23} = rate of uptake of P from plants by herbivores
y_{31} = rate of loss of P from herbivores to water

Fig. 1. Flow diagram for a very simple ecosystem composed of three components

Table 4. *The amounts and rates from Fig. 1 arranged in the format of Tables 1 and 2*

	Water	Component Plant	Herbivore
Amount	x_1	x_2	x_3
Inflow	a_1	0	0
Outflow	z_1	0	z_3
Transfers:			
Water	—	y_{12}	0
Plant	y_{21}	—	y_{23}
Herbivore	y_{31}	0	—

the water. These are a complete set of the data, and are summarized in Fig. 2. The reader will appreciate that if this were a real ecosystem with several hundred components an enormous amount of research just has been described.

As stated earlier, while this tells where the action is, it does not tell how it happens. We do not know how the system operates and cannot answer such questions as: What would happen if the inflow rate were doubled? What would happen if an herbivore with a greater tendency to emigrate were substituted? Will an herbivore that more efficiently utilizes plants tend to increase herbivore production and therefore reduce phosphorus outflow in the water? Another set of questions relates to the experimental design. How important are the various quantities? How accurately should each be estimated?

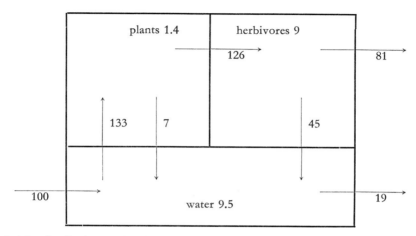

Fig. 2. The distribution (mg) and rates of movements (mg/day) of phosphorus in the three-component ecosystem after equilibration to a constant input rate of 100 mg/day (see Table 5)

To discover more about the system it has to be observed under varying conditions, through controlled experiment or natural variation. One approach is to run the system at different rates of phosphorus inflow, and observe the other nine quantities after equilibration has been achieved. Let us use the inflow rates of 25, 100, and 400 mg/day. (In the real world we should probably use more levels, but this is an imaginary system worked out without sampling error, and three are sufficient.) The results which might be obtained are shown in Table 5. All quantities increase with an increasing rate of flow, but not proportionally. Three quantities (the transfer from water to

Table 5. *Response of a three-component system to three different levels of input (a). Data taken after equilibration has been achieved (see Fig. 2)*

a_1	z_1	z_3	x_1	x_2	x_3	y_{12}	y_{21}	y_{23}	y_{31}
25	9	16	4.5	0.9	4	40.5	4.5	36	20
100	19	81	9.5	1.4	9	133	7	126	45
400	39	361	19.5	2.4	19	468	12	456	95

plants, the transfer from plants to herbivores and the outflow of herbivores) are very sensitive to the input rate, while the other six are much less sensitive.

Now we can proceed further with the analysis, even though this is but one of several kinds of treatments which might have been used. In real ecosystems, both local treatments affecting selected processes and overall treatments that affect the whole system are useful for eliciting changes (responses) in ecosystem processes.

Deriving Ecosystem Hypotheses

The next step is how the study of acquired data leads to the formulation of new concepts or *hypotheses*. Functions that relate the processes of the ecosystem to their most immediate causal factors need to be developed. This step occurs in all scientific work, and depends primarily on intuition or creativity on the part of the investigator. Since this particular problem was worked out backwards, beginning with functions and deriving the data, it is of interest to see how easily the reader can recover the functions given the data. Certainly a powerful influence is the "frame-of-reference", the past training or bias, with which the investigator works.

In this case it is too easy to write too many simple relations. For example, we can write:

$$z_1 = 2x_3 + 1 \quad \text{or} \quad z_1 = 4y_{21} - 9.$$

The best procedure, however, is to begin with proximate factors, and neither herbivore abundance nor the plant excretion rate have an obvious direct effect on the outflow rate of phosphorus in the water. The most obvious direct factor influencing z_1 is x_1, the amount of phosphorus in the water. Here we find that the simple relation $z_1 = 2x_1$ is an adequate and complete account for variations in z_1.

All of the rates have very simple functions that relate them to proximate factors (Table 6). The most simple are the negative exponentials of phosphorus outflow in the water, and in the return of phosphorus to water from plants and from herbivores. The two cases of active uptake or feeding (water to plants and plants to herbivores) are represented by the simple prey-predator notation of Lotka and Volterra. The emigration of herbivores is a simple expression of social interaction, the rate being proportional to the rate of contact between individuals. The functions for these rates are not defended as representative of the real world; they are used merely to produce a very simple system and to challenge the reader to consider other alternatives.

Table 6. *Functional relations of the six process rates in the simple three-component ecosystem (see Table 5)*

Numerical solution	Parameter functions	Parameter estimates
$z_1 = 2x_1$	$z_1 = c_1 x_1$	$c_1 = 2$
$z_3 = x_3^2$	$z_2 = c_2 x_3^2$	$c_2 = 1$
$y_{12} = 10x_1 x_2$	$y_{12} = c_3 x_1 x_2$	$c_3 = 10$
$y_{21} = 5x_2$	$y_{21} = c_4 x_2$	$c_4 = 5$
$y_{23} = 10x_2 x_3$	$y_{23} = c_5 x_2 x_3$	$c_5 = 10$
$y_{31} = 5x_3$	$y_{31} = c_6 x_3$	$c_6 = 5$

Since the numbers in the functions are only estimates and will vary from species to species, they should be represented by symbols, as shown on the middle column of the table. In this form we see that the system has been defined with six parameters $(c_1 \ldots c_6)$ in association with the ten variables. A real-world ecosystem with several hundred components will probably have thousands of parameters, since each process is itself a very complex function, certainly not as simple as those used here.

In Fig. 1 one can observe the arrows in and out of each component, and write the three differential equations for changes in the amount of phosphorus in the three components:

$$dx_1/dt = a_1 + y_{21} + y_{31} - y_{12} - z_1,$$
$$dx_2/dt = y_{12} - y_{21} - y_{23},$$
$$dx_3/dt = y_{23} - y_{31} - z_3.$$

Substituting the functions for these rates, produces a mathematical model for this system:

$$dx_1/dt = a_1 + c_4x_2 + c_6x_3 - c_3x_1x_2 - c_1x_1$$
$$dx_2/dt = c_3x_1x_2 - c_4x_2 - c_5x_2x_3$$
$$dx_3/dt = c_5x_2x_3 - c_6x_3 - c_2x_3^2.$$

Sensitivity Analysis

We are now ready to ask how accurately each of the six parameters should be estimated, since this will guide the continuing research on this system (assume that in the real world we are much less certain about the values of these parameters). For this problem we have techniques available in systems analysis collectively called sensitivity analysis. The current version of the model is set up in the computer, and then the value of each parameter in turn is varied upward and downward while the performance of the system is observed. If the system shows little response, it is not sensitive to the precise value of that parameter, which therefore need not be estimated with great accuracy. If the system or one important part of it responds strongly, the value of the parameter is important and must be estimated with precision. In this way future effort can be allocated to improve the accuracy of the sensitive parameters. Even in a system as simple as our example, the sensitivity of response varies considerably from one parameter to another.

Table 7. *Sensitivity analysis. The percentage error in predicted outflow of phosphorus in water* (z_1) *for an overestimation of one percent in each parameter. The relative sensitivity is obtained from the middle column by setting the largest error equal to 1.0*

Parameter	% error	Relative sensitivity
c_1	+0.900	1.00
c_2	−0.426	0.47
c_3	−0.900	1.00
c_4	+0.047	0.05
c_5	+0.853	0.95
c_6	0.000	0.00

This procedure of observing performance requires first a definition of the criterion for performance. If this were a potential system for removing phosphorus from water, our criterion could be the rate of outflow of phosphorus in water. The amount of change, after equilibration, in the rate of outflow of phosphorus in water (z_1 or c_1x_1) following a given small change in each of the six parameters of our model are shown in Table 7. For a change of one percent in the value of each parameter, the rate of outflow changes from zero to 0.9%. This means that, if any of the parameter estimations were in error by 1%, the predicted rate of phosphorus outflow would be in error by the indicated amount. Sensitivity is zero for the parameter, c_6. This implies that the rate of return of phosphorus to water from the herbivores, y_{31} or $c_3 x_3$, had no effect on the outflow of phosphorus in water, and would not need to be estimated at all! Conversely, the three parameters c_1, c_3, and c_5 are all very important, indicating that the majority of experimental effort should be expended in reducing their sampling errors. Moderate effort should be spent estimating c_2, but all that is needed for c_4 is the order of magnitude.

The relative sensitivities, obtained from dividing the second column by its largest entry, gives a good guideline for the allocation of effort in the estimation of these parameters. This assumes, of course, that we are interested only in the prediction of phosphorus outflow in water. Allocation then becomes an attempt to equalize the contributions from the several parameters to the final error of prediction.

A different criterion may produce different results. In ecosystem analysis, we do not have a single applied goal in mind, but wish to "understand ecosystems". In general, such a model would be exercised to predict the entire array of component amounts in the system. If this approach is applied to our simple system, the results are those of Table 8. Percentage responses of all three components are given for a

Table 8. *Sensitivity analysis. The percentage errors in predicted amounts (x) in the three components for an overestimation of 1% in each parameter. The sum is taken without respect to sign. The relative sensitivities are obtained from the sums by setting the largest equal to unity*

Parameter	Component			Sum	Relative sensitivity
	Water	Plant	Herbivore		
c_1	—0.100	—0.079	—0.106	0.285	0.14
c_2	—0.426	+0.354	—0.405	1.185	0.59
c_3	—0.900	+0.068	+0.106	1.074	0.53
c_4	+0.047	—0.004	—0.006	0.057	0.03
c_5	+0.853	—1.064	—0.100	2.017	1.00
c_6	0.000	+0.357	0.000	0.357	0.18

change of one percent in each parameter. One method (perhaps not the best) for summarizing the effects on the several components is to add up the responses without respect to sign. This reflects the summed percentage error of prediction for the system, and is shown on the fifth column. Dividing these by their largest entry yields the relative sensitivities shown in the last column.

Using the sum of the absolute component amounts as our criterion of sensitivity, we find that c_5, the parameter associated with herbivore grazing, is the most important, and should be estimated with the smallest percentage error. Parameters c_2, associated with herbivore outflow, and c_3, associated with plant uptake of phosphorus, are about half as important. The remaining three parameters are relatively unimportant, and need not be accurately estimated.

Of course, this model is not realistic, and its results should not be applied to any related real system. The intent here is to demonstrate that, even in a model this simple, sensitivity to error in parameter estimation varies greatly from one parameter to another. In more complex systems it can be expected that such variations will be at least as great, and that many parameters will not need be known with any great degree of precision. Using this knowledge in the allocation of further research will produce considerable savings in both time and effort. Thus, such a model is useful early in the research program.

Even a crude, tentative model can be useful as a tool for guidance in the allocation of further effort, but this approach has faults. Allocation of effort is based upon the assumption that the model is more or less correct. If it is wrong, the allocation of effort and future research direction also may be wrong. There is the possibility that the research group may become "model-bound", digging a deeper and deeper rut on the wrong road. (It is important, therefore, that critics and skeptics also be included.) Major disasters of this kind may be avoided by developing several studies in different places, each with its own effort toward synthesis and proceeding in a quasi-independent fashion, while at the same time maintaining compatibility between studies and providing for mutual evaluation and self-correction.

Assuming that our model is realistic enough to permit more detailed evaluation of the system, let us now return to a series of questions that was asked earlier concerning: What would happen if the inflow rate were doubled? Using our model, setting the inflow rate at 200 mg/day, and letting it run in the computer until the system stabilized, we would find that:

$$
\begin{array}{lll}
x_1 = 13.6 & z_1 = 27.3 & y_{21} = \quad 9.1 \\
x_2 = \quad 1.8 & z_3 = 172.7 & y_{23} = 238.4 \\
x_3 = 13.1 & y_{12} = 247.4 & y_{31} = \quad 65.7 \, .
\end{array}
$$

What would happen if an herbivore with a greater tendency to emigrate were substituted. Here we look at the sensitivity analyses (Table 7), and consider the effects of an increase in the parameter, c_2. The signs associated with the responses indicate the direction of change for an increase in the size of the parameter. Thus, Table 7 shows that the outflow of phosphorus in the water, z_2, would decrease. Table 8 suggests that the amount of phosphorus in the water inside the system also would decrease, while plant density would increase and herbivore density decrease.

Will an herbivore with a more voracious feeding habit tend to increase herbivore production and, therefore, reduce phosphorus outflow in the water? This involves an increase in the parameter, c_5. Table 8 indicates that phosphorus in plants and herbivores would *decrease*, while the phosphorus in water (and its outflow, Table 7) would *increase*. This effect may not be obvious, and is a good example of the value of using whole systems for drawing such conclusions. Further thought shows that the result is logical. If herbivores inflict greater plant damage, plants are less able to grow and take up phosphorus. This not only implies more phosphorus flowing out in water, but a lower plant density and, after equilibration, also a lower herbivore density. In food chain models it is generally found that, the more efficient in feeding the top trophic level becomes, the less common it becomes!

These questions and answers may help to explain what ecologists mean by "understanding" ecosystems. In part, it is being able to predict what would happen if some changes were induced in the system. It is more, however. After describing several systems, a reasonably large set of properties may be found which are common to many, if not all types of ecosystems, and which are *system* properties having little to do with the particular evolutionary adaptations of the species involved. If this should happen, then there is a discipline called the "principles of ecosystems", rather than a subject area which is merely a collection of studies of particular ecosystems. The following chapters on woodlands will take on significance reaching far beyond the particular cases they treat.

Postscript

Although a simple example was presented and analyzed as though it were set up in the computer, using simulation and various techniques of systems analysis, we were cheating. The model is so simple that it has a mathematical solution, making numerical methods unnecessary. After equilibration, the steady state values for the three components are:

$$c_2 c_3 x_3^2 + c_1 c_5 x_3 + c_1 c_4 = a_1 c_3$$

(solved for x_3 using the quadratic formula),

$$c_3 x_1 = c_4 + c_5 x_3,$$
$$c_5 x_2 = c_6 + c_2 x_3.$$

These can be differentiated for each variable (x) with respect to each parameter (c). For relative errors (percentage changes), an example would be: $(dx_1/x_1)/(dc_1/c_1)$. Simple mathematical solutions will probably not be possible for real ecosystems, and systems analysists will need to depend upon numerical methods using computers.

Suggested Reading

Forrester, J. W.: Industrial dynamics. Cambridge: M. I. T. Press 1961.
Holling, C. S.: The functional response of invertebrate predators to prey density. Mem. Entomol. Soc. Can. **48**, 1—85 (1966).
Watt, K. E. F.: Ecology and resource management. New York: McGraw-Hill Book Co. 1968.

Studying the Effects of Environmental Factors on Ecosystems

David W. Goodall

To study the effects of environmental factors on processes or parameters of the ecosystem as a whole is a formidable task. Since the environment is involved in, and a part of, the ecosystem, the problem cannot always be given any clear-cut interpretation. Cause and effect are mingled, often almost inextricably, and while the concurrent changes in biotic and abiotic factors can be described, it may hardly be possible to interpret the network of relations in terms of chains of causation.

The present chapter is concerned with parameters and functions of the ecosystem as a whole, such as productivity, rather than the individual species and populations of which it is composed. At present, very few observations are available which enable one to relate the behavior of the whole ecosystem to differences in the environment; this paucity of data is readily understandable in view of the difficulties involved in collecting them — particularly in such large-scale ecosystems as forests.

It might be thought, perhaps, that the large body of descriptive ecological data relating the occurrence of particular types of ecosystem (e.g., named associations) to particular environmental characters could be used as a basis for deducing the responses of the ecosystem to the environment. The chances of using such data seem extremely small. Because a community type occurs only in a particular combination of environmental conditions, it cannot be assumed that this combination is particularly favorable for it, in the sense of yielding optimal values of productivity or any other measure of community performance. The presence or absence of a particular type of ecosystem is the resultant of competitive and food-chain relations among the species of which it is composed, as well as other species in alternative ecosystems which might also be able to occupy the same area. Although the stable presence of an ecosystem in a particular locality implies that it is in some sense more viable there than possible alternative ecosystems, there are many ways in which this greater viability could come about.

If in place of measurements on the ecosystem as a whole we are content to look at more limited observations on particular biotic elements, more usable information is available. The growth of stem wood is one particular feature of forest ecosystems which has been widely studied. Its economic importance ensures that research is constantly in progress to relate this measure of growth (of certain, or all, tree species) to differences in the environment, either natural or artificially induced. If stem-wood growth could be shown in similar ecosystems to be closely correlated with productivity, this would be a valuable starting point for attempts to test the dependence of forest productivity (or other similar measures) on environmental variables. This correlation appeals intuitively, but needs careful test; if it can be shown to hold, then large amounts of published data are available for relating ecosystem parameters to environmental factors.

In so far as stem-wood growth is a parameter of value in the study of the ecosystem as a whole, the forest ecologist is particularly lucky in having available for many species a permanent record of past growth in tree rings. For wood growth, observations at a fixed point in time can do what, for other parameters, would require regularly repeated observations over many years. Consequently, the rewards from establishing a reliable relationship between wood growth and ecosystem parameters are all the greater.

Apart from wood growth, there is a large body of descriptive observations relating the quantity of particular species in forest ecosystems to different environmetal factors. Such observations can in principle contribute to the understanding of ecosystem responses to the environment; their interpretation, however, is often difficult. Although a species may be abundant in an ecosystem only where a particular factor is high, this cannot be taken as evidence that the species is responding to this factor — the effect may be indirect and mediated by responses of other competitive species, in which case a minor change in the biotic composition of the community might alter the response completely.

If observations on abundance of species in localities where environmental factors differ are difficult to interpret and of limited value, the same applies *a fortiori* to data where only the presence or absence of species has been noted. Observations of this sort can contribute only to estimates of the probability of encountering the species in samples of specified size and type, and then only if they are sufficiently numerous. Such estimates have very limited value as parameters of ecosystem functioning.

Besides observations on stem-wood formation and the very limited measurements of other ecosystem parameters, few existing data can contribute to analyses of the relation of ecosystem behavior to environmental factors. We must look to the collection of new data, and various approaches are possible.

Experimental Approach

Accepting the perhaps rather naive view of an ecosystem as the set of organisms occupying a certain volume of space during a certain period of time, plus the abiotic components affecting them (with the effect of an environmental factor represented by an imposed change of conditions), the ecosystem becomes susceptible to experimentation. One can divide it into replicate portions (or take replicate samples), some of these samples can be subjected to imposed change, others left without treatment, and differences subsequently developing can be ascribed to the imposed change, subject to the usual tests of statistical significance.

This approach is subject to rather serious practical limitations. For most ecosystems, the size of a representative sample is considerable, and the spread of influence to and from the periphery of any delimited area is such that very wide "buffer zones" would be required. The variability of natural ecosystems is very great, implying the need for numerous replicates in order to obtain conclusions with any reasonable degree of precision. Yet most measures of ecosystem function are so demanding in time and resources that a research team would find it difficult to repeat them in many different localities. For some purposes, an experimental design replicated in time rather than space may be acceptable. But in this case generalization beyond the particular sample of ecosystem used will be quite uncertain.

In forest ecosystems, with which we are primarily concerned in this volume, it is practicable to modify only certain environmental factors, and only in certain ways. Soil nutrients may be added to the surface, or by injection to deeper horizons, but nutrients cannot be removed. Soil moisture can be increased by surface or subsoil irrigation (with risk of disturbing soil structure) or decreased by covering the soil surface during precipitation. But changes to soil structure, mechanical properties, and moisture and nutrient retention, cannot be performed without such drastic operations as may destroy the original state of the ecosystem. Control of soil temperature *in situ* may be practicable, and also of the composition of the soil atmosphere. But the diffusion properties of soil for gases and water — so important for root development — are as difficult to control as mechanical properties.

Only very limited control is possible of the meteorological and micrometeorological conditions in a forest ecosystem. Any enclosure of the aerial portion of the system (even if practicable on the scale necessary) will affect every meteorological variable; and though establishment of any combination of these would then be possible in principle, resemblance of the system to the natural state would be little more than coincidental.

Even in cases where the experimental approach is practicable, interpretation of the results may be fraught with difficulty and uncertainty. An ecosystem is such an integrated object of study, that any imposed change is likely to have repercussions throughout the whole system. Consequently, any changes observed in the parameters measured may be indirect effects of the imposed change, mediated by a chain of influences on other environmental factors and components of the system.

The situation is quite different from that in agricultural experimentation, for instance, where the objective is to change the system in a particular way by a particular manipulation. In ecosystem studies the objective is to seek an understanding of the way in which the environmental factors interact to produce a given change, and experimental manipulations are merely tools for this purpose and not subjects of tests. If, then, an environmental manipulation is applied, it is essential not to treat it naively as a well-defined change in only the most responsive factor, but to monitor *all* the resulting changes. Furthermore, even though the manipulation may be identical in all replicate samples, the various secondary changes resulting from it will not be. Since the changes relevant to the observed effects on the ecosystem parameters may be exactly these secondary changes rather than those initially imposed by the experimenter, the value of replicated experiments is less in interpretative studies than in those where the manipulation is of direct interest. It may well be a better expenditure of effort if all replication is avoided in favor of different manipulations to give a wide range of relevant factor combinations.

Observational Approach

This leads us to an alternative to experimentation — namely, observations under a range of varying conditions provided by nature rather than imposed by the observer. We have here two alternatives: 1) study of differences between different samples of the ecosystem, observed at the same time but under different conditions; and 2) study of the same sample of the ecosystem at different times.

Observing different samples of the ecosystem at different locations, and hence under different conditions, can indeed give information about the response of ecosystem parameters to differences in environmental factors — but only if intrinsic differences in the samples can be taken into account. The floristic and faunistic composition of the samples certainly will not be identical; and the ecosystem parameters will be influenced by these differences as well as by differences in the factors of the environment. Statistical techniques may enable one to separate these effects, particularly if the form of the relationships is known. The technique of multiple regression is not well suited to this purpose, for it assumes that the "independent" variables are known and constant throughout the period of measurement. Though they may be known initially, few or none of the relevant variables will remain unchanged, and the values used in regression analysis will have to be samples at fixed and arbitrary times of these continuously changing variables. A distinction between dependent and independent variables is, in fact, far from realistic in the case of an ecosystem, where most variables are interrelated. Sophisticated application of the principle of path coefficients (Scott, 1966) might enable the changing state of the ecosystem itself to be taken into account in estimating the degree of dependence of ecosystem parameters on the environment. But to establish a network of path coefficients implies that the form and pattern of interrelationships is known, whereas in fact this may be one of the problems to be solved.

In the experimental approach, the degree of similarity of "replicate" samples needs consideration. The more similar they are the more precisely can one hope to estimate environmental effects; but, on the other hand, samples need to cover the whole range of composition over which generalization is intended. This implies that high precision and wide generalization are usually inconsistent aims.

The alternative approach is repeated observation on the same sample; although easier in practice, it has the drawback that it permits no generalization. Generalization becomes possible when similar observations through time are made at more than one site, and this may have advantages. Differences in the genetic composition of the various populations present can usually be eliminated, if comparisons are made between successive observations at the same site. For this reason among others, effects of environmental changes at a particular site may be subject to less random variation than comparisons between sites, and hence can be estimated more precisely — just as in split-plot designs for agricultural experimentation, where effects of treatments applied to sub-plots and their interactions with main-plot treatments can be estimated with increased precision, since large-scale random variation is excluded. The paradigm of this approach is the use of tree-ring measurements, where the single tree serves as the ecosystem sample for repeated but simultaneous observations.

As with contemporaneous comparisons, internal as well as external factors must be taken into account by some multiple-regression or path-coefficient procedure. Difficulties may arise from the fact that repeated observations at the same site are historically linked, and thus not independent. Extra observations may become necessary where historically linked states are being studied, which can be neglected without loss for contemporaneous comparisons.

Deductive Approach

All the types of observation discussed so far have assumed direct measurement of variables in intact ecosystems. An alternative is to deduce the values of ecosystem variables from other measurements made more easily. Admittedly deduced values can never have the certainty of observed values. But observed values may themselves be of value only as representing a whole population of other measurements that might be made. Consequently, the uncertainty attached to deduced values of ecosystem parameters is also associated with observed values, when they are regarded as estimates of population parameters, and confidence limits are generally required for estimates based on observation, just as much as on deduction.

Any ecosystem parameter (e.g., primary productivity) is the resultant of a large number of partial processes — rainfall and water penetration through the soil profile; root growth in the various horizons; water uptake; nutrient mobilization and uptake; shoot growth and leaf development; radiation adsorption and photosynthesis; translocation; use of translocate for reproduction; activities of herbivores, in turn affected by seasonal influences on their development and by predators; and many others. If enough is known about these partial processes, and about their dependence on environmental factors and on other constituents of the ecosystem, in principle one can use knowledge of these contributing factors, of the functions connecting them, and of the initial state of the ecosystem to estimate the value of the primary productivity or other ecosystem parameter. It will often be possible to set confidence limits around the deduced values of primary productivity, if information in the form of a covariance matrix is available concerning the penumbra of uncertainty surrounding each of these values used in the deductive process — whether it is intrinsic variation of values about a mean, or errors involved in a sampling process for estimating them.

Though use of a computer is not essential to this process of indirect, deductive estimation, it is excellently suited to it. The large set of differential equations involved can rarely be solved analytically, and it becomes necessary to solve the equations numerically over short time periods, or to replace them by difference equations. Iterative evaluation of the difference equations then enables one to obtain estimates of fluxes and changes in the system over longer periods. Thus a computer simulation can produce approximate solutions for systems too complex for explicit mathematical treatment.

The use of computer simulation to evaluate the effects of environmental changes can be illustrated by a model of a semi-arid shrub community grazed by sheep, which has been described in detail elsewhere (GOODALL, 1967, 1970). The data are imaginary in the sense that no particular concrete instance is in question, and that most of the parameter estimates are based on general experience rather than *ad hoc* measurement. The whole system, however, is realistic in most of its characteristics and in its behavior in the computer.

In brief, this model envisages the area over which the sheep are free to graze as divisible into several discrete portions varying in vegetation and soil properties, and in grazing pressure. Functions are developed expressing the relations of the various components of the ecosystem — the changes contributing to the soil water budget; plant growth; sheep movement and feeding habits; sheep reproduction, growth and mortality. These functions are set out in the Postscript. The initial state of the system

— composition of the vegetation in each part of the area; numbers, weight and age structure of the sheep population — and meteorological records for the area form the input for the computer program, which proceeds to calculate day by day the changes which the system will undergo. The results are reported at chosen intervals. It has been shown that this model can have practical value as a guide to management practices, such as varied stocking rates, or new watering points or fence lines, as well as contributing to the fundamental understanding of the ecosystem.

Suppose that we are interested in the total productivity of vegetation (gross growth in dry matter, before consumption by herbivores), the net change in biomass of the herbivores, and the effect of temperature upon these measures of ecosystem function. Clearly, this will be far from simple. Temperature changes will affect plant growth directly, but will also affect rates of evapotranspiration and hence soil moisture. Temperature changes also will affect sheep movements (at higher temperatures, they remain closer to water), and so will modify the grazing history of the plants, as well as the food intake by the animals, and thus have another indirect effect on current productivity. The temperature effect will depend on season, and will vary according to rainfall. We consequently must consider a particular time of year and the effects of a given temperature change under prevailing rainfall conditions.

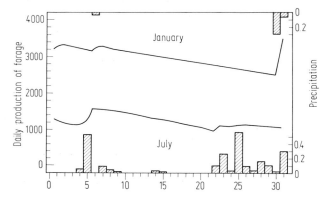

Fig. 1. Simulation by computer of precipitation (histogram) and changes in primary production (continuous graph) during January and July

Fig. 1 illustrates the daily course of forage production as simulated by the computer for two periods of one month in July (Australian winter — meteorological records for these comparisons are taken from Kalgoorlie) and January; in Table 1 are presented the figures for forage productivity and herbivore biomass as affected by a rise and fall of 5° C from the mean monthly temperature for these two months. In each case, the initial state of the ecosystem (quantity of forage of different species, soil moisture, stock numbers and weight) was the same, but ten different random samples of rainfall were taken from records for the month in question. Variations among the forage productivity and herbivore biomass estimates resulting from these variations in rainfall, the only stochastic element in an otherwise deterministic model, enabled confidence limits* to be established for the temperature dependence of the results.

---

* One may express reservations here, for the model is defective in that it does not take into account the undoubted correlation between rainfall and temperature.

Table 1. *Effect of temperature on processes in a computer model of a grazing ecosystem*

	Daily productivity of vegetation (g/ha)						Decrease in total live weight (kg) of 4000 sheep in one month	
	One day		Ten days		One month			
	Mean	SE	Mean	SE	Mean	SE	Mean	SE
July 6.3° C	1241.6	106.1	1449.5	108.0	1499.6	82.9	3389	35.5
July 11.3° C (mean)	1755.4	166.4	2161.7	158.1	2311.7	124.1	2999	51.8
July 16.3° C	2482.2	261.6	3068.2	250.6	3407.8	220.1	2504	64.8
Jan. 20.9° C	3092.8	225.6	2719.4	286.7	3352.9	389.9	2154	51.2
Jan. 25.9° C (mean)	4256.4	395.6	4011.2	433.3	5412.4	656.0	1375	46.6
Jan. 30.9° C	6158.2	602.4	5990.3	688.1	9390.8	1145.7	244	32.8

Mean productivity is expressed for periods of one day, ten days and one month over which these differences in mean temperatures operate, while decreases in herbivore biomass are expressed only as means for the month. The effect of the environmental variables increases with the period over which they operate, as would be expected with any exponential process such as growth.

A synthetic approach to estimation of environmental effects on ecosystem parameters does not enable one to avoid extensive observation or experimentation. On the contrary, it may even increase them, but the measurements shift to a different level. No longer is it necessary to make all observations on intact ecosystems in the field. It is not the response of the *whole* ecosystem to the *whole* environmental complex which is required, but responses of particular parts of it, or partial processes, to small groups of factors. The investigation thus becomes much more practicable to carry out on a small scale and over short periods of time. Some larger-scale tests may be necessary to ensure that no important factors or interrelations have been omitted, and to confirm that the computer model is giving satisfactory predictions; but these may well take the form of "spot checks" rather than a complete field study of the environmental effect in question.

The lack of clearcut meaning associated with the response of an ecosystem to environmental factors, when the environmental factors are themselves a part of the ecosystem, remains. This conceptual difficulty goes beyond the related practical difficulty that the various environmental factors are often linked in such a way that their effects are difficult to disentangle — for instance, sites differing in availability of soil nutrients are also likely to differ in soil moisture; a season of higher precipitation will usually have lower mean temperatures; radiation intensity is difficult to alter experimentally without also affecting temperature and humidity. If higher precipitation in a forest results in an increase in canopy density, which in turn reduces the wind speed within the forest and thus decreases the seed set of anemophilous herb species, is this to be regarded as an effect of the precipitation or of the wind speed? If the ecosystem were a fully discrete entity perhaps one could consider external abiotic factors differently from those internal to, and forming part of, the ecosystem. Ecosystems, in fact, rarely (or never) have clear-cut boundaries, and the degree of mutual influence between the ecosystem and its "exterior" declines gradually away from its nominal limits. In the example above, seed set will certainly be a function both of precipitation and of wind speed external to the forest. But neither of these variables is unaffected by the existence of the forest. We are in a situation where a distinction between dependent and independent variables cannot be upheld.

References

Goodall, D. W.: Computer simulation of changes in vegetation subject to grazing. J. Ind. Bot. Soc. **46**, 356—362 (1967).
— Simulating the grazing situation. In: Concepts and Models of Biomathematics: Simulation Techniques and Methods. (Heinmets, F. Ed.) (in press) 1970.
Scott, D.: Interpretation of ecological data by path analysis. Proc. N. Z. Ecol. Soc. **13**, 1—4 (1966).

Postscript

The Functions Used in the Sheep-Grazing Model

For each day, and each separate region, the soil moisture balance is calculated by computing input and output, the input taking the form of rainfall and surface flow (in flatter, lower-lying areas), the output evapotranspiration and surface flow. Rainfall is a random sample from the empirical distribution of daily rainfall for that month, different distributions being used according to whether or not the previous day had been dry. The functions connecting surface flow and evapotranspiration with daily rainfall are unsophisticated, and could undoubtedly be improved in the light of present-day hydrological knowledge. They are as follows:

$$R = a + bZ + cZ^2 \tag{1}$$

where R is surface flow (positive for run-off, negative for run-on) and Z is daily precipitation, but subject to the restrictions that the curve does not intersect the Z axis (i.e., R takes only one sign for any particular region) and that the positive slope does not exceed unity. Evaporation from an open water surface, V, is calculated by

$$V = a + bT + cT^2 \tag{2}$$

where T is the mean monthly temperature — a gross oversimplification on the broad scale, but an acceptable approximation within a limited region. The actual evapotranspiration, E, is given by

$$E = \max \{0,\ V(1 - e^{bS + cP + d})\} \tag{3}$$

where P is the proportion of ground covered vertically by vegetation, and S is the "available" soil moisture as a proportion of capacity — i.e., if W_1 is the moisture content at field capacity, W_0 the moisture content at the wilting point, and W_i the current moisture content, then

$$S = \frac{(W_i - W_0)}{(W_1 - W_0)} \tag{4}$$

Those parts of the plants that cannot be consumed by the livestock are ignored, and it is assumed that growth of the forage portions can be expressed in terms of a relative growth rate, which is a function of mean temperature and available soil moisture as defined above.

$$G_i = \max \left\{ 0, \left(a + \frac{b_i}{S + c_i} \right) \right\} \cdot (1 + d_i + f_i T)^T \tag{5}$$

where G_i is the relative growth rate (i.e., the daily increase in weight of the forage portion, divided by its initial weight) of the ith species, S and T have the same meanings as before, and the lower-case letters with subscripts are constants peculiar to the species.

It is assumed that distribution of consumption by the sheep over the fenced area is proportional to the product of two preference factors, one based on the location of each region in relation to water and to the boundary (proximity to each of which increases grazing pressure), the other on the density per unit area of species of differing palatability. The locality preference for the jth region, L_j, is given by

$$L_j = e^{a\,\Theta_j + bV A_j} \tag{6}$$

where Θ_j is the mean distance from the fenceline, A_j the mean distance from a water supply, and V is, as above, the evaporation rate (used as an index of the water stress on the animals). The forage preference factor F_j for the same region is given by

$$F_j = \sum_i A_i Q_{ij} \Big/ \sum_i A_i \sum_j Q_{ij} \tag{7}$$

where Q_{ij} is the mean quantity per unit area of the ith species in that region, and A_i expresses its relative palatability.

The amount of "effective" forage in the jth compartment, taking palatability and location into account, will then be

$$\chi_j = L_j F_j U_j \tag{8}$$

where U_j is the area of this compartment. If the total "effective" forage in the whole paddock is

$$D = \sum_j \chi_j \tag{9}$$

then the proportion of grazing time spent in the jth compartment, and of forage consumption taking place there, B_j, is given by

$$B_j = \frac{\chi_j}{D}. \tag{10}$$

The forage consumption by the sheep is taken to depend upon the amount of "effective forage" as defined above, and the consumption C per animal-day of grazing is then

$$C = a(1 - e^{bD}). \tag{11}$$

At higher stocking rates, the density of sheep might also have a direct effect on consumption, but at the very light stocking practiced in this country such an effect is likely to be negligible — that is, at any particular time the mean rate of forage consumption per animal will be almost the same, irrespective of whether a hundred or a thousand sheep are in the paddock.

The consumption per day in the jth region is then NCB_j, where N is the equivalent number of mature sheep in the paddock, immature individuals being weighted according to age. This consumption is distributed among the species according to their palatability. Let the m species be arranged in decreasing order of palatability. Then if Φ_{ij} expresses the ratio in the jth compartment of the weighted quantity of the ith species to all species of the same or less palatability, i.e.

$$\Phi_{ij} = \frac{A_i Q_{ij}}{\sum\limits_{k=i}^{m} A_k Q_{kj}} \tag{12}$$

the proportion Ψ_{ij} of consumption in this compartment which is taken from this species is given by

$$\ln \Psi_{ij} = a_i \ln \Phi_{ij} + b_i (\ln \Phi_{ij})^2 . \tag{13}$$

For the mth (least palatable) species,

$$\Psi_{mj} = 1 - \sum\limits_{k=1}^{m-1} \Psi_{kj} . \tag{14}$$

Use of time units of one day for this section of the calculations sometimes leads to impossible results, where one compartment or species is greatly preferred to others, but the quantity of forage available is small; provision is accordingly made for the time unit for calculation of grazing distribution to be decreased where more than one-tenth of the forage available in any compartment and species would be consumed.

For sheep of known age, the condition is often best expressed by their weight. This largely depends on forage consumption, weighted by factors expressing the nutritional value of the various species. If the nutritional value, X, of the forage eaten per mature animal is given by the expression

$$X = C \sum\limits_i K_i \sum\limits_j B_j \Psi_{ij} \tag{15}$$

where K_i is the nutritional value of the ith species, then the change in weight per animal, H, is

$$\Delta H = a + bX + cH + dHX . \tag{16}$$

The natural mortality, increasing as the condition of the flock deteriorates, is

$$M = a + be^{cH} . \tag{17}$$

For each of these expressions, different values of the parameters apply to the successive monthly cohorts of immature animals.

The reproductive rate per mature sheep per day, Ω, is also taken to be a function of condition as expressed by weight:

$$\Omega = a(1 - e^{bH + c}) \tag{18}$$

where the values of the constants vary with the calendar month.

This completes the set of functions required for the model in its present stage of development.

4

Phenology in Productivity Studies

H. Lieth

Phenology is a scientific field which exists on the borderline between floristics — ecology and meteorology, especially agrometeorology. The field is well established. Its concepts have been continuously applied and discussed for over a century, and its importance in agrometeorology is still increasing. Contemporary phenological studies have a variety of connections to many other scientific disciplines, a wide range of research goals and, therefore, different meanings to individual investigators. Phenology continues to undergo a metamorphosis with time as new applications and techniques are developed.

This chapter is not intended to be a complete review of phenology, although a modern English text would do much to promote this rather important aspect of ecosystem analysis. Probably the best introduction has been written by SCHNELLE (1955). Extensive treatments of the subject also are found in many textbooks and articles concerning ecology and meteorology, e.g., KÜHNELT (1965), WALTER (1960), SCHENNIKOW (1953, 1932), SHAW (1967), DAUBENMIRE (1948, 1968).

In this chapter, we are mainly concerned with the relation between phenology and productivity studies; other aspects of phenology are not emphasized, and interested readers are referred to the pertinent literature for additional information.

The Concept of Phenology

Phenology is generally described as the art of observing life cycle phases or activities of plants and animals in their temporal occurrence throughout the year. These studies permit a phenological calendar to be constructed and superimposed on the astronomic or civil calendar, such that the seasons of the year are not marked by calendar dates but rather by dated groups of phenological events. These events are much more meaningful in describing and explaining seasonal aspects of ecological phenomena. Such has been the application of scientists working in the temperate zones of the earth, for there it is apparent that the year has distinct seasons and that most of the observations in the life cycle of higher plants and animals can be related in one way or another to seasonal changes in the physical environment.

This application is characterized by two different approaches: the descriptive, observational viewpoint and the analytical viewpoint. Both applications have received wide practical use in floristic, ecological and meteorological studies. In most cases, descriptive phenological studies have been based upon morphological, anatomical, or behavioral characteristics which are easy to observe in nature. Furthermore, these studies have generally been restricted to higher plants or migrating vertebrates and insects.

Since the general concept of phenology historically was tailored to demonstrate and explain the more obvious seasonal aspects of higher plants and animals in the temperate zone, modifications in approach are necessary in cases where the lifetime of single individuals is shorter than the vegetation period or in geographical regions like the tropics where climatic changes occur in other than rhythmic, annual sequences. The first case might involve temporal characteristics of subsequent generations of the same population; in the second case, the familiar annual time schedule might be replaced by other periodic intervals.

Most classical phenological observations were basically qualitative; more recently, a quantitative approach has been emphasized (e.g., Morgen, 1949; Higgins, 1952; and Thornthwaite, 1952). This latter approach is called phenometry and is essentially the same as growth analysis, as we will see later. It is obvious that phenometry will have great impact on future productivity studies.

Historical Development

When Bliss and his associates published their recent appeal in BioScience (1967) about the organization of a descriptive phenological study group for the U. S. IBP (observational network and listing of species with desirable characteristics for phenological dating), they were initiating what Hoffman and Ihne (1884) and Linnaeus (1751) had done many years previously for Europe and Sweden, respectively. Another aspect of phenology (analytical), probably initiated by Reaumur (1735), has attempted to correlate the time required for agricultural crops to ripen with the

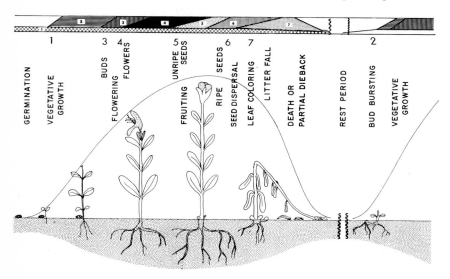

Fig. 1. The life cycle of a species and its use in phenology. A selection of growth and development phases (phenophases) in the typical life cycle of a flowering herbaceous plant. The patterned strip above the phenophases represents one method utilized to present phenological observations over a life cycle. The numerals in the various wedges correspond to the phenophases described in Figs. 3 and 8. This form of presentation has been described extensively by Schennikow (1932). The curved line indicates the accumulation of dry matter and symbolizes the phenometric approach applied to productivity studies as presented in Figs. 4 and 5

"heat sum" received during their growing period. The term "phenology", however, was not coined before the middle of the 19th century when suggested (1859—60) by CHARLES MORREN, a botanist from Lüttich (SCHNELLE, 1955; ABBE, 1950).

Hence, we see that the two main aspects of phenology, the descriptive and the analytical, have had a long and common history. Phenology was soon realized to be an important tool for agricultural and economic planning. Beginning with the early

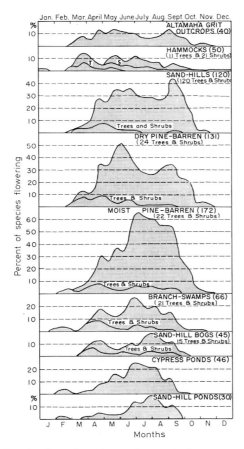

Fig. 2. Phenograms for the flowering phase of several plant communities in the Altamaha Grit (Georgia and Carolinas, USA) constructed from data collected by HARPER (1906). This figure permits comparison between the spring-fall bimodal curve of dry lands with the strongly pronounced summer maximum of wet-land communities. Values in parentheses denote the number of species comprising each phenogram

elaboration of seeding, flowering and harvesting calendars, phenology evolved voluminous services which resulted in impressive maps about the geographical patterns (isophenes) of certain phenological events. Phenology also developed into a tool which was able to use a selected number of widely distributed plants, together with a network of meteorological studies, to forecast a variety of environmental

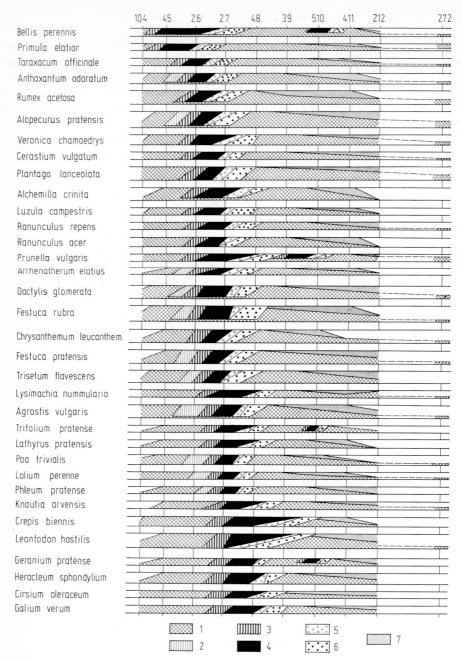

	10.4.	4.5.	2.6.	2.7.	4.8.	3.9.	5.10.	4.11.	2.12.	27.2.

Bellis perennis
Primula elatior
Taraxacum officinale
Anthoxantum odoratum

Rumex acetosa

Alopecurus pratensis

Veronica chamaedrys
Cerastium vulgatum
Plantago lanceolata

Alchemilla crinita
Luzula campestris
Ranunculus repens
Ranunculus acer
Prunella vulgaris
Arrhenatherum elatius

Dactylis glomerata

Festuca rubra

Chrysanthemum leucanthem.

Festuca pratensis

Trisetum flavescens

Lysimachia nummularia

Agrostis vulgaris

Trifolium pratense
Lathyrus pratensis
Poa trivialis

Lolium perenne
Phleum pratense
Knautia arvensis

Crepis biennis

Leontodon hastilis

Geranium pratense
Heracleum sphondylium
Cirsium oleraceum
Galium verum

1   2   3   4   5   6   7

Fig. 3. Phenological spectrum of an oatgrass association *(Arrhenatheretum elatioris)* in Poland taken from JANKOWSKA (1967). The phenodynamic strips are depicted in the same manner as in Fig. 1. The phenophase key is: 1) vegetative stage; 2) production of culms; 3) flower and inflorescence buds; 4) blossoming; 5) unripe seeds and fruits; 6) ripe seeds and dispersal; 7) yellowing of leaves. Only species with conspicuous phenophases with sharp temporal limits are suitable for phenological dating. For example, in *Leontodon hastilis* the start of flowering is appropriate, but the end of this phenophase is indistinct

properties or events. Quantitative relationships of the time between seeding and harvest of many crops, started by Reaumur (1735) and continued by several scientists
thereafter (Coutagne cited by Marie-Davy, 1883), finally developed the level of so-
phistication that enabled Thornthwaite (1952) to elaborate his phenological slide
rule. This last achievement was possible through detailed growth analyses of certain
crops — long a common practice in plant physiology — which was called "phenometry" by Kaempfert (1948) and Morgen (1949) in its adaptation to phenological
studies. Phenology is today a well-established tool in agrometeorology and ought to
become a common ground in IBP for discussion between ecologists and meteoro-
logists.

Phenological Observations in Productivity Studies

From discussion of the concepts of phenology and phenometry, it is obvious that
there are many aspects of productivity studies which can be categorized, predicted
and evaluated on the basis of phenological attributes. Growth rates, accumulation
and turnover of standing crop, and energy in their temporal aspect are but a few
examples. We need the appropriate date for many important biological events, e.g.,
beginning and end of the growing season (if any dormant period occurs), germination,

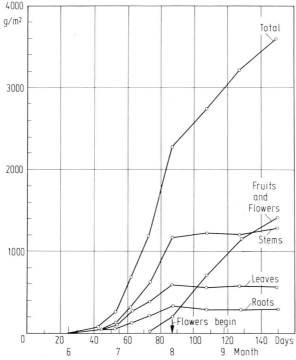

Fig. 4. A growth analysis study of sunflowers in Hohenheim, Germany (after Kreh, 1965).
Total dry matter accumulation is divided into plant parts which partly correspond with
phenophases (leaves, flowers and fruits). Such data are the basis for many production studies.
The abscissa denotes the number of days after sowing; the beginning of each calender month
of 1965 and opening of the first sunflower head are indicated. The ordinate axis gives dry
weight in g/m². Sunflowers were planted at a density of 8 plants per m²

seed set, mating, migration. All of these data involve phenology or phenometry, and it would be judicious to utilize the many important contributions of previous phenological studies in the analysis of ecosystems.

Terminology

To more easily interpret the phenological literature, it is necessary to understand the terms employed (Fig. 1). Each distinguishable phase within the life cycle of a species is called a "phenophase". Phenologists select for their observations the beginning or the end of phenophases — especially for organisms in which changes occur over a very short time period. The way in which the entire sequence of phenophases occurs throughout the year is called the "phenodynamic". Accounting for the percentage or absolute number of individuals of one species or number of species in a community entering a given phenophase allows one to construct "phenograms" (Fig. 2). Elaboration of the phenodynamic for each species in a community and their presentation in one comparative table is called the "phenological spectrum" (Fig. 3).

The investigation of quantitative changes within one phenophase (e.g., size or weight) is called "phenometry". This is demonstrated in Figure 1 by the curve which shows the fluctuation in dry matter of living vegetation during the growing season.

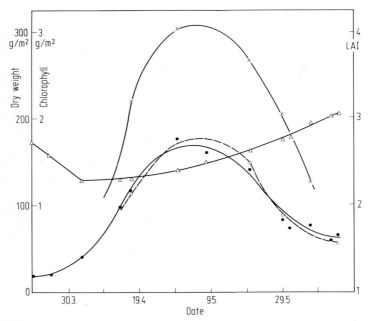

Fig. 5. Phenometric studies on an *Allium ursinum* stand in Hohenheim, Germany (after Medina and Lieth, 1964). Development of leaf area and chlorophyll content was measured concurrently with dry matter production. Abscissa — date; ordinate — dry matter in g/m², chlorophyll content in g/m², and leaf area index (LAI) in m/m². Symbols for the respective curves are: ● = aboveground dry matter, △ = belowground dry matter, ○ = chlorophyll content, × = photosynthetic surface. Such studies permit interpretation of the impact of certain morphological and physiological characters on the growth of plants. More data of this kind should be obtained in phenological studies

In phenometric studies, total dry matter is often divided into various categories (e.g., leaves, stems, roots, flowers and fruits) which often coincide with plant parts for which we distinguish different phenophases (Fig. 4); or other quantitative criteria can be measured, like chlorophyll content and leaf area index (Fig. 5).

The cartographic evaluation of phenological observations is usually done by drawing lines through geographic locations entering the same phenophase at the same time. These lines are referred to as "isophenes" (Fig. 6, see legend for further explanation).

Phenological Approaches to Productivity Studies

The Determination of the Growing Season

To conduct productivity studies in the field it is usually necessary to define both the beginning and end of the growing period in relation to the calendar year. The first green tillers in early spring or the bursting of leaf buds mark the beginning of the vegetation period much better than any given date of the astronomic calendar. Autumnal leaf coloring and abscission, on the other hand, mark the end of the vegetation period. Biological characteristics also demarcate periods of flowering,

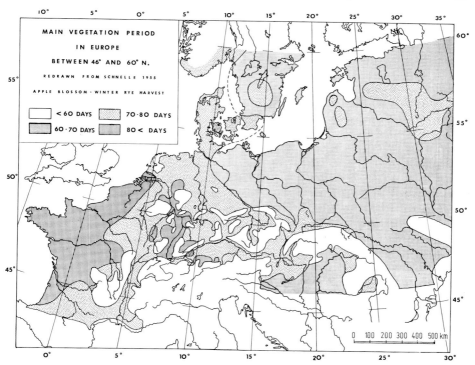

Fig. 6. Phenologically based map of the length of the main vegetation period in northern Europe (after SCHNELLE, 1955, map 4). The map is constructed from observations of two phenophases — apple blossom and winter rye harvest. The boundaries on this map, although they were derived originally from phenological observations, do not represent isophenes. Phenological determination of the vegetation period is one of the most useful tools for geographical evaluation of productivity

fruiting, and other important phenophases. We find information about these facts in the literature in great abundance. Although such statements are qualitative with regard to the productivity, they are already quantitative with regard to the cumulative effect of certain environmental factors on the organism's development. This is the essence of many agricultural phenological studies. The synthesized results can be presented in various ways, depending upon their ultimate application:

    a) in the form of tables or graphs covering one or more years for specific locations and individual species or populations,

    b) in the form of tables or graphs covering an entire year for each species of a plant community,

    c) as phenological profiles for spatial comparison of individual species or phenologically identifiable seasons,

    d) as phenological simultan maps for large-area comparisons of specific phenological events or combined extrapolations of several phenological characters.

Fig. 3 shows the phenological spectrum of a Polish meadow as modified from JANKOWSKA (1967). This same volume also contains a phenological spectrum of a deciduous forest by KAZMIERCZAKOWA (1967). Phenograms of the flowering times of higher plants in various communities of the Southeastern United States (after HARPER 1906) are presented in Fig. 2. Fig. 6 illustrates the construction of a phenological map based upon the main vegetation period in Central Europe (redrawn from SCHNELLE, 1955).

Life Cycles and Duration of Phases

Besides determination of the effective growing season, it is most important in productivity studies to know how different phenophases are related to each other as well as to changes in environmental factors. In most cases, development of a phenological spectrum (Fig. 3) yields an initial understanding of which species have concurrent phenophases and thus may be competitive. Another example of the importance of phenophase studies in primary production is the development of the herbaceous understory of temperate deciduous forests during early spring; here the species of two different strata (herbaceous and canopy) have complementary phenophases for vegetative growth. It also is obvious that the phenophases of herbaceous-feeding animals must be in phase with the development of their host plants. The birth dates of gazelles in the Cairo Zoo show close correlation to the rainy season (which determines food availability) in their country of origin (Fig. 7). Although phenological observations do not yield quantitative productivity estimates, they should be used conceptually in scheduling the timing for collection of data in ecosystem studies. This viewpoint leads then to another level of phenological studies: phenometry.

Phenometry and Growth Analysis

Phenometry is the quantitative analysis of the life cycle of an organism or certain specific phenophases and their correlation with environmental influences. Growth analysis is the quantitative study of the growth relations of organisms when environmental factors are held constant. Phenometry and growth analysis are very similar; only their goals are slightly different. Growth analysis is mainly concerned with physiological processes or plant properties; phenometry searches for environmental relationships, using plant growth as a microclimatic indicator.

The best analysis of growth is possible under controlled environmental conditions (WENT, 1957; EVANS, 1963). Growth chambers permit study of plant growth and development under well-defined conditions, with analysis of various environmental factors singly or in combination. But analysis of the productivity of an ecosystem is compounded by the inherent variability in growth rates of different species, each with potentially different growth habits and responses to changing environmental conditions. Under such conditions the phenometric approach is the only demonstrable solution which has proven practical, especially when the objectives of ecosystem research include analysis of time-dependent phenomena under environmental influence.

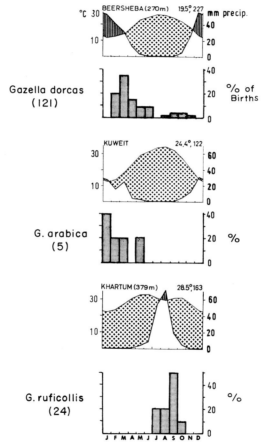

Fig. 7. The relation between phenophases of producers and consumers, i.e., the birth time of gazelle fawns in the Cairo Zoo in relation to the rainy season of their native country. The abscissa is demarcated in months; the ordinate axis for gazelle birth is percent of total birth- (in parentheses), while each division on the climate diagram represents 10° C in mean monthly temperature or 20 mm of rainfall. The heavier line is the rainfall distribution curve; the thinner line is the monthly mean temperature. Dotted areas indicate dry seasons; hatched areas mark humid seasons of the year. Mean annual temperature (° C) and cumulative precipitation are indicated in the upper right of each figure (data for gazelles after FLOWER 1942; climate diagrams from WALTER and LIETH, 1961—1967)

The next stage is the field experiment where the development of a single species is examined under natural climatic conditions. Such measurements have been done routinely for many years, e.g., Blackman and Black, 1959; Monsi and Saeki, 1953; Monsi, 1960. Agricultural and ecological literature are replete with results of growth analyses.

The gradual shift in emphasis from a growth chamber model to community analysis can be summarized as follows:

    a) growth chamber analysis: (plant growth) + (controlled environment) — (competition),

    b) field experiment: (plant growth) — (controlled environment) — (competition),

    c) community analysis: (plant growth) — (controlled environment) + (competition).

Conventionally, a) and parts of b) are classified as growth analysis, while c) and aspects of b) are phenometry. Figs. 4 and 5 show examples of the phenometric approach

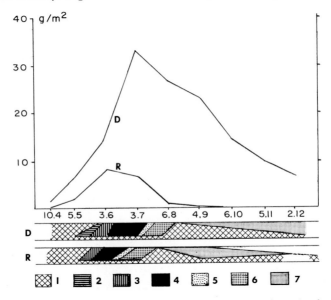

Fig. 8. Phenometric development curve and the phenodynamic strip for two species (D = *Dactylis glomerata* and R = *Rumex acetosa*) after Fig. 3 (from Jankowska, 1967). Abscissa, time of the vegetation period; ordinate, dry matter production in g/m². Phenophases 1—7 correspond with those in Figs. 1 and 3. The figure shows that, for these two species, maximum accumulation of biomass coincides with flowering

applied to the productivity of sunflowers and *Allium ursinum* in field plots in Hohenheim. In each case, total dry matter is subdivided into portions contributed by each major plant part. Each species component of a plant community can be studied in this same manner; then the wedge-shaped descriptions of phenological spectra can be replaced with more appropriate growth and development curves. Fig. 8 (redrawn from Jankowska, 1967) shows this for two species of the meadow depicted in Fig. 3. Similar data for at least the most important species of a given community provide the necessary quantitative information needed in food chain studies.

Correlation between Environmental Factors and Phenological Events

Most phenological studies imply correlation of the observed or measured facts with selected environmental factors measured independently. This correlation provides the basis for predicting certain important ecological events. Agricultural meteorology relies upon this phenological approach, generally to develop plant hardiness zones and to delineate crop zones for new varieties. In the execution of productivity studies, where instantaneous and cumulative production estimates and balances are required, relationships must be established between environmental stimuli and organismal responses. Delay of growth can result from abnormal cold or dry spells,

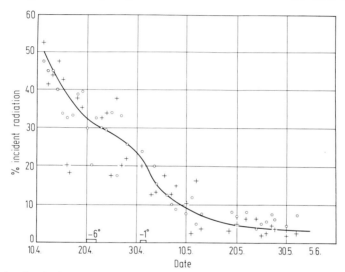

Fig. 9. Reduction in light intensity reaching the floor of a deciduous forest in Hohenheim, Germany, as a result of canopy leaf development (after LIETH, 1960; LIETH and ASHTON, 1961). The delay in leaf development due to two late frosts at the end of April is evidenced by irregularities in the exponential decline of light on the forest floor. The values above bars on the abscissa in late April give the minimum temperature (° C) for each frost period. Incident radiation (measured with continuously-recording light meters) reaching the forest floor is expressed as a percentage of that above the canopy (+ and ○ symbols represent replicate measurements)

and sudden increases in plant development often are correlated with favorable environmental conditions. Fig. 9 illustrates such an example in which two cold spells during early spring delayed leaf development of a forest canopy, as reflected in the decrease in light intensity reaching the forest floor.

The phenological and phenometric approaches become increasingly valuable when secondary producers are incorporated into the ecosystem analysis. The same environment that affects development of the food base often influences the life cycle of the animal consumer. Entomological literature contains studies demonstrating such relationships, and modern field studies of animal populations make consistent use of phenological techniques. Although WATT's (1968) book on quantitative ecology does not list the term phenology in the index, this book virtually overflows with phenological philosophy.

Parameters to be Measured

The first problem to be solved in any phenological study is what are the most important parameters to be observed. If the objectives of ecosystem analyses are descriptions of the cycle of organic matter, then the seasonal periodism in environmental variables and plant and animal populations should be examined. Pertinent variables with phenological dependencies are outlined below:

I. The Physical Environment
   a) light: including intensity, photoperiod, and spectral composition,
   b) temperature: daily, seasonal and annual fluctuations (mean, maximum and minimum), late and early frosts,
   c) water conditions: including rainfall patterns (amount, intensity and duration), vapor pressure deficit, and available soil moisture,
   d) nutrient levels: static and transient quantities of mineral nutrients as well as C, N, P and S, chemical pollutants.

II. Plants
   a) all life cycle phases (phenophases): germination and/or bud bursting, leaf development, flowering time, fruiting time, fruit or seed dispersal, plant or leaf death, litterfall,
   b) phenometric analyses: leaf area index, chlorophyll content, photosynthetic activity, respiration rates, organic productivity, energy and nutrient composition.

III. Animals
   a) all life cycle phases: birth, growth and development, mating, emigration, immigration, and death,
   b) feeding periods and food habits,
   c) phenometric analyses comparable to those of plants.

The Information Gained

Obviously, the application of phenological principles to productivity studies will necessitate greater field labor and overall effort. Therefore, initially it may be necessary to limit phenometric analyses to the most important species in an ecosystem. Such expenditures, however, eventually will be repaid in increased predictability of our environmental models. If the objectives of an ecosystem study are simply biomass turnover, one might question the need for such detailed measures of productivity as fluctuations in plant growth rates during the growing season. The importance of such measurements, however, cannot be overestimated for the following reasons.

First, harvest at the end of a growing season yields only an estimate of net production. The differences between net and gross production — those quantities consumed, respired or otherwise lost from the system — can often be calculated only from differences of short-term field measurements obtained during phenometric studies. Secondly, when we correlate primary productivity with solar radiation to determine ecological efficiencies, it is important to know whether the logarithmic growth phase occurred during a hot, dry period or under optimal growing conditions. Finally, when relating the activities of animal consumers to their physical environment and food supply, detailed information is needed on the amount of food available, and at what time and under what environmental conditions it was present.

Table 1. *Information from phenological and phenometric analyses useful in ecosystem productivity studies*

Floristic

Duration of growth and vegetation period
Absolute length of each life cycle and the comparison of phenophases among different species
Appropriate bases for calculation of ecological efficiencies
Quantitative data on the growth and development of each species
Production and turnover of biomass throughout the year

Faunistic

Migration patterns of various animals
Quantitative data about the development of populations in time and space
Relations of population growth to food resources

Environmental

Correlations with macroclimatic variables
Correlations with microclimatic variables
Environmental influences resulting from combined climatic and edaphic characteristics

Biospheric

Phenological maps
Phenological profiles
Relations (predictive) between various phenological events and environmental conditions

---

These points may elucidate why seasonal fluctuations in productivity and consumption are as important to understanding production processes in an ecosystem as is the net annual organic-matter balance. The phenologic-phenometric approach is in many cases the only one which will yield values for time-dependent variables required for system modeling. In Table 1 are summarized the synthetic and predictive capabilities of phenology and phenometry in the analyses of ecosystems. Some of the generalizations which can be synthesized from phenological and phenometric interpretations have been demonstrated earlier. One more example may demonstrate the validity of the phenological approach for understanding geophysical and biological cycles of materials.

In 1963, BOLIN and KEELING published their investigations on the seasonal fluctuations in atmospheric CO_2 content over the earth's surface. Their map suggested that photosynthetic activity and distribution of vegetation had a great impact upon this fluctuation. CZEPLAK (1966) substantiated this inference by using the world productivity map of LIETH (1965) (see map inside the front cover), a simple model of seasonal release and uptake of CO_2 by vegetation, and a first approximation of the length of vegetation periods through climate diagrams (WALTER and LIETH 1961—67). The results of CZEPLAK's calculations are compared with the measurements of BOLIN and KEELING in Fig. 10. Comparison of the two isopleth diagrams shows reasonable agreement, although the phenological model was not completely satisfactory.

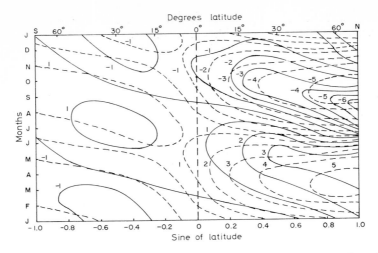

Fig. 10. Seasonal variation in CO_2 concentration of the atmosphere as a function of latidude and its relation to seasonal photosynthetic activity of vegetation and long-distance diffusion. Lines in the diagram are isoplethes, deviations from the latitudinal mean for each month. Solid lines are isoplethes according to the measurements of BOLIN and KEELING (1963); the dashed lines are computed values for impact of vegetation (after CZEPLAK, 1966). Numerical values in the figure give deviations from the model in ppm CO_2

Phenological Modeling

Since phenology is closely related to meteorology, a field which utilizes many mathematical techniques, it is not surprising that many mathematical interpretations and extrapolations have been made of phenological data. Such an example is COU-TAGNE's "law of growth" cited by ABBE (1905). COUTAGNE developed the formula

$$v = a \cdot e^{-\left(\frac{x-c}{n}\right)^2}$$

for the momentary rate of plant development, where v = optimum development of a plant (in a given climate), a = the coefficient of development ($1/a$ = longevity), e = the natural log base, x = plant temperature, c = the temperature for which the most rapid development of the plant is obtained, and n = a coefficient which describes the sensitivity of the plant to temperature. Integration of this equation yields the total growth time of the plant at any given temperature, x. The modification of the logistic equation is evident here and in later models developed by MONSI and SAEKI (1953), KIRA et al. (1956), LIETH (1963), von BERTALLANFFY (1951), and WATT (1968). THORNTHWAITE (1952) developed COUTAGNE's model into an outstanding application of phenometric modeling, the crop slide rule, by which the civil calendar can be converted into a climatic (phenological) calendar predicting the growth performance of different agricultural crops. The real value of such phenometric approaches to productivity and turnover studies is the predictive capability of models under varying environmental conditions.

An example of graphic modeling in meteorological phenology is given by SCHNELLE (1955) who attempted to enlarge the scope of phenologically-computed

development times as calculated from meteorological events. BERG (1940) had previously defined the continentality of an area by the simple equation

$$K = \frac{C + M}{M}$$

where K = continentality, C = endurance time of continental air masses, and M =
endurance time of maritime air masses. SCHNELLE (1955), in a profile through Central
Europe, correlated percent continentality with phenological vegetation periods

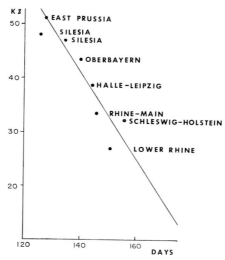

Fig. 11. Correlation between the continentality of a climate and the length of the vegetation
period (after SCHNELLE, 1955). Length of the vegetation period (days) is defined by the time
span between flowering of the snowdrop *(Galanthus nivalis)* and the harvest of winter rye
(Secale cereale). Vegetation periods calculated from this relationship average a little over
2 months longer than those obtained from Fig. 6. Pilot research in each region is necessary
to demonstrate which plants and phenophases are the best indicators for the vegetation
period (for further explanation see the text)

(similar to Fig. 6) of the number of days between the flowering of *Galanthus nivalis*
and the harvest of winter rye. Fig. 11 demonstrates the linear relationship between
vegetation period and continentality. This relationship, however, has only limited
value in the analysis of ecosystem processes.

A first practical phenological model for geophysical productivity research is the
mathematical formulation used by CZEPLAK (1966) to calculate the sink and source of
CO_2 for the world. He used for the Northern Hemisphere (about 50° N) a simple
cosine function which is phased for a maximum source in January and a maximum
sink in July. The maximum values in atmospheric CO_2 for this region, however,
have been reported in April and May while the minimum values occur during
September (BOLIN and KEELING, 1963; CZEPLAK, 1966). If the cumulative release and
uptake of CO_2 for a productive land surface in this region is calculated phenometri-
cally, a function is obtained similar to curve 3 in Fig. 12a. The shape of this curve
is much more similar to the annual fluctuation of atmospheric CO_2 concentration for

this region as measured by Bolin and Keeling (1963) (Fig. 12b). This similarity suggests that their assumption is correct and that the land vegetation is the primary factor influencing this fluctuation.

In the future, graphical interpretations will have to be replaced by stochastic computer models to synthesize a complexity of variables and give rapid results. But such models are no better than the data upon which they are based. More accurate and extensive data will be required than are presently available. Productivity studies within

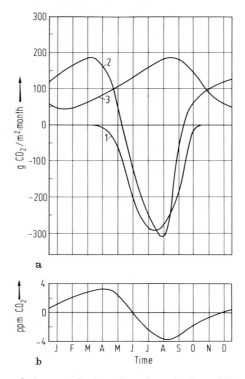

Fig. 12. The influence of photosynthesis and total respiration of biota upon carbon dioxide fluctuations in the atmosphere around 50° N latitude is presented as an example of graphic modeling of phenometric data. a) shows the sequence of steps in graphic calculation of the cumulative change in atmospheric CO_2 input and output during the year. The ordinate axis represents grams of CO_2 per m^2 taken up or released by the biota; negative values refer to uptake, while positive represent release. Curve 1 describes photosynthetic activity (Lieth, 1962) and curve 3 represents total respiration (after Lieth, 1963) — both for central Europe. Curve 2 shows the cumulative impact of biota on the CO_2 of the atmosphere throughout the year. The curve was constructed by arbitrarily beginning in January and calculating cumulative totals of released and consumed CO_2 (cumulative values from curves 1 and 3). The resultant curve was then rebalanced to effect zero net annual change (equal areas above and below the horizontal zero line). — b) This diagram illustrates the annual observed CO_2 fluctuation in ppm about the mean latitudinal CO_2 concentration at 54° N (taken from Czeplak, 1966). The predicted curve from biological measurements (part a) agrees quite well in general form, although maxima and minima in both figures are about 15 days out of phase. Quantitative comparison between curve 2 in a) and the curve in b) will be possible only when calculations can be made of the air mass volume and its dilution factor for CO_2 exchanged with the landscape

the IBP framework hopefully will provide us with these data. The analysis of an ecosystem without consideration of phenological principles should be untenable to analytical ecologists.

Acknowledgements

I wish to acknowledge the help of Mrs. SUSAN HENDERSON and Miss JUNE FOUSHEE in preparing this manuscript.

References

ABBE, C.: A first report on the relations between climates and crops. U. S. Department of Agriculture, Weather Bureau, Government Printing Office, Washington, D. C., Bull. No. 36, W. B. No. 342, 386 p. (1905).

BERG, H.: Die Kontinentalität Europas und ihre Änderung, 1928/37 gegen 1888/97. Ann. Hydr. **68**, 124 (1940).

BLACKMAN, G. E., and J. N. BLACK: Physiological and ecological studies in the analysis of plant environment. XI. A further assessment of the influence of shading on the growth of different species in the vegetative phase. Ann. Bot. **23**, 51—63 (1959).

BLISS, L. C.: Phenology program of the IBP. BioScience **17**, 712—714 (1967).

BOLIN, B., and C. D. KEELING: Large-scale atmospheric mixing as deduced from the seasonal and meridional variations of carbon dioxide. J. Geophys. Res. **68**, 3899—3920 (1963).

CZEPLAK, G.: Modellrechnung über den meridionalen turbulenten Großaustausch von Kohlendioxyd und Ozon in der Atmosphäre. Meteorologisch-Geophysikalisches Institut, Mainz: Johannes Gutenberg-Universität, 72 p. June 1966.

DAUBENMIRE, R.: Plant communities: A textbook of plant synecology. New York: Harper and Row 1968.

— Plants and environment: A textbook of plant autecology. New York: John Wiley and Sons 1948.

EVANS, L. T. (Ed.): Environmental control of plant growth. New York: Academic Press 1963.

FLOWER, S. S.: Notes on the recent mammals of Egypt, with a list of the species recorded from that Kingdom. Proc. Zool. Soc. Lond. **1932**, 369—450 (1932).

HARPER, R.: A phytogeographical sketch of the Altamaha Grit region of the Coastal Plain of Georgia. Ann. N. Y. Acad. Sci. 17, 1—414 (1906).

HIGGINS, J. J.: Instructions for making phenological observations of garden peas. Johns Hopk. Univ. Lab. Climatol. Publ. Climatol. **5**, 1—11 (1952).

HOFFMANN, H., and E. IHNE: Phänologischer Aufruf, Gießen 1884.

JANKOWSKA, K.: Sezonowe zmiany roslinnosci i produkcja pierwotna netto w placie taki *Arrhenatheretum elatioris*. Zaklad Ochrony Przyrody Polskiej. Akademii Nauk, Studia Naturae — Seria A **1**, 153—173 (1967).

KAEMFERT, W.: Zur Phänometrie. Wetter und Klima. **1**, 40—52 (1948).

KAZMIERCZAKOWA, R.: Runo lasu bukowego *Fagetum carpaticum*, jego fenologia i ekologia produkcji pierwotnej. Zaklad Ochrony Przyrody Polskiej Akademii Nauk, Studia Naturae — Seria A, **1**, 95—114 (1967).

KIRA, T., H. OGAWA, K. HOZUMI, H. KOYAMA, and K. YODA: Inst. Polytech., Osaka City, Univ., Ser. D., Biol. **7**, 1—14 (1956).

KREH, R.: Untsuchungen über den Aufbau und die Stoffproduktion eines Sonnenblumen-bestandes. Zulassungsarbeit zur wissenschaftlichen Prüfung für das höhere Lehramt, Tech. Hochsch. Stuttgart, 1965.

KÜHNELT, W.: Grundriß der Ökologie unter besonderer Berücksichtigung der Tierwelt. pp. 81—87. Jena: VEB Gustav Fischer 1965.

LIETH, H.: Die Stoffproduktion der Pflanzendecke, 156 pp. Stuttgart: Gustav Fischer 1962.

— Über den Lichtkompensationspunkt der Landpflanzen I und II. Planta **54**, 530—576 (1960).

— Mathematische Probleme in der Biologie. I—IV, Praxis der Mathematik, **2—4**. Köln: Aulis Verlag Deubner u. Co. (1960—1963).

LIETH, H.: The role of vegetation in the carbon dioxide content of the atmosphere. J. Geophys. Res. **68**, 3887—3898 (1963).

— Versuch einer Kartographischen Darstellung der Produktivität der Pflanzendecke auf der Erde, pp. 72—80. In: Geographisches Taschenbuch. 1964/65. Wiesbaden: Franz Steiner 1965.

—, and D. H. ASHTON: The light compensation points of some herbaceous plants inside and outside deciduous woods in Germany, Canad. J. Bot. **39**, 1255—1259 (1961).

LINNAEUS, C.: Philosophia Botanica. Stockholm: Kiesewetter 1751.

MEDINA, E., and H. LIETH: Die Beziehungenz wischen Chlorophyllgehalt assimilierender Fläche und Trockensubstanzproduktion in einigen Pflanzengemeinschaften. Beitr. Biol. Pflanzen **40**, 451—494 (1964).

MONSI, M.: Dry matter production in plants. I. Schemata of dry matter production. Bot. Mag. Tokyo **73**, 81—90 (1960).

—, and T. SAEKI: Über den Lichtfaktor in den Pflanzengesellschaften und seine Bedeutung für die Stoffproduktion. Jap. J. Bot. **14**, 22—52 (1953).

MORGEN, A.: Phänometrie des Flächenwachstums lebender Pflanzenblätter. Wetter u. Klima **2**, 1—15 (1949).

REAUMUR, R. A. F. DE: Observations de thermometre, faites a Paris pendant l'annee 1735, comparees avec celles qui ont ete faites sons la ligne, a l'Isle de France, a Alger, et en quelques-unes de nos isles de l'Amerique. Mem. Paris Acad. Sci. **1735**, 545 ff. (1735).

SCHENNIKOW, A. P.: Pflanzenökologie, 380 pp. Berlin: Deutscher Bauernverlag 1953.

— Phänologische Spektra der Pflanzengesellschaften, pp. 251—266. In: Handbuch der biologischen Arbeitsmethoden. (E. ABDERHALDEN, ED.) Abtlg. XI. **6,2**, 251—266 (1932).

SCHNELLE, F.: Pflanzen-Phänologie. (Probleme der Bioklimatologie Bd. III). Leipzig: Akad. Verlagsges. 1955.

SHAW, R. H. (Ed.): Ground level climatology. Washington, D. C.: AAAS Publications 1967.

THORNTHWAITE, C. W.: Climate in relation to planting and irrigation of vegetable crops, pp. 290—295. In: Proc. VIII th Gen. Assem. — XVII th Congr. Internat. Geographical Union, Washington 1952.

VON BERTALLANFFY, L.: Theoretische Biologie. Bern: A. Francke 1951.

WALTER, H.: Grundlagen der Pflanzenverbreitung, pp. 103—140. In: Einführung in die Phytologie, Bd. III. Stuttgart: Eugen Ulmer 1960.

—, and H. LIETH: Klimadiagramm-Weltatlas. Jena: VEB Gustav Fischer 1961—1967.

WATT, K. E. F.: Ecology and resource management: A quantitative approach. New York: McGraw-Hill Book Co. 1968.

WENT, F. W.: The experimental control of plant growth. Waltham, Massachusetts: Chronica Botanica Co. 1957.

5

Biomass and Productivity Models of Forest Canopies

H. A. I. Madgwick

Knowledge of the amount, distribution and functioning of foliage within canopies is important in understanding the productivity of plant communities because photosynthesis forms the basis for energy capture and dry matter growth. Most forest trees maintain their productive structure through systems of branches which, in themselves, contain a considerable fraction of dry matter produced.

Models of canopy productivity or function are necessary if we are to fully understand forest growth. Thus, we know from seedling studies that the total dry matter production of young trees over the growing season may be drastically affected by the relative distribution of photosynthate to leaves, stems and roots. The distribution of photosynthate within forest stands and methods of manipulating this distribution to maximize harvests are largely unknown.

The purpose of this paper is to review information on forest canopies with emphasis on models underlying the estimation of canopy biomass and productivity.

Estimating Biomass

Two main methods have been used as bases for estimating stand biomass of foliage and branches. In several studies it has been assumed that the tree having "mean dimensions" for the stand also has mean canopy weight. "Mean dimensions" have variously been interpreted as including either mean diameter or mean basal area and, in some cases, mean height. Others have sampled a number of trees in the stand, and used regressions of logarithm canopy weight ($\log W$) on logarithm stem diameter breast height ($\log D$) for estimating weights of branches and leaves on remaining trees, i. e.,

$$\log W = a + b \log D \tag{1}$$

where a and b are constants (Kittredge, 1944)*.

In recent years the use of average trees has come under more or less vigorous attack. A number of writers have noted the mounting evidence that trees which are average for one characteristic such as basal area, tend not to be average in other respects such as canopy weights (Ovington and Madgwick, 1959; Madgwick, 1963; Baskerville, 1965; Attiwill, 1966; Attiwill and Ovington, 1967). The same authors, however, tend to accept the logarithmic regressions without critical evaluation.

---

* All logarithmic transformations in this paper are to base e.

Use of the Mean Tree

On small plots of *Pinus densiflora* and *Cryptomeria japonica* Satoo (1966) found that the trees of mean basal area gave stand values of foliage mass ranging from 96 to 102 percent of those based on clear felling. More recently, Ovington, Forrest and Armstrong (1968) tested the mean tree approach by clear felling a one-hundred-tree stand of *Pinus radiata* and weighing each tree. They found that using the average weight of at least five trees with basal areas closest to the mean for the stand yielded stand estimates of branch and foliage biomass within ± 7 percent of the values obtained by weighing all trees.

Although it appears that "mean" trees may prove a reasonable method for estimating canopy biomass, this method has disadvantages. An understanding of the functioning of the forest canopy as a photosynthesizing machine involves a knowledge of the changing distribution of foliage among trees of different class from dominant to suppressed as the stand develops. Obtaining information for mean trees alone yields none of the information necessary. Furthermore, there is no basis for testing whether the particular stand under study departs from the underlying assumption that trees of average dimensions are also of average weight.

Use of Logarithmic Regressions

The use of logarithmic regressions of the form of Eq. (1) for estimating canopy weights would appear to overcome the objections raised concerning the use of mean trees. However, such regressions should not be accepted without question.

In several instances, sample trees from a number of different stands have been combined to calculate regression constants for estimating canopy weights from tree dimensions (e.g., Storey, Fons and Sauer, 1955; Storey and Pong, 1957; Cable, 1958; Loomis, Phares and Crosby, 1966). The validity of combining data for a number of stands is supported by Baskerville (1965) who found that stand density from 1,700 to 12,000 stems per hectare had no effect on the relationship between logarithm diameter and logarithm canopy weights of *Abies balsamifera*. In contrast Satoo (1962, 1965) and Shinozaki et al. (1964a) found significant differences in the regression constants between stands of each of a number of species including *Pinus densiflora*, *Betula* spp. and *Cryptomeria japonica*. Between-stand differences are also obvious in regressions both for *Pinus banksiana* and for *Pinus strobus* plotted by Kittredge (1944). Furthermore, Satoo (1962) in a study of four *Pinus densiflora* stands found that applying the regression for one stand to another could give estimated foliage biomass from —17 to 179 percent of the values derived from regressions for the stands themselves.

Satoo (1966) has emphasized that the values of the constants in Eq. (1) for any species are dependent on competition within the stand. Thus, equations relating logarithm canopy weight and logarithm diameter based on sampling trees from a variety of stands may be inaccurate unless some measure of competition also can be included.

Shinozaki and co-workers (1964a, b) have theoretically justified and empirically demonstrated the value of a logarithmic relationship between diameter at the base of live crown and either branch or foliage weight. Their conclusions are supported by the work of Storey, Fons and Sauer (1955), Storey and Pong (1957) and Loomis,

PHARES and CROSBY (1966), who found that logarithm crown weight was more closely related to logarithm diameter at base of crown than to logarithm dbh. The concept of using diameter at base of live crown is similar to that of using diameter at the base of each live branch suggested by CUMMINGS (1941) and used by ROTHACHER, BLOW and POTTS (1954) and ATTIWILL (1962). Diameter at base of live crown is more expensive to measure accurately, and the question of finding biologically sound hypotheses for relating leaf or branch mass to more easily measured tree dimensions and plot conditions requires investigation.

Fitted regressions using the logarithmic transformation of weight yield an estimate of the geometric, rather than arithmetic, mean weight for trees of a given diameter. The geometric mean underestimates the arithmetic mean, the degree of bias being affected by the statistical distribution of canopy weights (FINNEY, 1941; PATTERSON, 1966). Assuming that the logarithm weights are normally distributed, FINNEY (1941) has shown that an efficient estimator of the population arithmetic mean would be

$$e^{\bar{x} + 1/_2 s^2} \left[1 - \frac{s^2 (s^2 + 2)}{4n} + \frac{s^4 (3s^4 + 44s^2 + 84)}{96 n^2} + \cdots \cdots \cdots \right]$$

where $\bar{x}$ and s^2 are the estimated mean and variance of the transformed data and n is the number of observations. Terms involving higher values of n can probably be ignored in estimating canopy weights. No published estimates of tree weights based on logarithmic regressions have been found which reportedly take into account the above bias.

An additional problem in estimating canopy weight using regressions arises from the alternative assumptions which may be made in calculating regression coefficients. LIU et al. (1966) illustrate the effects of acknowledging measurement errors in the regressor variable which result in larger absolute values of slope. The consequent effect on estimated stand weights appears not to have been studied.

Tests of stand estimates based on logarithmic regressions against clear felling have been reported by SATOO (1966). In each of three stands the logarithmic equations overestimated stand foliage biomass by between one and nine percent. Incorporating an estimate of bias due to the use of logarithmic transformations would increase the degree of overestimate.

Estimates of stand canopy mass based on logarithmic regressions are usually heavier than those based on trees of mean basal area (OVINGTON and MADGWICK, 1959; BASKERVILLE, 1965; ATTIWILL, 1966; SATOO, 1966). It is usually assumed that the difference between the two methods results from biased underestimates using trees of mean basal area. But the difference also could be due to overestimates using the regression technique.

Error Estimates

Published estimates of stand weights do not include error terms. Similarly, few authors present information on the accuracy of predictions of individual tree weight from estimating equations. Unpublished data for separate regressions for eighteen sets of data for *Pinus resinosa* and *P. virginiana* indicate a standard error of estimate of the logarithm of needle or branch weight of about 0.30 with branch weight rather less accurately predictable than foliage. Thus the 95 percent confidence interval of the logarithm of individual tree foliage or branch weights would be ± 0.60.

Decoding from logarithms requires taking the antilog of 0.60 and converting the addition or subtraction to multiplication or division. Thus the 95 percent confidence interval ranges from 55 to 180 percent of the estimated mean. Regressions based on combining data from a number of different stands have been shown to be less accurate except where diameter base of live crown was used as a regressor variable. LOOMIS, PHARES and CROSBY (1966) working with *Pinus echinata* from a number of stands found standard errors of estimate of logarithm foliage and branch weights to be 0.463 and 0.665, respectively, when diameter breast height was used as regressor variable. Basing estimates on diameter base of live crown reduced the standard errors to 0.143 and 0.244 for foliage and branch-wood, respectively. These error estimates are in logarithmic units, so that using a method which halves the standard error represents relatively greater improvement in the accuracy of arithmetic predictions.

Estimating Productivity
Foliage

The annual production of foliage has been estimated both from values of standing biomass and litter fall. Estimates from standing biomass usually assume that annual production is equal to the weight of one-year-old leaves on the trees at the end of the growing season. Such an assumption overlooks any decline in the weight of individual leaves at the end of the growing season, losses due to insect feeding and, where long-period estimates are of interest, year to year fluctuations in foliage production (BRAY and GORHAM, 1964). Thus, using the weight of leaves on trees at the end of the growing season to calculate annual foliage production will tend to lead to biased underestimates. Foliage production estimates from litter traps involve similar assumptions, and also are dependent on the type of litter trap employed (KERR, personal communication).

Branches

Branch biomass reflects a complex of factors including annual production, death and decay rates. Thus, accurately estimating branch productivity of stands is very difficult.

Branch production may be estimated by combining bole diameter growth measurements with regressions of branch weights on bole diameter. This method assumes that regressions are unaffected by changing tree age and, for periods of a few years, may give reasonable estimates of the net change in branch mass but neglects loss of dead branches, so underestimating total production.

More realistic estimates of branch productivity may be obtained from detailed subsampling of branches within the canopy (WHITTAKER, 1965). Such an approach may be illustrated with detailed branch data for twenty *Pinus sylvestris* of OVINGTON and MADGWICK (1959). The arithmetic average weights of each branch whorl for five different tree diameter classes are plotted in Fig. 1. There is a similar pattern of branch growth for each tree class, with a rapid, almost logarithmic phase for the first four years. The whorls of the smallest diameter class (Class 5) appear to have a slightly faster initial growth rate but this is an artifact produced by the gradual slowing of growth of the leaders of these trees as they become suppressed. After the fourth year the growth rate drops remarkably and between the seventh and twelfth years branch weight varies relatively little. As expected, within the stand the larger trees have the greatest branch mass on any particular whorl.

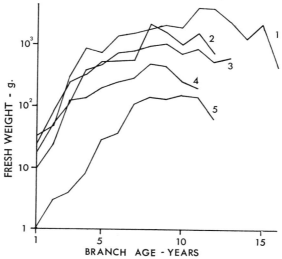

Fig. 1. The arithmetic mean weights of branch whorls based on four trees of five different dbh classes in a plantation of *Pinus sylvestris*. The number near each line refers to the relative diameter of the trees which ranged from 16.2 cm (labeled 1) near the maximum for the stand to 6.5 cm (labeled 5) near the minimum

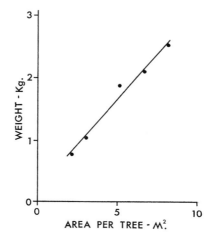

Fig. 2. The annual production of branch wood on trees of mean basal area as related to the inverse of stand density. Each point represents an average of three trees per stand

Since relatively little death of branches occurs in the upper whorls, the growth curves in Fig. 1 may be used to estimate annual increments of branch weight within the stand. Thus the annual weight increment of a particular branch whorl can be estimated by subtracting the weight of the next youngest whorl. It follows that the

4*

average weight of the branch whorl in the stable plateau region (whorls 7 to 12 in the present case) is an estimate of the total annual branch production of the tree, if steady state conditions are approximated.

Between stands, the estimated weight of branch production in trees of average diameter is markedly affected by density (Fig. 2). Average land area per tree (the inverse of density) has been used as a measure of crowding. From these limited data branch production per hectare was independent of density and averaged 3.1 metric tons per hectare fresh weight. Partial data on moisture content suggests a dry weight production of 1.4 metric tons per hectare. The effect of age on this relationship is not known.

The analysis of branch production for these *Pinus sylvestris* is simple because their canopies had a regular structure and the sample trees had had a relatively uniform, undisturbed growth. Severe weather conditions and defoliation by insects can drastically affect the initial development and subsequent growth of branches. Such effects will have to be considered in the construction of more general models.

Canopy Models

One important goal of the study of forest canopies is an understanding of the photosynthetic process within the forest. Foresters have been interested in the relative contribution of different parts of the tree canopy to the growth of stems (LADEFOGED, 1946; MOLLER, 1960; HALL, 1965). Most such studies have been based on pruning various fractions of the live crown with less regard to the amount of foliage removed.

A realistic production model of a forest canopy must incorporate information on canopy morphology, physiology and environment, which together regulate the quantity of foliage in the canopy and determine the photosynthetic capacity of the canopy. SAEKI (1960) proposed a simple model involving the quantity and orientation of leaves on the stand, the relationship between light intensity and photosynthesis and light distribution within the canopy. An empirical test of SAEKI's model by PEARCE, BROWN and BLASER (1967) indicated agreement within about 10 percent between predicted and measured rates of photosynthesis. DE WIT (1965) has extended SAEKI's model and provided Fortran programs for simulating photosynthesis of plant canopies using digital computers.

MADGWICK (1968) has proposed a model to explain the distribution of foliage within *Pinus resinosa* crowns. The weight of needles on a new branch whorl is directly related to the height growth of the tree. Under unrestricted conditions, the weight of needles developing on the same whorl in subsequent years increases exponentially at an empirically determined rate of 160 percent. However, under forest conditions shading increases with subsequent canopy development. So, as the branch whorl ages, growth falls short of that in unrestricted conditions. Individual needle development is reduced in proportion to shading by the foliage mass above until the weight of an individual needle at the base of the canopy weighs only about 43 percent of individual needles near the tree apex. Tree nutrition affects needle longevity. On potassium deficient trees nearly all needles died within the first three years growth. On trees adequately supplied with potassium needle life was considerably extended and many needles lasted five years.

Mathematical modeling of canopy growth cannot be divorced from the more general problem of growth of the whole plant. BLACKMAN (1919) and BRIGGS, KIDD and WEST (1920) postulated simple growth models of plants in which growth rate was expressed as the change in weight per unit plant weight and leaf amount, respectively. Numerous studies have been based on these early models and have led to many useful ideas concerning factors affecting plant growth (WATSON, 1968).

Recent work by NEWHOUSE (1968) suggests a more complex model for plant growth with several, interacting, factors contributing to growth. NEWHOUSE not only included the "unit leaf assimilation rate" of BRIGGS et al. (1920) but also noted the effects of seed size, differences in the pattern of distribution of photosynthate within the plant, and leaf area: weight ratios, on plant growth. From a study of seedling growth of 14 hardwood species he concluded that distribution of photosynthate within the plant and leaf area: weight ratios varied systematically throughout the growing season and among species. These factors can be combined in a set of mathematical equations to simulate plant growth but variations in net assimilation rates, apparently related to weather conditions, are so large that agreement between simulated and actual growth is low.

Conclusion

While the development of mathematical models of plant growth are currently in their infancy important advantages of this type of research can be seen. Model plants may be manipulated to suggest ways of increasing productivity so providing basic information for the "genetic engineering" of plants. Secondly, the effects of environmental conditions on plant growth can be simulated so leading to a clearer view of critical periods of plant development and suggesting improved cultural practices to increase productivity. Finally, a conceptual framework is provided for further study of the ecology and physiology of plant growth.

References

ATTIWILL, P. M.: Estimating branch dry weight and leaf area from measurements of branch girth in Eucalyptus. For. Sci. 8, 132—141 (1962).
— A method for estimating crown weight in *Eucalyptus*, and some implications of relationships between crown weight and stem diameter. Ecology 47, 795—804 (1966).
—, and J. D. OVINGTON: Determination of forest biomass. For. Sci. 14, 13—15 (1968).
BASKERVILLE, G. L.: Estimation of dry weight of tree components and total standing crop in conifer stands. Ecology 46, 867—869 (1965).
BLACKMAN, V. H.: The compound interest law and plant growth. Ann. Bot. 33, 353—360 (1919).
BRAY, J. R., and E. GORHAM: Litter production in forests of the world. Adv. Ecol. Res. 2, 101—157 (1964).
BRIGGS, G. E., F. KIDD, and C. WEST: A quantitative analysis of plant growth. Ann. Appl. Biol. 7, 103—123 and 202—223 (1920).
CABLE, D. R.: Estimating surface area of ponderosa pine foliage in central Arizona. For. Sci. 45—49 (1958).
CUMMINGS, W. H.: A method for sampling the foliage of a silver maple tree. J. For. 39, 382—384 (1941).
FINNEY, D. J.: On the distribution of a variate whose logarithm is normally distributed. J. Roy. Stat. Soc. Ser. B 7, 155—161 (1941).

Hall, G. S.: Wood increment and crown distribution relationships in red pine. For. Sci. 11, 438—448 (1965).

Kittredge, J.: Estimation of the amount of foliage on trees and stands. J. For. 42, 905—912. (1944).

Ladefoged, L.: The productive importance of the individual parts of the crown in spruce. Det. forstl. Forsögsv. Dan. 16, 365—400 (1946).

Liu, T. K., R. T. Odell, W. C. Etter, and T. H. Thornburn: A comparison of clay contents determined by hydrometer and pipette method using reduced major axis analysis. Soil. Sci. Soc. Amer. Proc. 30, 665—669 (1966).

Loomis, R. M., R. E. Phares, and J. S. Crosby: Estimating foliage and branchwood quantities in shortleaf pine. For. Sci. 12, 30—39 (1966).

Madgwick, H. A. I.: Nutrition research: some problems of the total tree approach. Soil. Sci. Soc. Amer. Proc. 27, 598—600 (1963).

— Some factors affecting the vertical distribution of foliage in pine canopies, pp. 233—245. In: Primary productivity and mineral cycling in natural ecosystems (H. E. Young, Ed.). Orono: Univ. Maine Press 1968.

Newhouse, M. E.: Some physiological factors affecting seedling growth of hardwoods. Ph. D. Thesis, Virginia Polytechnic Institute, Blacksburg, Virginia, pp. 69 (1968).

Ovington, J. D., W. G. Forrest, and J. S. Armstrong: Tree biomass estimation, pp. 4—31. In: Symposium on primary productivity and mineral cycling in natural ecosystems, (Young, H. E., Ed.) Orono: Univ. Maine Press 1968.

—, and H. A. I. Madgwick: Distribution of organic matter and plant nutrients in a plantation of Scots pine. For. Sci. 5, 344—355 (1959).

Patterson, R. L.: Difficulties involved in the estimation of a population mean using transformed sample data. Technometrics 8, 535—537 (1966).

Pearce, R. B., R. H. Brown, and R. E. Blaser: Photosynthesis in plant communities as influenced by leaf angle. Crop. Sci. 7, 321—324 (1967).

Rothacher, J. S., F. W. Blow, and S. M. Potts: Estimating the quantity of tree foliage in oak stands in the Tennessee valley. J. For. 52, 169—173 (1954).

Saeki, T.: Interrelationships between leaf amount, light distribution and total photosynthesis in a plant community. Bot. Mag. Tokyo 73, 55—63 (1960).

Satoo, T.: Notes on Kittredge method of estimation of amount of leaves of forest stand. J. Jap. For. Soc. 44, 267—272 (1962).

— Production and distribution of dry matter in forest ecosystems. Misc. Inform. Tokyo Univ. For. 16, 1—15 (1965).

Shinozaki, K., K. Yoda, K. Hozumi, and T. Kira: A quantitative analysis of plant form — the pipe model theory. I. Basic analysis. Jap. J. Ecol. 14, 97—105 (1964a).

— — — — A quantitative analysis of plant form — the pipe theory. II. Further evidence of the theory and its application in forest ecology. Jap. J. Ecol. 14, 133—139 (1964b).

Storey, T. G., W. L. Fons, and F. M. Sauer: Crown characteristics of several coniferous species. U. S. For. Ser. Div. Fire Res., AFSWP-416, 1—95 (1955).

—, and W. Y. Pong: Crown characteristics of several hardwood tress species. U. S. For. Serv., Div. Fire Res. AFSWP-968, 1—36 (1957).

Watson, D. J.: A prospect of crop physiology. Ann. Appl. Biol. 62, 1—9 (1968).

Whittaker, R. H.: Branch dimensions and estimation of branch production. Ecology 46, 365—370 (1965).

Wit, C. T. de: Photosynthesis of leaf canopies. Versl. Landbouwk Onderz. Ned. No. 663, 1—57 (1965).

6

A Synthesis of Studies by the Harvest Method: Primary Production Relations in the Temperate Deciduous Forests of Japan

TAISITIROO SATOO

In Japan many studies on primary forest production utilizing the harvest method have been published since 1952 which include various forest types ranging from boreal conifer forests of northern Japan to tropical forests of Cambodia and Thailand, and from pine groves of the seashore to the creeping pine carpets of alpine timberline. These studies, which present data on more than 300 stands were reviewed recently by KIRA and SHIDEI (1967) and recorded in a recent IBP handbook (NEWBOULD, 1967). SATOO (1968b) also has synthesized the existing data on woodlands of *Pinus densiflora*. Since most of these investigations have been in the interest of silviculture, data for broadleaved deciduous forests, which are of less commercial value, are not often compared with coniferous forests. Almost all of them lack information on root and undergrowth biomass, and data for primary production are very limited. Because of steady progress in methodology, these data are not always based on the same method, even for the same investigator(s). This makes it difficult to compare the existing data, and, if we select the data by method, data obtained by one and the same method are too scarce to be synthesized. Thus, the contents of this synthesis are inevitably of more or less tentative nature; yet, it is hoped that this summary may be of use as a starting point for more advanced studies.

Temperate Deciduous Forests of Japan

According to HONDA (1912), the temperate deciduous forest zone occupies the northern half of Honshu and southern half of Hokkaido, and is limited roughly by the isothermal lines of 6 and 13° C for the mean annual temperature. In the southern half of Japan this type of forest occupies higher elevations. The climax of this zone is represented by beech, *Fagus crenata*, forests, but contains profuse deciduous broadleaved species and sometimes is mixed with conifers. Birches are among the important pioneer species.

Estimation of Biomass and Production

Estimation of biomass by the harvest method is the basic procedure of studies of production relations. Since methods of estimation are described in detail in the recent IBP handbook (NEWBOULD, 1967), I shall review only those methods commonly used in Japan. Clear-cutting and measuring all trees in a reasonably large area of forest is laborious and is neither practical nor permitted under many circumstances. Also, sampling of too small an area of a forest may result in biased and erroneous estimates for the total stand. Clear-cutting is normally used for comparison with and standardization of other indirect methods of estimating biomass. Estimation of biomass by clear-cutting the trees of a given area (W) is described by:

$$W = \Sigma w \tag{1}$$

where w is the value for each harvested tree. Most studies have been made by harvesting a large number of sample trees, weighing them, and converting such values as quantity of stem, branches and leaves into a unit ground area basis with the aid of a stand table. The procedure consists of two steps: (1) harvesting and measuring the sample trees and (2) constructing a stand table for conversion of data to a ground area basis. It is necessary that both series of measurements are not only accurate but easy and practical.

Many methods of estimating the volume of timber in a forest, which are described in textbooks of forest mensuration, could be applied, but three have been generally used. One method is harvesting trees of mean cross-sectional area for the stand, and either multiplying the mean values for the sample trees ($\overline{w}$) times the number of trees per unit ground area of the site (N), or multiplying the sum of the values for the sample trees (w) times the ratio of basal area (G) to the sum of cross-sectional area of sample trees (g). Thus,

$$W = N \cdot \overline{w} \tag{2}$$

or

$$W = \Sigma w \left(\frac{G}{\Sigma g} \right) \tag{3}$$

where W is the value per unit ground area. This method is simple, but it is not always possible to secure a sufficient number of average trees. A variation of this approach is to stratify trees in the study area into size groups.

The second method of estimating biomass is the random harvesting of sample trees of varying size classes which represent the whole range of tree diameters in the study area. Multiplying the sum of the values for the harvested sample trees (w) times the ratio of basal area of the study area (G) to sum of cross-sectional area of sample trees (g) yields W. This method also is described by Eq. (3).

The third method involves harvesting trees of varying size sufficient to cover the whole range of tree diameters in the study area and determining the equation representing the relation between a dimension of tree and the values for the sample trees. Then the production values for each size class (w') are calculated with the equation, multiplied times the number of trees per unit area of corresponding size class (n) and the products are summed:

$$W = \Sigma (w'n) . \tag{4}$$

When weights of whole trees, or parts thereof, and a stem dimension of individual trees are plotted on double logarithmic coordinates, the trends are linear and can be represented by simple regression equations. These relations can be used for indirect estimation of biomass of trees of each size class. This method is called allometry (KIRA et al., 1960). For a tree dimension, stem diameter at breast height (d) or the square of the diameter multiplied against height (d^2h) generally has been used; d^2h is a substitute for stem volume. When diameter is used as the dimension,

$$w' = ad^b \tag{5}$$

where a and b are constants. This method is an application of a classical method for estimation of stand stem volume known as BERKHOUT's equation (PRODAN, 1965), and was proposed by KITTREDGE (1944) for the estimation of leaf biomass. This method also is used for branch and stem biomass, increment of stem and branch production, etc. (SATOO et al., 1956). Limitations of this method have been discussed elsewhere (SATOO, 1962, 1966). In the case of leaf biomass, the constants of Eq. (5) change systematically with intensity of competition (SATOO, 1962).

While the construction of a stand table is comparatively easy, securing the necessary regression data can be an exceedingly laborious task. Since the constants of Eq. (5) differ among stands (SATOO, 1962, 1966; KIRA and SHIDEI, 1967), tree height (h) can be introduced as a second variable. The equation is then rewritten as:

$$w' = a(d^2h)^b . \qquad (6)$$

This is an application of a method for estimation of stand stem volume known as SPURR's equation (PRODAN, 1965). By introducing stem height (h), stem, branch and root (but not leaf) dry weight of trees from forests of widely different types can be expressed by one type of regression equation (OGAWA et al., 1965; KIRA and SHIDEI, 1967). Although Eq. (6) is excellent in describing the relations obtained from harvested trees, this equation is not necessarily as good for estimating stand biomass. Exact measurement of tree height in forests is very laborious and often difficult. If h is calculated from the relation between d and h as determined from cut trees or from the measurement of parts of standing trees (OGAWA et al., 1965), Eq. (6) is nothing but a function of d, though not always of the same form as Eq. (5), and does not serve to introduce another parameter. In the case of leaf biomass, constants of Eq. (6) also change with the intensity of competition (TADAKI, 1966).

Since the constants of Eq. (5) and (6) differ among stands for leaf biomass, stem diameter at the joint of the lowest living branch (or cross-sectional area of the stem) has been proposed as a more significant predictive variable (YAMAOKA, 1958; SHINOZAKI et al., 1964; KIRA and SHIDEI, 1967). Using this approach, leaf biomass of sample trees from forests of widely different types has been described by the same regression equation (KIRA and SHIDEI, 1967). This method is good in theory; it is difficult, if not impossible, in practice. The difficulty arises in construction of the stand table, another important step in the estimation. To measure stem diameter at the joint of the lowest living branch for every standing tree in the study area is usually impossible. If the diameter at this specific position of the stem is not measured directly and is only estimated from the relation to dbh on sample trees (YAMAOKA, 1958), then this method is nothing but the original equation using dbh. The relation between dbh and stem diameter at the lowest living branch is dependent on taper of the bole and differs stand by stand, causing a serious error (SATOO and SENDA, 1966). However, this method may be useful in the continuous estimation of leaf biomass of fixed trees in nondestructive sampling.

An example of the comparison of estimates by these methods with clear-cutting is given in Table 1. Regardless of the method, relative error of estimation was generally less than 5 percent, although methods using different tree size classes, instead of average trees, appeared to offer some advantage. Total net production is estimated by summing the production of each component of the forest. Stem production of the

Table 1. *Comparison of estimates of biomass by different methods for a 22-year-old stand of Betula ermanii (source: 17)*

Method	Stem metric t/ha	Branch metric t/ha	Leaf metric t/ha	Above- ground[a] metric t/ha	Leaf area ha/ha	Increment of stem metric t/ha/year
Clear cutting	50.83	8.92	2.77	62.52	5.55	5.28
Average trees:						
cross-sectional area	*51.61*	7.71	*2.69*	*62.01*	4.53	4.67
number of trees	*50.71*	7.58	*2.65*	*60.91*	4.45	4.61
Trees of different size:						
cross-sectional area	*49.78*	*8.80*	*2.75*	*61.08*	5.87	*5.05*
ad[b]	*51.11*	*8.49*	*2.77*	*64.15*	5.68	*5.09*
a(d²h)[b]	*51.55*	8.33	*2.73*	*63.90*	5.82	*5.09*

Note: Figures *in italics* are within 5% of relative error: the difference from the value of clear-cutting divided by the value of clear-cutting.

[a] Calculated from the aboveground biomass of each tree, not the sum of the first three columns.

tree layer is determined from stem analysis as volume and converted into dry weight with bulk density. Increment of branch growth sometimes has been determined by branch volume analysis (similar to the stem analysis) from subsamples, or by multiplying the biomass of branch size class by the relative growth rate of cross-sectional area of each size class. Some investigators have assumed that the relative branch growth rate is the same as that of stem. Other investigators have assumed that the allometric relation between stem dimensions and biomass does not change over the period of a few years, and "estimate" increment in branch or whole aboveground parts from the difference between biomass determined with the same allometric equation and stem dimensions at the harvest and one or two years before.

However, not only are these assumption unsupported by evidence, but there are some evidences against them (Whittaker, 1965; Ando et al., 1968). Although there are some published figures on net production based on such assumptions, these figures have not been used in this synthesis. Estimated production of leaves is approximated by the existing leaf biomass of deciduous species, but these estimates fail to account for loss during the growing season. We still do not have proper methods to estimate root production, and mere assumptions sometimes are made. Production values obtained for sample trees are converted into a unit ground area basis comparable to aboveground biomass.

Thus, there are many sources of bias in the estimation of production. Harvest usually is made at the end of a growing season, but many leaves already have been shed (Fig. 8). The same situation applies to twigs. In many cases, flowers, fruits, seeds, leaf scales and such minor components are neglected and cause underestimation (Ovington, 1963). Grazing by insects also is usually neglected, although it is common for large amounts of insect frass to be found in litter traps in apparently healthy woodlands. All these factors cause underestimation of production. It is necessary to compensate for these biases by collecting litter and frass during a growing season before the harvest.

The crown of harvested trees is usually separated into layers or strata (Fig. 9) and processed and weighed separately, as the characters of leaves and their environment differ gradually from the top to bottom of the crown.

Estimation of biomass and production by the shrub layer is made in the same way as for the tree layer. Estimation of biomass and production of grass and the herbaceous layer is made by harvesting subsamples in the study area, as in the studies of grassland communities.

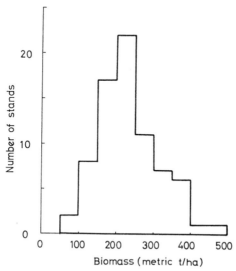

Fig. 1. Biomass of the tree layer of mature beech *(Fagus crenata)* forests in Japan (sources: 2, 12, 13, 14, 18). Multiply values by 100 for g/m^2 or metric tons/km²

For conversion of respiration losses by nonphotosynthetic tree tissues into unit ground area bases, which is necessary for estimation of gross production, the allometric method is also useful (KIMURA, 1960; YODA et al., 1965).

Total Biomass

Biomass of the tree layer of mature forests of *Fagus crenata* is shown in Fig. 1. The majority of mature beech forests had tree-layer biomasses between 150 and 250 metric

Table 2. *Distribution of biomass among components of pioneer forests in Hokkaido, Japan (Betula platyphylla, B. ermanii, B. maximowicziana, and Populus davidiana).* Sources: 5, 15, 17, 20

	Biomass metric t/ha	Leaf %	Branch %	Stem %	Under- growth %	Number of stands sampled
mean	47— 65	4.7	11.7	77.5	6.1	2
range		4.3—5.1	9.7—13.8	76.4—78.5	3.4—8.8	
mean	104—158	1.8	10.5	85.7	2.0	6
range		1.6—1.9	6.7—14.0	80.1—90.5	0.2—4.3	

tons per hectare. However, the total biomass itself is not a measure of productivity in perennial plant communities such as forests. Biomass is rather a function of age, because stem biomass which is accumulated year by year constitutes an increasingly greater part of total community biomass (Table 2, Fig. 2). Table 2 shows the distribution of biomass among the components of pioneer forest communities. In Fig. 2,

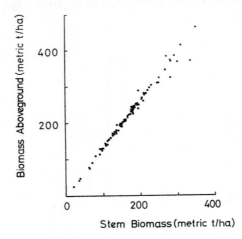

Fig. 2. The relation between stem biomass and total aboveground biomass of beech forests *(Fagus crenata)* (sources: 2, 12, 13, 14, 18)

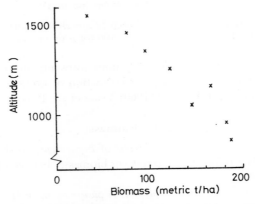

Fig. 3. Total aboveground biomass change in beech forests *(Fagus crenata)* along altitudinal series (modified from Yamada, 1955)

total biomass of the tree layer of forests of *Fagus crenata* is plotted against stem biomass. Total aboveground biomass of the tree layer shows a linear increase with increasing stem biomass. Total biomass averages 1.3 times stem biomass. Yamada (1955) followed the change along altitudinal series of growing stock of climax forests of *Fagus crenata* in central Japan. His mean values of the stem volume at each altitude were converted into dry weight with the bulk density of wood and then multiplied by 1.3 to obtain total biomass. Total aboveground biomass decreased with altitude (Fig. 3).

Leaf Biomass

While stem biomass is a function of age and branch biomass also is affected by age and stand density (SATOO, 1968b), leaf biomass of closed stands does not appear to be

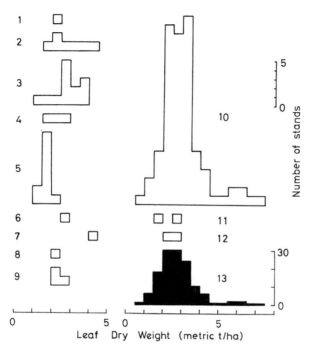

Fig. 4. Leaf biomass of the tree layer of deciduous broadleaved forests in Japan (individual species and total stand values). The numbers in the figure correspond to the numbers below. *1 Populus davidiana; 2 Betula platyphylla; 3 B. ermanii; 4 B. maximowicziana; 5 Alnus inokumai; 6 A. hirsuta; 7 A. sieboldiana; 8 Quercus serrata; 9 Q. crispula; 10 Fagus crenata; 11 Zelkowa serrata; 12 Fraxinus mandshurica; 13* all species. (sources: 1, 2, 3, 4, 5, 6, 8, 9, 10, 11, 12, 13, 14, 15, 17, 18, 20)

influenced by many factors (MÖLLER, 1945). SATOO (1966, 1967) reported leaf biomass of closed stands of *Cryptomeria japonica* to be influenced only by site quality. Leaf biomass can not increase infinitely with forest growth, and there must be a limit specific to species or groups of ecologically similar species (SATOO, 1966). Existing data of leaf biomass of the tree layer of Japanese deciduous broadleaved forests are graphically presented in Fig. 4 and 5. Leaf dry weight per hectare ranged between 1 and 5 metric tons and many forests were between 1.5 and 3.5 metric tons. The leaf area per unit area of ground, or leaf area index, ranged between 2 and 5.5. In Fig. 6, dry weight of leaves per unit ground area is compared for birch and beech forests of Europe and Japan. Differences in leaf biomass between European and Japanese species are minor. For other genera of deciduous trees, we have not yet

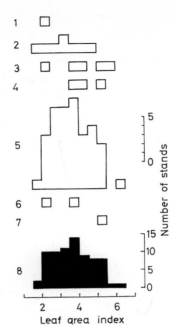

Fig. 5. Leaf area index (LAI, leaf area per unit ground area) of the tree layer of deciduous broadleaved forests in Japan. The numbers in the figure correspond to the numbers below. *1 Populus davidiana; 2 Betula platyphylla; 3 B. ermanii; 4 B. maximowicziana; 5 Fagus crenata; 6 Zelkowa serrata; 7 Fraxinus mandshurica; 8* all species (sources: 8, 9, 10, 11, 12, 14, 15, 16, 17, 18, 20)

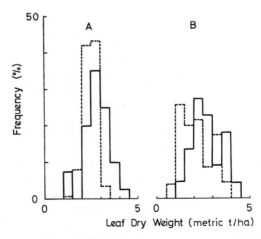

Fig. 6. Leaf biomass of beech and birch forests of Europe and Japan. A beech; B birch (solid lines, Japan; broken lines, Europe). Sources: beech, Europe (166 stands), 23; beech, Japan (40 stands), 14; birch, Europe (23 stands), 21, 22, 24, 25, 26, 27, 28; birch, Japan (21 stands), 1, 11, 15, 16, 17, 20

accumulated enough data for such comparison. Mean values of existing data on leaf dry weight for various types of forests in Japan are illustrated in Table 3. It may be seen that deciduous forests have less leaf biomass than do evergreen forests.

Table 3. *Leaf biomass of pure stands in Japan* (metric t/ha)

		Species	Number of stands sampled	Leaf biomass
Evergreen	coniferous	*Abies* spp.[a]	9	14.9
		Chamaecyparis obtusa	7	13.2
		Pinus thunbergii	12	9.9
		Pinus densiflora	108	6.6
	broadleaved	*Camellia japonica*	4	8.6
		Castanopsis cuspidata	11	8.5
Deciduous	coniferous	*Larix leptolepis*	15	3.9
	broadleaved	*Fagus crenata*	86	3.1
		Betula ermanii	10	2.7
		Betula spp.[b]	21	2.7

[a] *Abies sachalinensis, A. veitchii*, and a mixture of *A. veitchii* and *A. mariesii*, altogether.
[b] *Betula ermanii, B. platyphylla*, and *B. maximowicziana*, altogether.

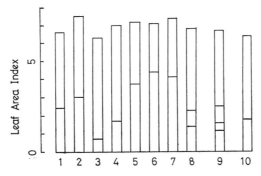

Fig. 7. Leaf area index of all layers of deciduous forest communities (1—8) and other types of forests (9, 10) in Japan. Each section of the column of the figure represents each layer, top to bottom: crown of the tree layer to the ground vegetation. The numbers in the figure correspond to the numbers below. *1 Betula platyphylla + Sasa apoiensis*, source: 20; *2 B. platyphylla + S. paniculata*, source: 15; *3 B. ermanii + S. paniculata*, source: 20; *4, 5, 6 B. ermanii + S. kurilensis*, source: 16; *7 B. ermanii and B. platyphylla + S. kurilensis*, source: 16; *8 Fagus crenata* + many species, source: 18; *9 Larix leptolepis* + many species, source: 19; *10 Cinnamomum camphora* + many species, source: 7. *1—7* and *10* tree layer + ground vegetation; *8* tree layer + shrub layer + ground vegetation; *9* tree layer + understory of hardwood tree layer + shrub layer + ground vegetation

Few data exist on leaf biomass of temperate deciduous forest communities which include undergrowth; data available are presented in Fig. 7 and compared with some other forest types of Japan. Biomass of undergrowth was negligible in the total biomass of forest communities (Table 2), but leaf area of undergrowth is a significant part of total leaf area. While leaf area of the layer varies stand by stand, the leaf area index of the whole community is within a very narrow range of 7.

As previously reported (Satoo, 1968b), in plots of *Pinus densiflora* leaf biomass nearly as large as that of mature woodlands can be attained even at very young ages, if sufficient plant density occurs. This is also true for deciduous forest species. Tadaki and Shidei (1960) reported leaf biomasses as large as 435 g/m² of ground surface (4.35 metric tons/ha) in three-year-old plots of *Ulmus parvifolia*. Kawahara et al.

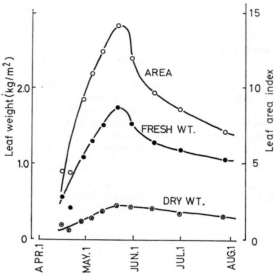

Fig. 8. Seasonal change of leaf biomass and leaf area index in experimental plots of *Ulmus parvifolia* (source:10)

(1968) also reported leaf biomass as large as 453 g/m² (4.53 metric tons/ha) in three-year-old plots of *Camptotheca cuminata*. The seasonal change of leaf biomass in experimental plots of *Ulmus parvifolia* is shown in Fig. 8. After flushing in the spring, leaf biomass increased until late May and then decreased gradually until autumn leaf-fall. However, in spite of its importance, no data are yet available for the seasonal change of leaf biomass in mature deciduous forests of substantial size.

Vertical Distribution of Biomass

Vertical distribution of leaf biomass and extinction of light through foliage determines the productive, photosynthetic structure of the forest. According to Monsi and Saeki (1953), relative light intensity (I/I_0) at a certain depth within the forest canopy dependent upon the leaf area above it ($\overline{F}$, m²/m²), obeying the Beer-Lambert law, and described as

$$I/I_0 = e^{-k\overline{F}} \tag{7}$$

where k is light extinction coefficient for each species, and I and I_0 are light intensities at a given depth of canopy and above the canopy, respectively. This relation, which was originally developed for grass and herbaceous communities, also applies to the forest canopy, and leaf weight can be used instead of leaf area (Satoo et al., 1955). With productive structure and light-photosynthesis curves, Monsi and Saeki (1953) calculated the gross production of communities. The coefficient k can be approximated if both the relative light intensity below the crown canopy and leaf biomass are known. KIMURA (1960) applied this method to an evergreen broadleaved forest in southern Japan, and obtained fairly good agreement between the estimate and the harvest value. NOMOTO (1964) also estimated the primary productivity of beech forests in Japan using this method. The pattern of vertical distribution of leaf biomass in the tree layer is influenced by stand density (or the intensity of competition)

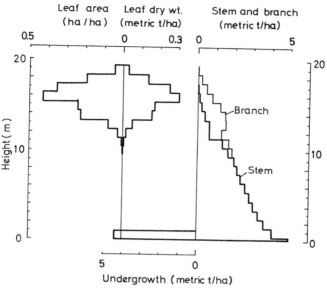

Fig. 9. Vertical distribution of leaf area and leaf, stem and branch biomass in a stand of *Betula platyphylla* (source: 20)

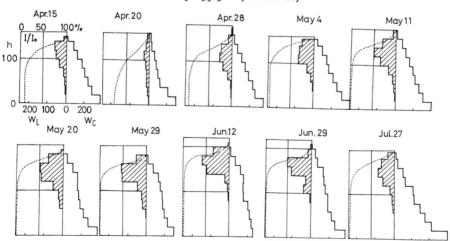

Fig. 10. Seasonal change in the productive structure of experimental plots of *Ulmus parvifolia* (source: 10). h height (cm); W_l leaf biomass (g/m²); W_c woody organ biomass (g/cm²); I/I_o (dotted lines) light intensity at a given height (I) relative to the light intensity above the stand (I_o)

although total leaf biomass is unaffected (SATOO, 1968b). An example of vertical distribution of biomass in a *Betula platyphylla* forest is given in Fig. 9. The light extinction coefficient, k, was 0.60 for leaf dry weight and 0.31 for leaf area. Seasonal changes in the photosynthetic structure of small stands of *Ulmus parvifolia* (TADAKI and SHIDEI 1960) are shown in Fig. 10. Initially leaves develop in all strata, but afterwards they fall from the lower strata.

Net Biomass Change

The total biomass itself is not a measure of production in natural forested ecosystems, because it consists of many plant constituents of different ages. Using harvest methods, net production as estimated here by summing the measured biomass

Table 4. *Some examples of net aboveground production estimates*[a] *for temperate deciduous broadleaved forests of Japan* (metric t/ha/year)

Species	Stem	Branch	Leaf	Tree total	Under-growth	Stand total	source
Betula maximowicziana	2.9	1.1	2.2	6.2	0.2	6.3	17
Betula maximowicziana	3.7	1.0	2.6	7.2	0.3	7.5	17
Betula maximowicziana	2.8	1.0	1.8	5.5	0.5	6.1	17
Populus davidiana	5.6	0.9	2.2	8.7	3.6	12.4	8
Fagus crenata	2.5	1.8	3.0	7.3	—[b]	—[b]	18

[a] Some estimates here and in later tables might be increased if leaf consumption and litter decomposition were known for the periods of biomass measurement.
[b] No data available.

Table 5. *Net aboveground production* (metric t/ha) *of tree layers of some adjacent natural deciduous broadleaved forests and conifer plantations in the same altitudinal zone*

Species	Age (yr)	Net production	Source
Betula maximowicziana	47	6.2	17
Betula maximowicziana	47	7.2	17
Betula maximowicziana	47	5.5	17
Populus davidiana	40	8.7	8
Picea abies	47	12.4	17
Picea abies	46	11.7	17
Picea abies	46	11.4	17
Picea abies	45	7.3	17
Abies sachalinensis	26	13.0	17
Larix leptolepis	21	16.5	17

changes occurring in each component for the year may be underestimates, because there were no corrections for unmeasured losses, e.g., early leaf drop or insect consumption. Although much is known about the biomass of temperate deciduous forests of Japan, there have been no studies estimating gross production. Many data exist on increment growth of stems and leaves, but few data are available for branch increments and nothing is known about roots. The available estimates on the net production of temperate deciduous forest of Japan are summarized in Table 4. There are no substantial differences between stands in these approximations of total net production of temperate deciduous forest of Japan are summarized in Table 4. There are no substantial differences between stands in these approximations of total net production. Additional estimates of net production of conifer plantations within several kilometers of the deciduous forests and within the same altitudinal zone are presented in Table 5. Net production by deciduous broadleaved forests was lower than that of conifer plantations, except for the one *Picea abies* plantation on poor soil. Annual net production estimates per hectare for potatoes (8.5 metric tons) and for sugar beets (11 metric tons) were available for the same region. Considering the possible underestimation of forest production due to the exclusion of root biomass, net production of deciduous forest may not be too dissimilar from the net production of these agricultural crops. OVINGTON (1956) presented a similar comparison with English data, giving similar conclusions.

Distribution of Net Primary Production

Distribution of the aboveground net primary production in Table 4 is shown in Table 6. Distribution ratios differed among forests, but generally less than one half of

Table 6. *Percentage distribution of net primary production among components of forest ecosystems* (sources: 8, 17, 18)

Species	Within tree layer			Within ecosystem	
	Leaf	Branch	Stem	Tree	Undergrowth
Populus davidiana	24.8	10.1	64.3	70.7	29.3
Betula maximowicziana	35.2	17.7	47.1	97.5	2.5
Betula maximowicziana	35.9	13.6	50.6	96.1	3.9
Betula maximowicziana	31.7	17.1	51.1	91.3	8.7
Fagus crenata	41.1	24.7	34.2	—a	—a

a No data available.

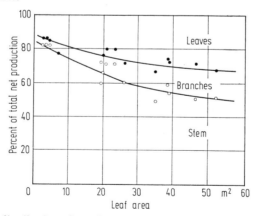

Fig. 11. Percentage distribution of net aboveground production in leaves, branches and stem of *Populus davidiana* trees as a function of total leaf area (source: 8). The percent composition is represented by the distances between lines

the dry matter production was distributed in trunks which are harvested in forestry. An example of the pattern of distribution of net production in individual *Populus davidiana* trees within a forest is shown in Fig. 11. Trees with larger leaf areas had a greater proportion of total net production deployed in canopy structure. The same trend was observed in a woodland of *Pinus densiflora* (Satoo, 1968b), but a converse relationship occurred in a plantation of *Cinnamomum camphora* (Satoo, 1968a). More information is needed to adequately describe this distribution phenomenon.

Primary Production Efficiencies
Efficiency of Total Net Leaf Production

Since total net production is a product of photosynthesis, it can be divided into leaf area and the production efficiency,

$$\bar{P} = \bar{F} \times \bar{P}/\bar{F} \tag{8}$$

or

$$P = F \times P/F \tag{9}$$

where $\bar{P}$ and P are total net production per unit ground area and per tree, and $\bar{F}$ and F are leaf area per unit ground area and per tree, respectively. The quotients $\bar{P}/\bar{F}$ and P/F (net production efficiency or net assimilation rate) are derived values useful in comparing the productive potential of different forests. Figure 12 shows the relation between net aboveground production and both the area and production efficiency of

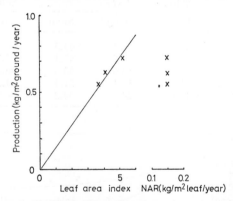

Fig. 12. Net aboveground production of stands of *Betula maximowicziana* in relation to leaf area index and "unit leaf area" net assimilation rate (NAR) (source: 17)

leaves in 47-year-old stands of *Betula maximowicziana*. Net production of the stands was linearly proportional to leaf area, while it was independent of the production efficiency of leaves. This relation was also the same among individual trees of these stands (Fig. 13). The relations between the net production and leaf area are described as

$$\bar{P}_{(kg/m^2ground/year)} = 0.145\ \bar{F}_{(m^2/m^2ground)}$$

for the stands, and

$$P_{(kg/tree/year)} = 0.148\ F_{(m^2/tree)}$$

for individual trees. The production efficiency of leaves or net assimilation rate, $\bar{P}/\bar{F}$ and P/F, was similar in both cases — 0.145 kg/m²leaf/year for stands and 0.148 kg/m² leaf/year for individual trees. These values are somewhat lower than the value of net assimilation rate reported for *Betula verrucosa* in England (OVINGTON and MADGWICK, 1959). Annual net assimilation rate per kg leaf biomass was 2.91 kg (mean of the three stands). Annual net assimilation rate of the beech stand shown in Table 4 was 0.162 kg/m² leaf or 2.43 kg/kg leaf dry weight.

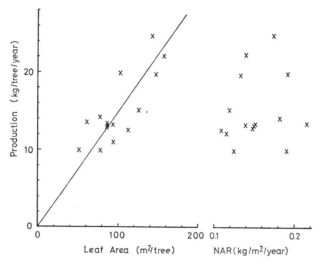

Fig. 13. Net aboveground production by *Betula maximowicziana* trees within stands in relation to leaf area and net assimilation rate (NAR) (source: 17)

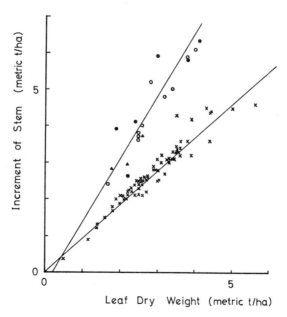

Fig. 14. Production of stemwood of stands in relation to leaf biomass: ○ *Betula ermanii*, ● *B. platyphylla*, ▲ *B. maximowicziana*, × *Fagus crenata* (sources: birch: 1, 11, 17, 20; beech: 2, 14, 18)

Efficiency of Stemwood Production by Leaves

Very few data are available for total net production by forests, while there are abundant data on production of stemwood and leaf biomass. The relation between leaf biomass and stemwood production of beech (*Fagus crenata*) and birch (*Betula platyphylla, B. ermanii,* and *B. maximowicziana*) is shown in Fig. 14. In both cases, stemwood production (P_s) is proportional to leaf biomass of the stand (W_l). These relations are described as

for birch, and
$$P_{s \text{ (metric t/ha ground/year)}} = 1.67 \ W_{l \text{ (metric t/ha ground)}}$$

$$P_{s \text{ (metric t/ha ground/year)}} = 0.951 \ W_{l \text{ (metric t/ha ground)}}$$

for beech. With equivalent leaf biomass, birch forests produce more stemwood than do beech forests.

Acknowledgements

I acknowledge the generosity of Dr. Y. Oshima, Mr. K. Maruyama and Mr. K. Morita who allowed use of their unpublished data. I also thank Professor A. Kumura for advice in converting the yield data of agricultural crops into dry matter. Contributions from JIBP-PT No. 29.

Sources of Data

I. Japanese Sources:

(1) Akai and Asada, 1955
(3) Ikushima, 1964
(5) Kato et al., 1966
(7) Satoo, 1968a
(9) Satoo et al., 1959
(11) Tadaki et al., 1966
(13) Yuasa and Shidei, 1966
(15) Morita (unpublished data)
(17) Satoo (unpublished data)
(19) JPTF-66-Koiwai[a] (unpublished data)

(2) Asada et al., 1966
(4) Kan et al., 1964
(6) Ohmasa and Mori, 1937
(8) Satoo et al., 1956
(10) Tadaki and Shidei, 1960
(12) Yamada and Maruyama, 1963
(14) Maruyama (unpublished data)
(16) Oshima (unpublished data)
(18) JPTF-66-Ashu[a] (unpublished data)
(20) (unpublished data)[b]

II. European Sources:

(21) Jokela and Jlänen, 1956
(23) Möller, 1945
(25) Nordfors[c]
(27) Smirnova and Gorodentzewa, 1958

(22) Knudsen and Mauritz Hanson[c]
(24) Mork[c]
(26) Ovington and Madgwick, 1959
(28) Tamm[d]

[a] Field work for discussion and training in methods held by the Woodland Working Group of Japanese PT in 1966 at Ashu (Kyoto) and Koiwai Farm (Iwate), respectively.
[b] Joint studies on forest productivity (Hokkaido, Tokyo, Kyoto and Osaka-City Universities).
[c] In Möller, 1945.
[d] In Ovington, 1962.

References

Akai, T., and S. Asada: Productivity in forests of *Betula ermanii*. Trans. 75th Mtg., Jap. Forest. Soc. (1964) 349—351 (1965)**.

Ando, T. et al.: Studies on the system of density control of sugi *(Cryptomeria japonica)* stand. Bull. Govt. Forest Expt. Sta. **209**, 1—76 (1968)*.

Asada, S., T. Akai, and T. Nozasa: Productivity of beech forest in northern Sinsyu. Trans. 76th Mtg. (1965), Jap. Forest. Soc. 151—153 (1966)**.

Honda, S.: Forest Zone of Japan (revised Ed.) Tokyo: Miura Syoten 1912**.

Ikushima, I.: Production structure of woody communities, pp. 106—125. In: Rept. Sci. Survey, Tanzawa Mts., Yokohama: Kanagawaken 1964**.

Jokela, E., and J. Yläinen: Koividoiden lehtisadon määrästä. Metsätal. Aikakaus **4**, 131—132 (1956).

Kan, M. et al.: On productivity of evergreen broadleaved forests. 2. On forests of *Castanopsis cuspidata*. Trans. 74th Mtg. (1963), Jap. Forest. Soc. 171—172 (1964)**.

Kato, R., K. Segawa, and S. Oba: The growth of *Alnus inokumae* in the young stands of various density. Ann. Rept. Tohoku Branch, Govt. Forest Expt. Sta. **7**, 81—99 (1966)**.

Kawahara, T. et al.: Movement of nutrients in a model stand of *Camptotheca acuminata* Dence. J. Jap. Forest. Soc. **50**, 125—134 (1968)*.

Kimura, M.: Primary production of warm temperate laurel forest in southern part of Osumi Peninsula, Kyushu, Japan. Misc. Rept. Res. Inst. Nat. Resources **52/53**, 36—47 (1960).

Kira, T., and T. Shidei: Primary production and turnover of organic matter in different forest ecosystems of the western Pacific. Jap. J. Ecol. **17**, 70—87 (1967).

— et al.: Estimation of standing crop, pp. 45—64. In: Studies on the Productivity of the Forest. I. Essential Needle-Leaved Forests of Hokkaido, (Shidei, T., Ed.) Tokyo: Kokusaku Pulp. Co. 1960**.

Kittredge, J.: Estimation of amount of foliage of trees and shrubs. J. Forest. **42**, 905—912 (1944).

Möller, C. M.: Untersuchungen über Laubmenge, Stoffverlust und Stoffproduktion des Waldes. Forstl. Forsøgsv. Danmark **17**, 1—287 (1945).

Monsi, M., and T. Saeki: Über den Lichtfaktor in den Pflanzengesellschaften und seine Bedeutung für die Stoffproduktion. Jap. J. Botany **14**, 22—52 (1953).

Newbould, P. J.: Methods for estimating the primary production of forests. IBP Handbook 2, Oxford: Blackwell 1967.

Nomoto, N.: Primary productivity of beech forest in Japan. Jap. J. Botany **18**, 385—421 (1964).

Ogawa, H. et al.: Comparative ecological studies on three main types of forest vegetation in Thailand. 2. Plant biomass. Nature & Life in Southeast Asia **4**, 49—80 (1965).

Ohmasa, M., and K. Mori: The amount and decomposition of the leaf litter of the foresttrees of Japan. Bull. Forest Expt. Sta. Imperial Household **3**, 39—103 (1937)*.

Ovington, J. D.: The form, weights and productivity of tree species grown in closed stand. New Phytol. **55**, 289—304 (1956).

— Quantitative ecology and woodland ecosystem concept, pp. 103—191. In: Advances in ecological research, Vol. 1, (Cragg, J. B., Ed.) New York-London: Academic Press 1962.

— Flower and seed production, a source of error in estimating woodland production, energy flow and mineral cycling. Oikos **14**, 148—153 (1963).

—, and H. A. I. Madgwick: The growth and composition of natural stands of birch. 1. Dry matter production. Plant and Soil **10**, 271—283 (1959).

Prodan, M.: Holzmeßlehre, 644 S. Frankfurt a. M.: J. D. Sauerländer's Verlag 1965.

Satoo, T.: Notes on Kittredge's method of estimation of amount of leaves of forest stand. J. Jap. Forest. Soc. **44**, 267—272 (1962).

— Production and distribution of dry matter in forest ecosystems. Misc. Inform. Tokyo Univ. Forests **16**, 1—15 (1966).

— Efficiency and quantity of leaves of closed stands of *Cryptomeria japonica* as influenced by site quality. 14th IUFRO Congress, Munich. Papers **2**, 396—404 (1967).

— Primary production and distribution of produced matter in a plantation of *Cinnamomum camphora*. Bull. Tokyo Univ. Forests **64**, 241—275 (1968a).

Satoo, T.: Primary production relations in woodlands of *Pinus densiflora*, pp. 52—80. In: Symposium on Primary Productivity and Mineral Cycling in Natural Ecosystems, (Young, H. E., Ed.). Orono: University of Maine Press 1968b.

—, R. Kunugi, and A. Kumekawa: Materials for the studies of growth in stands. 3. Amount of leaves and production of wood in an aspen second growth in Hokkaido. Bull. Tokyo Univ. Forests **52**, 33—51 (1956)*.

—, K. Nakamura, and M. Senda: Materials for the studies of growth in stands. 1. Young stands of Japanese red pine of various density. Bull. Tokyo Univ. Forests **48**, 65—90 (1955)*.

—, K. Negisi, and M. Senda: Materials for the studies of growth in stands. 5. Amount of leaves and growth in plantations of *Zelkowa serrata* applied with crown thinning. Bull. Tokyo Univ. Forests **55**, 101—123 (1959)*.

—, and M. Senda: Materials for the studies of growth in stands. 6. Biomass, dry matter production, and efficiency of leaves in a young *Cryptomeria* plantation. Bull. Tokyo Univ. Forests **62**, 117—146 (1966)*.

Shinozaki, K. et al.: A quantitative analysis of plant form: the pipe model theory. Jap. J. Ecol. **14**, 97—105, 133—139 (1964).

Smirnova, K. M., and G. A. Gorodentseva: The consumption and rotation of nutritive elements in birch woods. Byull. M. O-va ISP. Prirod. Otd. Biol. **63**, 135—145 (1958)***.

Tadaki, Y.: Some discussions of the leaf biomass of forest stands and trees. Bull. Govt. Forest Expt. Sta. **184**, 135—161 (1966).

—, and T. Shidei: Studies on production structure of forests. 1. The seasonal variation of leaf amount and the dry matter production of deciduous sapling stand *(Ulmus parvifolia)*. J. Jap. Forest. Soc. **42**, 427—434 (1960)*.

— et al.: Studies on productive structure of forest. 2. Estimation of standing crop and some analysis on productivity of young birch stand *(Betula platyphylla)*. J. Jap. Forest. Soc. **43**, 19—26 (1961)*.

Whittaker, R. H.: Branch dimensions and estimation of branch production. Ecology **46**, 365—370 (1965).

Yamada, M: Forest physiographical studies on the microtopographical analysis. Tokyo: Rinya Kosaikai 1965*.

—, and K. Maruyama: Quantitative ecological studies on natural beech forest. Trans. 72nd Mtg. (1962). Jap. Forest. Soc. 245—248 (1963)**.

Yamaoka, Y.: The total transpiration from a forest. Trans. Amer. Geophys. Union **39**, 266—272 (1958).

Yoda, K. et al.: Estimation of the total amount of respiration in woody organs of trees and forest communities. J. Biol. Osaka City Univ. **16**, 15—26 (1965).

Yuasa, Y., and T. Shidei: Productive structure and productivity of beech forest in Ashu. Trans. 76th Mtg. (1965). Jap. Forest. Soc. 153—155 (1966)**.

---

     * In Japanese with English summary.
   ** In Japanese only.
 *** In Russian with English summary.

7

Metabolism of Terrestrial Ecosystems by Gas Exchange Techniques: The Brookhaven Approach

G. M. Woodwell and D. B. Botkin

There is something rejuvenating in the tacit but progressive acceptance of F. E. Clement's classical assertion that the community is an organism (Clements, 1928), a concept that was hardly original with him but under his pen proved almost destructively provocative, at least in America. Certainly in considering the metabolism of segments of the earth's surface we recognize that there is at least an analogy between the community and an organism. Just as measurements of metabolism integrate function in a plant, so does metabolism integrate complex processes in arrays of organisms as diverse as forest, city, lake and ocean. And when we see the products of the metabolism of civilization becoming so dense as to affect life, as they are in too many places right now, the arguments for detailed knowledge of the metabolism of ecosystems become compelling.

The most conspicuous objective in study of metabolism is measurement of total respiration because it is a measure of the total amount of life. Photosynthesis is also important because it is the source of all the energy driving natural ecosystems. As simple as these two objectives appear, their measurement has proved a major challenge. The problems have been especially great in forests for the very reason that makes forests an important subject: they are among the most complex of terrestrial ecosystems.

We discuss: first, the relationships between primary production, respiration and photosynthesis in terrestrial ecosystems; second, ways of solving the equations that define the relationships between net and gross production and respiration; and, third, the contributions that measurements of gas exchange can make to the solution of these equations. Our discussion is drawn primarily from our own experience, extending over several years, with gas exchange techniques in a intensively studied oak-pine forest at Brookhaven, Long Island, New York, where we have made a special effort to develop the techniques.

The Production Equations

The relationships between photosynthesis and respiration of the plant community and of the ecosystem are clarified by referring to our analogy with a single green plant. Its net increase in weight during a year (NPP or net primary production) is equal to the total amount of photosynthesis it carried out (GP or gross production) less its respiration (Rs). We can write this:

$$NPP = GP - Rs. \tag{1}$$

The same formula can be applied without change to the entire community of green plants in a terrestrial ecosystem. Measurement of the NPP of the plant community, however, is a much more difficult task than it is for a single plant. With slight modification the formula can also be applied at the ecosystem level, with another tenfold jump in the complexity of measurement. The modifications involve recognition that the net increment of organic matter to an ecosystem is less than the NPP by an amount equal to the consumption (Rs_H) by heterotrophs. Thus, for the ecosystem we can write:

$$NEP = GP - (Rs_A + Rs_H) \qquad (2)$$

where NEP is net ecosystem production, the annual increment of biomass to the ecosystem; Rs_A is respiration of the autotrophs, the respiration of the plant community; Rs_H is respiration of the heterotrophs including the decay organisms; and $Rs_A + Rs_H = Rs_E$, the respiration of the ecosystem; $NEP = NPP - Rs_H$. It is a first objective in study of the function of terrestrial ecosystems to solve these equations (Woodwell and Whittaker, 1968; Whittaker and Woodwell, 1969).

Methods for Measurement of the Metabolism of Ecosystems

The two general categories of techniques available for such measurements are mutually dependent. The harvest technique with its various modifications offers direct measurements of NPP only (Woodwell and Bourdeau, 1962; Whittaker and Woodwell, 1968; Woodwell and Whittaker, 1968; Newbould, 1967; Odum, 1960; Lieth, 1962). A variety of applications of measurements of gas exchange in conjunction with other measurements offer at least the possibility for measurement of Rs_A and Rs_H, direct measurement of Rs_E, a second estimate of NPP, and even an estimate of NEP. GP can probably not be measured directly due to the problems of measuring respiration during photosynthesis (Zelitch, 1964). The other measurements have proved considerably more difficult than would appear at first. The techniques for measurement of community metabolism can be summarized in somewhat simplified form as (Woodwell and Whittaker, 1968):

1. Harvest Techniques (NPP and details of structure)
   a) Plots,
   b) Dimension Analysis.

2. Exchange of Gases
   a) Large Enclosures (NEP, Rs_E),
   b) Flux or Aerodynamic Techniques (NEP, Rs_E),
   c) Small Enclosures (NPP, NEP, $Rs_E \sim Rs_A$, Rs_H and other details).

Dimension analysis, applied to forests, is probably the single most versatile harvest technique, providing direct estimates of standing crop, net production, surface area of various parts and a host of other details of structure of the plant community. Many of the details are basic to use of gas-exchange techniques, especially for the small chamber approaches. In any case it is clearly not possible to complete the production equations on the basis of the harvest technique alone since the technique applies only to the plant community. Supplementary data include a measure of litter fall, an estimate of the rate of decay of humus and litter, and a measure of total respiration obtainable only by gas-exchange techniques. A summary of the biomass of the plant community of the Brookhaven oak-pine forest appears in Table 1. The summary was adapted from Table 3 of WHITTAKER and WOODWELL (1969).

Table. *Biomass of the Brookhaven oak-pine forest, Long Island. Adapted from* WHITTAKER *and* WOODWELL (1969)

	Stem, wood and bark	Branch live wood and bark	Branch dead wood and bark	Current twigs	Leaves	Fruits	Flowers	Above-ground total	Total roots	Full total
				Biomass, dry g/m²						
Quercus coccinea	1964	705	110	26	164	7	1	2967	1335	4302
Q. alba	1280	345	19	16	109	8	1	1788	1485	3272
Pinus rigida	808	262	35	9	116	5	—	1235	315	1550
Quercus velutina	271	98	16	4	23	1	—	413	190	603
Tree-stratum total	4323	1410	180	55	412	21	2	6403	3325	9728
Quercus ilicifolia	6.2	2.9	0.9	0.2	1.6	—	—	11.8	50.9	62.7
Gaylussacia baccata	47.4	23.1	4.4	4.5	15.6	0.6	0.2	95.8	160.9	256.7
Vaccinium vacillans	5.1	5.1	0.4	1.9	7.6	1.0	0.2	21.3	66.8	88.1
V. angustifolium	—	3.0	0.5	0.6	1.8	—	—	5.9	12.3	18.2
Other shrub species[a]	4.0	2.3	0.3	0.2	1.4	—	—	8.2	14.6	22.8
Tree sprouts[b]	8.0	4.0	0.4	0.4	2.7	—	—	15.5	—	15.5
Shrub-stratum total	70.7	40.4	6.9	7.8	30.7	1.6	0.4	158.5	305.5	464.0
Total woody plants	4393	1450	187	63	443	23	2.4	6561	3631	10192

[a] Current twig clippings (dry g/m²) for minor shrub species: *Myrica pensylvanica* 0.750, *Kalmia angustifolia* 0.624, *Pyrus arbutifolia* (L.) L. F. 0.103, *Rubus* sp. 0.048, *Lyonia mariana* (L.)D. Don 0.024, *Amelanchier stolonifera* Wieg. 0.016, *Vaccinium corymbosum* (L.) 0.015, *Comptonia peregrina* (L.) Coult. 0.002, *Smilax rotundifolia* 0.0002.

[b] Current twig clippings (dry g/m²) for tree sprouts and seedlings not reaching 1 cm breast height: *Quercus coccinea* 1.57, *Q. alba* 1.04, *Q. prinoides* Willd. 0.266, *Pinus rigida* 0.030, *Sassafras albidum* 0.018.

Methods using gas-exchange are all based on measurement of CO_2, usually by infrared absorption. It is possible to make individual measurements within 1 minute or less and with an accuracy of ± 1 ppm. While the thought of measuring O_2 concentration simultaneously is attractive and the measurements would be important for some purposes, to our knowledge no equipment is available that can make field measurements of O_2 concentrations with an accuracy in the range of a few ppm against a background of 20% O_2, which is what is required. The choice among the three approaches to gas-exchange measurements hinges in part on objectives and in part on practicality. All the techniques depend on measurement of a difference in CO_2

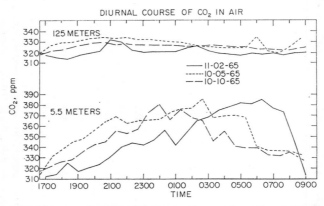

Fig. 1. Carbon dioxide concentration of the atmosphere at Brookhaven National Laboratory during 3 days. Heights are in meters above the ground. Absolute concentrations are within ± 1 ppm of a standard from the National Bureau of Standards

concentration of air that has passed over a plant or through an ecosystem, an objective that has proved more elusive than it appeared at first. Accuracy depends not only on the gas analyzer but also on the stability of the CO_2 concentration in the ambient air. Difficulties increase if the CO_2 concentration fluctuates as it often does and if the differences measured are small as they usually are. The CO_2 concentration of air within a few meters of the ground fluctuates greatly as shown by Fig. 1 in which are plotted records of CO_2 concentration at 125 m above the ground and at 5.5 m at Brookhaven, New York over several days.

It is clear that changes of more than 50 ppm sometimes occur in minutes. Against such a background of variation in CO_2 concentration it is extremely difficult, even impossible, to make accurate measurements with ambient air as the air supply, since most chambers have a half-time for flushing that is large in comparison with the rates of change in CO_2 concentration in the ambient air. Although this difficulty can be solved, even for large chambers, solution requires a 2-5 fold increase in technology and has not yet appeared worth the cost. The same problem plagues flux measurements, recently summarized for the IBP so ably by Baumgartner (1968), forcing an expansion of technology there to as much as 10 times what has been possible to date, to make that technique useful.

This set of considerations, coupled with recognition of the heat-load/flow-rate dilemma of assimilation chambers, led the first author several years ago to settle on the "small chamber" approach as the most versatile and most promising. This approach offers reasonable solutions to the technical problems of fluctuating CO_2 concentrations, heat-loads, low differential concentrations of CO_2, and replication. Large chambers that enclose a segment of land area present similar technical problems on a larger scale. They have no advantage of simplicity and provide in return only the possibility of measurement of respiration (Rs_E) and net photosynthesis (NEP) of the ecosystem. The harvest or even the flux techniques should appear preferable in the long run for such measurements.

Our objective is to show what can be done at present, to clarify some of the objectives and to contribute to the simplification of the techniques. We have focussed necessarily on our own experience and on the Brookhaven Forest because it has sustained the brunt of our efforts over the years. We discuss in particular the small chamber technique and one modification of flux measurements.

Our solution to the fluctuating CO_2 concentration problem, still not simple and still not perfect, has been to use many small chambers fed by a single air supply system. Concentrations of CO_2 in the supply air have been buffered slightly by passing all the air through a large room where it is mixed continuously before entering the air-supply manifold. Flow rates to the chambers have been measured with flow meters and controlled by pressure-reducing valves. To reduce further the errors due to fluctuating CO_2 concentrations in the air supplied to the chambers, all comparisons of CO_2 concentration from experimental chambers have been with the exhaust from dummy chambers that are the same size and have the same flushing time.

MEASUREMENT OF CO_2 EXCHANGE OF FOREST

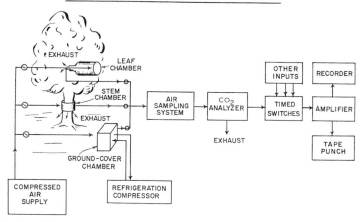

Fig. 2. The Brookhaven small-chamber system for monitoring CO_2 exchange rates and various environmental factors

The small-chamber system we have used was designed around three types of chambers and an automatic data-recording system (Fig. 2). The chambers were the simplest we have been able to design. The leaf chamber was a cylindrical polyvinyl-chloride bag, supported by a light, rigid, aluminum-alloy frame and gathered tightly around a branch. In our application it contained one twig and its leaves. Air was supplied at about 10 l/min, blown against the bottom of the bag and exhausted to the atmosphere through the neck of the bag which was gathered tightly around the branch. Air temperature was monitored by an aspirated thermocouple in a white-painted tube in the neck of the bag. The CO_2 content of the aspirating air was recorded once every 30 min. A very detailed study of temperature in the chambers showed that leaf and air temperatures very rarely differ by more than 2° C, although the temperature may rise in the chamber as much as 10° C.

Stem chambers have been operated on a similar principle, enclosing the bark around a segment of tree stem. The chambers were painted with an opaque reflective paint and, although air temperature rose somewhat, subsurface bark temperatures followed those on shaded portions of the tree stem.

We have approached the problem of measurement of gas exchange of the ground cover and soil surface in several ways. For a contribution to measurement of total respiration of the ecosystem we have used large, compressor-cooled chambers covering 0.25 m² of ground surface and standing 1 m high. This size is sufficient to include both the ground cover and low shrubs of the Brookhaven Forest. The bottom edge of the chamber is a knife edge that penetrates the humus layer and reduces the possibility of leakage. Chambers have not been pressurized. Temperature has followed the temperature of the ambient air, regulated by a special thermostat.

A recording system, now entering its third modification, records data from each of 18 stations once each 30 min. Data are printed by a teletype machine and punched simultaneously on tape for computer handling. Manifolds and patch panels in the field provide rapid switching from one set of experiments to another. Computer programs have been written in a general form that allows constants to be changed easily.

Total Respiration of the Brookhaven Forest: Temperature Inversions — A Simplified Flux Technique

We have made three different direct estimates of Rs_E. The earliest estimate was based on the rate of accumulation of CO_2 during temperature inversions, which are frequent on Long Island and useful for our purposes because the terrain is forested and flat over several square miles, providing a long "fetch" and reducing the importance of cold air drainage to a minimum.

The pattern of CO_2 accumulation during one of these inversions is shown in Fig. 3 (Woodwell and Dykeman, 1966). The slope of the CO_2 concentration curve is presumably a measure of the rate of respiration of the vegetation at that height. Integrating these slopes over the full depth of the forest gives the rate of CO_2 production for the forest. We have done this for 41 temperature inversions during one year. Relating the rate of Rs_E to mean daily temperature produced two curves best expressed as simple linear regressions (Fig. 4). While the data are scattered more than one might wish due to the inherent crudeness of the technique, it seems clear that there is an annual course of Rs_E, that it is related to temperature, and that there are two rates, one for the dormant season and one for summer. Applying the curve of mean monthly temperatures to these regressions gives an integrated sum of Rs_E for the year of about 3500 g of CO_2. If 1 g of CO_2 represents 0.614 g of dry matter, which is an approximation when carbohydrate ($C_6H_{10}O_5$) is burned, then the 3500 g of CO_2 is the equivalent of 2104 g of dry matter, the first estimate of Rs_E.

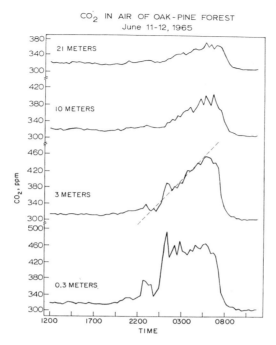

Fig. 3. CO_2 concentrations at several heights in the Brookhaven Forest during a temperature inversion. The slope of the curve showing the increase in CO_2 concentration during the inversion is an index of the rate of respiration of the forest at that height

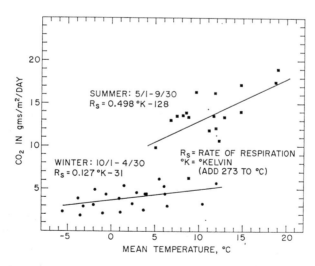

Fig. 4. Respiration rates of the Brookhaven oak-pine forest plotted against mean temperature (the mean of the average temperature at 10 m and 3 m, measured every 15 minutes during the inversion). Rs = rate of respiration in grams of CO_2 per square meter per day (WOOD-WELL and DYKEMAN, 1966)

Respiration by Small Chamber Technique

Total respiration of the ecosystem by the small chamber technique involves summing the data from the three types of chambers. Data from leaf and stem chambers must be taken by species and summed for all. To arrive at a figure for the ecosystem an estimate of total leaf and bark surface must be available. At Brookhaven we have the advantage of the detail offered by an elaborate application of dimension analysis in five separate samples (WHITTAKER and WOODWELL, 1969). Total surface of stems and branches is 1.5 m²/m² of land (WHITTAKER and WOODWELL, 1967).

Two applications of the small chamber technique are possible: direct measurement of the annual course of respiration, integrating under the annual curve; or, establishment of the relationship between the rate of Rs and some, preferably controlling, variable such as temperature. This latter curve can be used to construct the annual curve from weather records. Both techniques have been used here.

Respiration of Stems: Annual Curve

The curve expressing rate of CO_2 evolution by stems of trees through the year can be constructed by plotting daily means, averaged over 24 hours, for the series of measurements available to date, a series spanning 4 years. The curve for scarlet oak

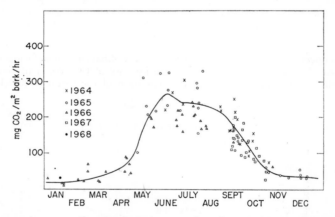

Fig. 5. The annual course of respiration of stems of scarlet oak *(Quercus coccinea)* trees based on measurement over 4 years. Each point is the mean of approximately 50 measurements over a 24-hour-period

(Querus coccinea MUENCH.) (Fig. 5) shows a maximum in mid-June, similar to maxima reported for several species by JOHANNSEN (1933). For convenience we have expressed all our data as mg of CO_2 evolved per m² of bark surface. The curve drawn here is a somewhat freely drawn average of the data spanning the 4 years, the data representing each year deviating appreciably from this average. In 1966, for instance, rates of CO_2 evolution during mid-summer were consistently lower than in other years due possibly to a severe drought.

Integrating under this curve and multiplying the sum by the total area of stems and branches of scarlet oak in the forest, a number available from dimension analysis, provides an estimate of the mean annual total CO_2 production by scarlet oak. Summing with similar estimates for white oak and pitch pine, provides a total for the tree stratum of the forest, estimated as equivalent to 950 g of dry matter per m² per year.

Respiration of Stems: Respiration/Temperature Method

Alternatively, it is possible to establish a correlation, or a series of correlations, between Rs and temperature during the year as shown in Fig. 6 (WOODWELL and WHITTAKER, 1968). Here individual observations made over single days are plotted against absolute temperature. Each day's observation can be expressed most simply by a simple linear regression, and the series of regressions through the year forms a complex curve as shown. The progression forms an ellipsoidal pattern starting at the lower left in winter, rising clockwise to a peak in late summer, and declining abruptly in fall to the mid-winter low. The pattern is a refinement of the pattern shown earlier by Rs_E determined by the temperature inversion technique (Fig. 4).

It would be convenient if all woody species followed this pattern, but the diversity of nature is such as to spare us even this simplification. To be sure, white oak (*Quercus alba* L.) follows a pattern similar in broad outline to that shown for scarlet oak. Pitch pine (*Pinus rigida* MILL.), however, at least in the data available at present, is quite different. Rates rise slowly in early spring and abruptly in early summer, declining slowly in fall. This generates an ellipsoidal pattern similar to that of scarlet oak but it runs counterclockwise, implying an annual course of respiration skewed to the right of the curve shown for scarlet oak (Fig. 5).

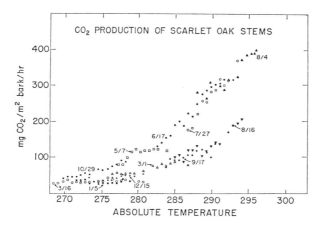

Fig. 6. CO_2 production by stems of scarlet oak as a function of temperature during the course of a year. Each point is the mean of approximately 50 measurements made over a 24 hour period

Using these graphs and daily mean temperatures based on 15-year records at Brookhaven we have estimated the total CO_2 production by stems and branches as equivalent to 1242 grams of dry matter per m², providing a second estimate, this one slightly higher than that based on the annual curves alone.

Respiration of Soil, Shrubs and Ground Cover

The other major contribution to Rs_E is CO_2 evolved by the soil surface, low shrubs and ground cover. This figure, of course, includes the respiration of roots. It does not include losses of CO_2 in water percolating into the ground water table, assumed to be small.

Application of the same annual curve technique indicates a total consumption of organic matter in the soil and by vegetation less than 1 m tall equal to 1045 grams of dry matter per m² annually.

Total Respiration of the Ecosystem by the Small Chamber Estimates

Total respiration of the ecosystem is the sum of the respiration of leaves and twigs, bark of stems and branches, and the low shrubs, ground cover and soil surface. Total respiration of the leaves and twigs cannot be measured in the field because there is no practical way of monitoring respiration during photosynthesis. In our data net photosynthesis is confounded with total respiration of leaves and twigs, introducing a small bias toward lower total respiration for the forest. Thus, the estimate of Rs_E by the small chamber technique reported here is based on summing of estimates of CO_2 production by stems and branches, plus the observations from ground cover and soil. The sums are 1995 grams dry weight/m²/year and 2287 depending on whether the annual curve or the estimates based on the Rs/temp. correlations are used.

Net Photosynthesis

Measurement of net photosynthesis presents greater problems. If we plot rates of net photosynthesis, averaged over 24 hours, for the total number of observations of

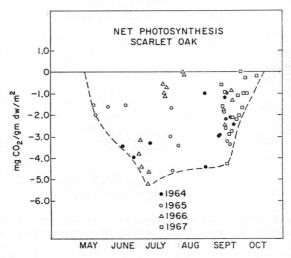

Fig. 7. Net daily photosynthesis (24 hour means) for scarlet oak leaves during the course of a year

scarlet oak available, the annual pattern appears as shown in Fig. 7. Even in midsummer there are days when net photosynthesis is low, at least for some leaves. For simplicity in our discussion, we have estimated the maximum that photosynthesis could be for all three species. The sum of all species, plus an increment of 336 grams dry weight/m² from the ground cover chambers, indicates a net annual maximum photosynthesis of 3453 grams dry weight/m². To correct this to GP we would have to add an unmeasured and still unmeasurable increment of respiration of leaves during lighted periods plus dark respiration. This approach using annual curves constructed from direct measurements, holds little promise in measurement of NPP in our view, being much more cumbersome and probably less accurate than other estimates based on the distribution of light (see SAEKI, 1960, among others).

Discussion

The details available from the chamber technique allow several further insights and measurements useful in solving the production equations. Their usefulness is, however, heavily dependent on supplementary measurements. For instance, Rs_A cannot be measured directly due to the confounding of respiration of roots and respiration of the surface organic horizons, which is of course part of Rs_A. An estimate of the rate of decay of the litter and humus has been possible in the Irradiated Forest at Brookhaven using the gradient of radiation damage (WOODWELL and MARPLES, 1968). This work suggests that decay consumes about 621 g d.w./m²/year. Since we know from the measurements with small chambers that respiration of the soil surface and vegetation 1 m and less tall accounts for an estimated total of 1045 g/m²/year, then the respiration of roots of higher plants and shoots of plants 1 m or less tall consumes 1045 — 621 = 424 g/m²/year. The confounding of roots and shoots is unavoidable. Despite the confounding the figure is useful in that it, added to the mean estimate of respiration for bark and branches (1096), gives an estimated 1520 g/m²/year for Rs_A.

NPP has been measured by dimension analysis as 1195 g/m²/year (WHITTAKER and WOODWELL, 1969), making possible an estimate of GP, not otherwise measured, of 2715 g/m²/year. Applying this figure to calculate NEP indicates that the ecosystem is adding an increment of 2715 — 2127 = 588 g/m²/year. The previous estimate based on dimension analysis was 542 g/m²/year (WHITTAKER and WOODWELL, 1969).

What is important here is not the fact that these estimates differ by as much as 20%, but that the gas-exchange technique has made important contributions to the solution of the production equations. The contributions have been through measurements of respiration, especially total respiration of the ecosystem. They have specifically not been direct estimates of NPP or NEP, at least in the comparatively simple and direct applications reported here. Nor are these estimates in any sense final; they are preliminary data obtained in the course of development of techniques. Because a wide diversity of data are available describing various attributes of the Brookhaven forest, the small chamber technique has been especially appropriate, enriching the overall study with still further details and allowing several independent estimates of the same attributes of the ecosystem.

It seems amply clear, however, that even in a center such as Brookhaven, where technical assistance is available, the monitoring and replication needed to develop annual curves of metabolism by direct measurement are so difficult, subject to so many failures of equipment, as to be an unreasonable objective. Alternatively it is possible at least for Rs of stems and branches to develop correlations with temperature

6*

and season that are clear and allow calculation of the annual curve on the basis of monitored data from as few as 12 days, a very great reduction in labor. It is not to be assumed, however, that measurements of one tree, casually selected, will suffice. Rates of respiration vary among trees by factors of 2x or more with growth rate and other factors. It is possible, nonetheless, for purposes outlined here to select "average" trees that integrate the variation in growth rates and offer a reasonable estimate of rates of CO_2 production. It is better, of course, to measure the variations if such measurements are possible. Similarly, we have shown elsewhere that rates of photosynthesis, measured also on carefully selected trees in carefully selected places, can be correlated with temperature and light on the basis of few measurements.

Despite these important simplifications and despite the conspicuous and noteworthy success of REINERS (1968) in a similar attempt, we have been unable yet to define any acceptable simplifying correlation expressing CO_2 production by the soil surface. These rates are clearly dependent on season; they are also heavily dependent on soil moisture availability, often doubling within minutes of a heavy rain and remaining high for many hours. At the moment there is for us no alternative to attempting to define the annual course as we have done.

There is no question of the importance of the contribution of gas-exchange techniques in study of the metabolism of terrestrial ecosystems. There is every reason to pursue both the small chamber approaches, seeking to simplify them further, and the flux techniques, which will ultimately be simplified to the point of offering short-term estimates of both respiration and NEP. Neither is a panacea; both are extraordinarily expensive in equipment, time and effort. They are adjunct to, but not yet a replacement for, harvest techniques.

Acknowledgements

Research carried out at Brookhaven National Laboratory under the auspices of the United States Atomic Energy Commission.

References

BAUMGARTNER, A.: Meteorological approach to the exchange of CO_2 between the atmosphere and vegetation, particularly forest stands. Presented at the IBP Workshop on Tree Photosynthesis. Ottawa, Canada (1968).

CLEMENTS, F. W.: Plant succession and indicators New York: Wilson Co. 1928.

JOHANSSON, N.: The relation between the respiration of the tree stem and its growth. Svenska Skogsvardsforen Tidskr. 31, 53—134 (1933).

LIETH, H. (Hrsg.): Die Stoffproduktion der Pflanzendecke. Stuttgart: G. Fischer 1962.

NEWBOULD, P. J.: Methods for estimating the primary production of forests. IBP Handbook No. 2. Oxford: Blackwell Sci. Publ. 1967.

ODUM, EUGENE P.: Organic production and turnover in old field succession. Ecology 41, 34—49 (1960).

REINERS, W. A.: Carbon dioxide evolution from the floor of three Minnesota forests. Ecology 48, 471—483 (1968).

SAEKI, T.: Interrelationships between leaf amount, light distribution and total photosynthesis in a plant community. Botan. Mag. Tokyo 73, 55—63 (1960).

WHITTAKER, R. H., and G. M. WOODWELL: Surface area relations of woody plants and forest communities. Amer. J. Botany 54, 931—939 (1967).

— — Dimension and production relations of trees and shrubs in the Brookhaven Forest, New York. J. Ecol. 56, 1—25 (1968).

— — Structure, production and diversity of the oak-pine forest at Brookhaven, New York. J. Ecol. 57, 155—174 (1969).

Woodwell, G. M., and P. F. Bourdeau: Measurement of dry-matter production of the plant cover, pp. 283—289. In: Methodology of plant eco-physiology, (F. E. Eckardt Ed.). Montpellier, France: UNESCO 1962.

—, and W. R. Dykeman: Respiration of a forest measured by CO_2 accumulation during temperature inversions. Science **154**, 1031—1034 (1966).

—, and T. G. Marples: The influence of chronic gamma irradiation on production and decay of litter and humus in an oak-pine forest. Ecology **49**, 456—465 (1968).

—, and R. H. Whittaker: Primary production in terrestrial communities. Amer. Zool. **8**, 19—30 (1968).

Zelitch, I.: Organic acids and respiration in photosynthetic tissues. Ann. Rev. Plant. Physiol. **15**, 121—142 (1964).

8

Insect Influences on the Forest Canopy

RUDOLPH T. FRANKLIN

The forest environment, its species diversity and structural complexity, provides many niches for exploitation by numerous undisturbed populations of insects. The ecological interactions which are involved may include only those between insect and tree; usually, however, they include a complex series of reactions between many kinds of organisms and physical factors in the environment. Any syntheses of this subject area draw heavily upon the field of forest entomology. The ultimate objective of forest entomology is to discover how to prevent insects from causing economic damage to forests. The literature, therefore, emphasizes pest species. Quantitative data may be available for the pest species, noneconomic species receive considerably less attention.

The need for fundamental study aimed at the physiological and biological aspects of insects in the forest environment is recognized, but such study is essentially of long-term nature. Population fluctuations may be violent. The injury caused to trees by the attack of forest insects may be expressed in loss of increment, degradation of the wood, or in the deterioration of killed standing timber. The full extent of the biological and ecological impact of insects upon the forest canopy is not known. Past and present research has failed to account fully for all of the effects of insect activity on productivity. This is in part due to 1) limitations in knowledge of the insects and their effects on host trees and 2) the particular ecosystem of which they and the host tree are a part. Shortcomings in our methods of study also contribute to the lack of knowledge. Many effects of insect activity cannot be measured in quantitative terms. The many interactions between insect and host are difficult to describe completely — much less measure.

Insect-Plant Relations

Classification of Feeding Habits

The most important foliage feeding insects belong to the orders Hymenoptera, Coleoptera, and the Lepidoptera. Species of the Orthoptera, Diptera, and Hemiptera also feed upon foliage and several other orders regularly occur in the forest canopy. Other groups of insects such as the bark beetles (Scolytidae), while not feeding there, may have dramatic effect upon the canopy by killing trees as a result of their bark-feeding habits.

In general, foliage feeding is classified into three groups: leaf chewing, skeletonizing, and leaf mining. Some species of insects skeletonize, or eat all but the vascular portions of the leaves during part of their growth, or are leaf miners, feeding between the upper and lower epidermal layers, but become leaf chewers during later stages. Recent textbooks in forest entomology, ANDERSON (1960) and GRAHAM and KNIGHT (1965), have attempted to avoid specifically treating the vast number of insects that feed upon trees by treating them as groups that feed on specific parts of the tree. Thus, the Coleoptera, Lepidoptera, and Hymenoptera that are defoliators can be treated as such without delving into specific classification and description.

A discussion of insects that have an effect upon the forest canopy must consider many groups beside the defoliators (Fig. 1). The inner-bark-boring insects may kill trees, cause breakage, or introduce pathogens. Sap-sucking insects may cause enlarged growth (galls), foliage disturbances such as chlorosis and deformations, and

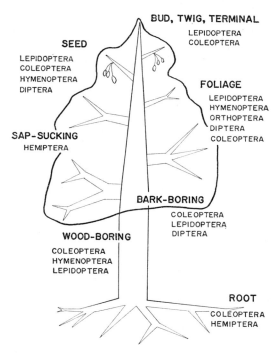

Fig. 1. Categories of insect feeding relationships, and the main orders of insects comprising each category, which have a significant impact upon the primary production of forest trees

disease transmission. Bud and twig-feeders have direct effect upon the foliage; seedling and root-feeding insects may have long-term affects upon regeneration and succession. Cone and seed consumption which, of course, occur in the canopy may reduce reproduction. Each category may include all or several of the orders of insects, but even the categories do not have defined limits because the same species may occur in more than one category during its life history.

Primary and secondary are terms that have been used to classify insects in relation to their hosts. A primary insect is capable of completing its development in a healthy tree while a secondary is not. Defining a healthy tree then becomes a problem. For the purpose of this discussion the term primary will be reserved for those insects that feed on growing plant parts throughout their life history. These species are the defoliators, sap-suckers, and shoot-feeders. Primary species vary in their requirements for living tissue. Some thrive best on young vigorous tissue; others prefer mature tissue and feed only on old foliage. Some of the sawflies (Hymenoptera) feed on pine needles that are one or two years old.

Factors Influencing Population Size

The kind of food that an insect needs determines its effect on productivity and the vulnerability of different tree species or different individuals within a species. The amout of favorable food, and its distribution within a forest stand, strongly influences population levels. As soon as the supply of food suitable for full development of the feeding stage becomes inadequate, loss through dispersal increases and there is a drop in the mean fecundity of the surviving adults. The physiology of both trees and insects is involved in understanding the reciprocal effects of the tree on the insect and the insect on the tree.

MORRIS (1967) found significantly lower survival of *Hyphantria cunea* Drury larvae on late-season foliage, and the fecundity of the moths was considerably decreased. When the filial generation was subjected to nutritional stress, the influence of paternal food was apparent throughout the larval, pupal and adult stages. However, when the filial generation had favorable food there was no significant difference in survival rates. Elm spanworm, *Ennomos subsignarius* (HUBNER), moths produce more eggs when the larvae are fed pignut hickory, *Carya glabra* (MILL.), and the pupal weights are higher than when the larvae are fed northern red oak, *Quercus rubra* L., and white oak, *Q. alba* L. (DROOZ, 1965). FINNEGAN (1965) found pine needle miner, *Exoteleia pinifoliella* (CHAMB.), used empty mines made by larvae of previous generations for oviposition sites. Thus, previously infested trees tended to become more infested while nearby healthy trees remained relatively uninfested. The spread of this insect to uninfected foliage is brought about by either migration of young larvae or wind dispersal of dead needles containing eggs. *Neodiprion nigroscutum* MIDDLETON sawfly adults when offered a choice deposited more eggs on jack pine, *Pinus banksiana* LAMB., than on either red pine, *P. resinosa* AIT., or eastern white pine, *P. strobus* L. Larvae had higher cocooning success, cocoons were larger, and adult emergence success was highest on jack pine (BECKER and BENJAMIN, 1967). Age of the foliage consumed by spruce budworm, *Choristoneura fumiferana* (CLEM.), may affect fecundity when larvae are forced to feed on old foliage with the resulting moths producing fewer eggs. Severe defoliation also reduces the availability of suitable oviposition sites (MORRIS, 1963). Different types of individuals among the progeny of the western tent caterpillar, *Malacosoma pluviale* (DYAR), are concentrated in different parts of the egg mass. The most agile progeny come from some of the first eggs laid, and the least viable are among the last deposited. There is evidence that this serial arrangement stems from unequal partitioning of the maternal food reserves during egg production. The differences in feeding rate and food capacity displayed by the different types of females during their own larval stage affect the proportions of the various types of progeny per egg mass as well as the viability of consecutive groups of eggs within the mass (WELLINGTON, 1965).

MARTIN (1966) observed the faunal changes in the crown stratum of four red pine stands for four years. The stands were 5, 15, 25 and 35 years old. During the establishment stage (0—5 years) the ecological conditions are essentially those of an old field

(Table 1). At 15 years the crown closes and the monoculture stage begins. The transition then is to young forest. Many of the arthropods in the establishment stage were replaced by other species in the transition stage (Table 2). All of the primary feeders disappeared in the transition stage. The original coccinellid fauna was reduced from six species to three. The aphids, neuropterans and a number of spiders increased in the transitional stage. In the Hymenoptera, the species of ants in the transition stage were different from those in the monoculture and young-forest stages.

Table 1. *Arthropod fauna in the crown stratum during the establishment stage of a red pine community, Kirkwood Township, Ontario, 1960—1964* (from Martin, 1966)

Species or group	Percent of trees occupied				
	1960	1961	1962	1963	1964
Sparganothis sulfureana (Lepidoptera)	5.0	79.0	43.0	12.1	43.5
Rhyacionia adana (Lepidoptera)	3.0	3.4	6.0	9.4	13.4
Pissodes approximatus (Coleoptera)	2.0	4.0	9.0	0	0
Pineus strobi (Homoptera)	0	6.9	10.7	21.2	15.9
Cinara spp. (Homoptera)	0	0	15.0	14.8	61.2
Schizolachnus pini-radiatae (Homoptera)	0	0	2.7	16.8	35.9
Magdalis perforatus (Coleoptera)	0	0	0.6	0	2.1
Coccinellidae (Coleoptera)	0	0	3.8	4.4	10.5
Araneida	0	0	2.6	8.0	24.6
Elateridae (Coleoptera)	0	0	0.8	1.2	12.8
Neuroptera	0	0	0	30.9	26.6
Miscellaneous Hemiptera	0	0	0	0	10.5

The Collembola showed the tendency to rise rather steadily to a peak and then decline rapidly. This type of change might be expected in a movement from a simple to a complex community. However, in this case there is little reason to suspect increased complexity from the standpoint of species represented, since the number and kinds of species were about the same in all three stages. The food relations of the arthropod fauna, calculated from the feeding habits of the families represented, showed some interesting differences among the three stages of the community. The percentage of predators and parasites in the total population remained remarkably stable throughout the three successional stages, and although their density in relation to crown size decreased steadily with increasing stand age, a ratio of one secondary feeder to three primary feeders was maintained. The relatively large number of transient species, such as the psyllids, many flies and beetles, that did not feed in the crown either on foliage or as predators and parasites was particularly interesting. This group undoubtedly served as a source of food for many predators and decreased pressure on the primary feeders. The density of transients in relation to primary feeders increased from approximately one-quarter of the prey population in the transition stage to one-half in the young-forest stage.

Insect Dispersal

Dispersal of insect, in addition to flight of adults, may occur in all stages of insect development. Brown (1965) reported the transport of large numbers of moths of the forest tent caterpillar, *Malacosoma disstria* Hubner, by the turbulent air associated with a cold front for a distance of at least 300 miles in 12 hours. Maple leaf cutter, *Paraclemensia acerifolia* (FITCH), larvae were found to drop from the foliage, and would

Table 2. *Percentage composition of the arthropod fauna in the crown stratum of three stages in the development of a red pine community, Kirkwood Township, Ontario, 1961—1964 (from* MARTIN, *1966)*

Order	Transition stage				Monoculture stage				Young-forest stage			
	1961	1962	1963	1964	1961	1962	1963	1964	1961	1962	1963	1964
Hemiptera-Homoptera	33.6	67.4	40.5	21.2	22.0	45.2	13.6	9.8	16.1	59.5	26.8	20.7
Acarina	12.3	8.2	22.1	46.8	29.3	18.6	31.2	44.0	12.9	7.3	17.0	17.9
Diptera	29.9	7.7	19.7	14.5	21.9	11.6	13.1	10.0	20.5	9.5	19.8	16.3
Collembola	3.3	4.0	2.0	2.6	10.6	9.7	10.6	9.5	24.2	12.0	14.0	24.6
Araneida	8.3	4.6	4.9	3.9	7.0	3.5	3.5	3.0	6.5	2.6	3.4	3.4
Hymenoptera	5.9	3.4	4.4	3.1	3.0	3.7	4.3	2.8	4.3	2.4	4.7	3.9
Coleoptera	3.6	2.4	2.9	2.7	2.6	2.4	2.3	1.7	6.0	2.0	3.7	1.9
Psocoptera	0.2	0.2	0.7	2.1	1.3	2.0	19.3	17.1	4.5	1.5	7.4	6.8
Neuroptera	0.6	0.8	0.7	1.6	0.3	2.0	0.7	1.3	0.9	2.0	1.5	1.7
Lepidoptera	1.2	0.5	0.6	0.7	0.3	0.5	0.7	0.5	0.8	0.5	1.2	1.1
Thysanoptera	0.5	0.3	0.3	0.3	0.3	0.1	0.1	0.2	1.5	0.2	0.2	0.8
Trichoptera	0	0	0	0	0.6	0	0.1	0	1.2	0	0.1	0
Others	0.6	0.5	1.2	0.5	0.8	0.7	0.5	0.1	0.6	0.5	0.2	0.9
	100.0	100.0	100.0	100.0	100.0	100.0	100.0	100.0	100.0	100.0	100.0	100.0

re-establish themselves on foliage if trees were close enough together or sapling undergrowth was fairly dense. These results suggest that intra- and inter-tree dispersal, especially in heavy infestations, may be more extensive than was heretofore suspected (Ross, 1962). The transfer of larvae from tree to tree also depends upon air movement. Morris (1963) found that small spruce budworm larvae, *Choristoneura fumiferana* (Clem.), often drop on silk threads, which break near the point of attachment to the twig if air currents blow the larvae about. Therefore, more dispersal occurred in open stands permitting free movement of air. Dispersal is greater in stands with severe infestation than in stands with only light or moderate infestation. Lambert and Franklin (1967) found that motile nymphs of the balsam woolly aphid, *Adelges piceae* (Ratz.), dropped from infested Fraser fir and by wind transport could start new infestations at considerable distances from the old host tree.

Recent work with pine bark beetles (McCambridge, 1967; Wood and Vité, 1961) has shown that attraction produced by initial attacking beetles concentrates the activity of the remaining population into a mass attack. The resulting spot kill has a devastating influence upon the forest canopy.

Effects of Insect Feeding

Except for a relatively few forest pests, most leaf-eating insects occur in comparatively small numbers. Insect populations are variable, and with the great amount of foliage on a healthy tree, probably cause slight effect upon the growth of the tree. Reliable measurements of insect populations and effect upon growth are almost impossible to obtain.

Defoliation injures trees by reducing photosynthesis, by interfering with transpiration and translocation of food. In conifers, future photosynthesis is affected by the absence of foliage for 1, 2, or 3 years, and changes in future foliage production by changes in shoot and needle growth. Defoliation may cancel competitive advantage in respect to the environment. To the extent that they sometimes thin overdense stands or kill decadent and suppressed trees insects may actually increase productivity of a forest site (see Chapter 9).

Using radiocesium, Crossley (1966) estimated *Chrysomela knabi* Brown larval food consumption at 7 to 16 mg dry of plant per larva per day in field areas. Laboratory measurements of food consumption rates, by comparison, were about 9 to 10 mg per larva per day for larvae of similar size. The first instar larvae of *Chrysomela crotchi* Brown feed gregariously in compact clusters on the same leaf on which the egg mass was laid. Feeding results in skeletonizing. Second and third-instar larvae tend to be less gregarious. The upper and lower epidermal layers of the leaf, small veins and some larger veins are consumed. Adults leave only the mid-rib and the main lateral veins. The adults have a fall and a spring feeding period. During the fall 22 first-year adults averaged 9.68 cm² of leaf surface per individual, and in a 2-week period in the spring averaged approximately 0.06 cm² per individual (Smereka, 1965).

The leaf-miner *Phytomyza lanati* Spencer feeds in the uniform depth of the palisade layer. By measuring the surface of the mine an accurate volume of food intake by each instar may be obtained. A ratio of approximately 1:5:30 was estimated to be the rate of tunnelling for each instar. This indicates that the rate, as well as the amount of feeding, increased greatly with each molt. The third instar not only eats 40 times as much as the first instar, but its rate of food consumption is 30 times greater (TAUBER and TAUBER, 1968).

REICHLE and CROSSLEY (1967) found that insects consumed an average of 5.6% of the total leaf area of a *Liriodendron tulipifera* L. forest. BRAY (1964) found the percentage leaf utilization by primary consumers of *Acer* (maple), *Fagus* (beech), and *Fraxinus* (ash) canopies to range between 5.0 and 7.7% in forests in southern Canada.

Pine sawflies have a variety of effects upon young and old trees. The larvae of the European pine sawfly, *Neodiprion sertifer* (GEOFFROY), injure their hosts by old-growth defoliation, debarking, and current shoot girdling. Defoliation is undoubtedly responsible for nearly all the growth loss. Both height growth and radial growth are inversely related to sawfly population levels, but radial growth is affected most, especially at higher sawfly population levels. The extent of growth loss is a function of the amount and rate of foliage removed. Even at very high insect population levels defoliation seldom exceeds 90% when old and new foliage are considered together. Since new foliage continues to grow after the sawflies leave the host, the trees "recover" later in the season (WILSON, 1966). Defoliation by the first generation of the sawfly *Neodiprion rugifrons* MIDDLETON is slight but the second generation larvae feed on foliage of all ages often completely defoliating the trees late in the season, completely defoliating the trees late in the season. Completely defoliated trees die, and trees retaining less than 10% of the total crown foliage often die the following spring. Terminal growth in trees with less than 10% total crown defoliation and undamaged terminal shoots is not significantly reduced (WILKINSON et al., 1966). The effects of defoliation by *N. swainei* MIDD. on jack pine are less when old foliage is fed upon than when new and old foliage is consumed (O'NEIL, 1962).

BLAIS (1965) recorded the last six outbreaks of spruce budworm in Quebec through radial-growth studies. The intervals between the six budworm outbreaks in the Laurentide Park have been 44, 60, 26, 76, and 37 years. BLAIS speculated that the outbreaks were not as severe as in other areas of eastern Canada because of marginal conditions, especially weather, for development of budworm populations.

MILLER (1966) studied the black-headed budworm *Acleris variana* (FERN), in eastern Canada during the peak of an outbreak. Outbreaks tend to occur every 10 to 15 years in maturing stands with a high content of balsam fir. The black-headed budworm may cause 2 years of severe defoliation but stands will survive this rate of attack without top-killing or a marked reduction in radial increment. There is probably an association between weather and population release. The maximum larval density at the peak of the outbreak was approximately 200 larvae per 0.93 m^2 of branch surface area. This resulted in 100% defoliation of the current foliage for two successive years. The impact of the black-headed budworm in hemlock-fir forests in coastal British Columbia is apparently more severe than in spruce-fir forests of eastern Canada.

Defoliation by the forest tent caterpillar, *Malacosoma disstria* HBN., showed a number of effects in the aspen forests Minnesota (DUNCAN and HODSON, 1958) (Table 3). Heavier intensities of defoliation reduced average leaf size and induced tip-clustering of leaves during the year after defoliation.

Aspen showed nearly complete recovery in the second year following the last defoliation. There was no evidence to show that defoliation alone caused aspen mortality. Although defoliation by the forest tent caterpillar significantly reduces the

Table 3. *Comparison of growth during and following defoliation by the forest tent caterpillar in aspen stands in Minnesota* (from DUNCAN and HODSON, 1958)

Defoliation history	Average percent of normal[a] growth occurring during year of last defoliation	Average percent of normal[a] growth occurring during the first year following defoliation
Heavy, one year	28	78
Light after one year of heavy	35	100
Heavy, two years	13	84
Light after two years of heavy	17	84

[a] Growth during the five years preceding defoliation.

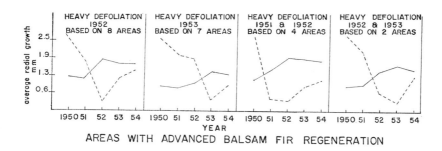

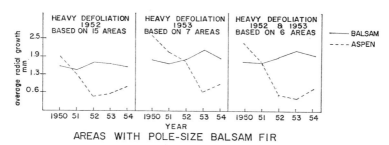

Fig. 2. Relationship between aspen and balsam radial growth during different years of heavy defoliation of the aspen by the forest tent caterpillar, for two size classes of balsams (from DUNCAN and HODSON, 1958)

growth rate of aspen, both height and radial growth of understory balsam fir was greater during this period (Fig. 2).

WICKMANN (1963) found two main effects in white spruce from feeding of the Douglas-fir tussock moth, *Hemerocampa pseudotsugata* McD. Mortality occurred from defoliation alone and from the attack of inner bark boring beetles on trees weakened by defoliation. In addition to mortality and growth loss, top-killing was also an important impact. Defoliation had an immediate effect on radial growth. Red oak, severely defoliated by the winter moth, died back from the top and produced numerous epicormic branches (EMBREE, 1967); these trees usually died during dormancy.

Trees can continue to increase their foliage during the following years, if initial defoliation was no more than 30%. When defoliation was complete, the number of leaves the following year was reduced by as much as 40%.

GIESE and BENJAMIN (1964) studied the entomosociology of the sugar maple complex in relation to maple blight in Wisconsin. A complex ecological relationship exists between a series of leaf rollers and the maple webworm. The maple webworm, whose defoliation is apparently the cause of maple blight, is absolutely dependent upon the rolled leaves for oviposition sites. The leaf roller, *Sparganothis acerivorana* MACKAY, was the initial defoliator in the complex. Sugar maple was the main host; red, silver, and mountain maples also were fed upon. Twenty primary parasitic Hymenoptera were reared from this leaf roller. Some spiders and a bug also were involved in the population. The second leaf roller to appear, *Acleris chalybeana* (FERNALD), had 21 associated Hymenoptera. Eleven of these also occurred on *S. acerivorana*. A third leaf roller, *Gracilaria* sp., also contributed rolled leaves and had a number of parasites. The maple webworm, *Tetralopha asperatella* (CLEMENS), attacks a variety of hardwoods including sugar, red and mountain maple. Eggs of the webworm were found only on the surface at the edge of the roll on partially rolled leaves. An egg parasite, several parasites, lacewings and bugs are associated with the webworm. Several parasites and predators are common to all of the defoliators. The list of defoliating insects occurring on sugar maple in Florence County, Wisconsin consists of 2 orders, Lepidoptera and Coleoptera, with 20 species in 13 families. In addition, 6 sucking forms including aphids, scales, and tree hoppers were collected.

Cone insects of grand fir destroyed an average of 60 seeds per cone. Indirect losses were estimated resulting from seeds deprived of nutrients, fusion of the seed to the scale, and damage to the seed by predacious larvae seeking prey. Seed chalcids were the most important pests (HEDLIN, 1967). COULSON and FRANKLIN (1968) found 14.3% of the second year cones of shortleaf pine, *Pinus echinata* MILL., damaged by coneworms, *Dioryctria* spp., and Cecidomyiidae. First year cones and other cone bearing surfaces were relatively uninjured.

Experimental Approaches

Artificial Defoliation

The artificial removal of foliage of different ages from trees is a means of assessing the primary effects of defoliation. KULMAN (1965) found the average shoot growth of red pine to be reduced when old needles were clipped from the tree. Trees which had new needles clipped showed significantly less shoot growth than trees which had only the old needles removed. LINZON (1958) found that the removal of any year's foliage had an effect upon height growth. Diameter growth was severely reduced in those trees that had their one-year-old foliage removed in combination with another year's foliage. All ages of foliage are of value to white pine but 1-year-old foliage may be the most important. O'NEIL (1962) manually defoliated young jack pine. The removal of 2-year-old or 3-year-old foliage had no appreciable effect on tree growth, but removal of both reduced height growth. Height growth also was reduced by the removal of 1-year-old foliage. Current foliage was essential for the maintenance of normal height, diameter, and shoot growth. Removal of current foliage induced high bud mortality, production of profuse adventitious growth, and a reduction of the rate of shoot elongation. Complete defoliation resulted in tree death shortly thereafter.

KULMAN (1965 b) found disbudding of sugar maple, *Acer saccharum* MARSH, killed 70% of the previous year's shoot growth and 10% of the 2-year-old growth. Terminal function was not assumed by any of the adventitious shoots. Removal of all terminal buds caused 65% reduction in terminal shoot length. Removal of only the distal terminal bud caused the same pattern of damage but growth reduction was less severe. Red maple, *A. rubrum* L., was much less sensitive to disbudding.

IVES and NAIRN (1966) subjected young tamarack, *Larix laricina* (DU ROI), to four successive years of different amounts of artificial defoliation. Fifty percent mortality occurred after two years of complete defoliation, but lesser amounts caused no mortality within four years. Foliage production, stem increment and possibly root growth were reduced by as little as 25% defoliation.

Methods of Population Assessment Sampling

The sampling universe for insects influencing canopy production may be divided into three categories: crown — including foliage, buds, shoots, branches, and seeds and cones; stem — inner bark, soil and litter, and roots; and free air. The feeding stages of the insects are found on plant parts that serve as food. Sampling a population at a specific time other than during the feeding period requires consideration of the life cycle and habits of the insect. Most sampling and population studies of forest insects have indicated aggregated or clumped distribution (WATERS, 1962). The statistical elements contributing to aggregation include the form and size of the sampling unit, sampling intensity, superimposition of independent random distributions, and compounding of random and logarithmic series. The biological elements include aggregation in response to physical factors of the environment; aggregation in response to host plants or parts thereof; reproductive behavior — issuance from egg masses or clumped eggs, with limited movement of subsequent stages; intraspecific mutual attraction, e.g., gregariousness and mating — including courtship; interspecific competition, e.g., restriction of the organism to subareal zones of the habitat, and modification of behavior by parasites and disease.

The measurement of insect populations in the forest presents many difficult and unique problems and no simple, rapid techniques of direct sampling have yet been found that will establish population means with the precision required in intensive studies.

MORRIS (1955) considered spruce budworm sampling units and a method of converting the basic unit to population per acre (0.4 hectare). In stands between 35 and 55 years of age branch surface per hectare is increasing at the mean rate of 3.4% per year. Therefore, even if the population per m^2 remains constant in such stands over a period of 10 years, absolute population increases by 34%. Foliage quantity per hectare probably reaches its peak near age 55, and then starts to decline. The mean annual decrement up to age 105 is about 2% per year. The absolute insect population also decreases at this rate if population per unit of branch surface remains constant. The basic unit of population related to tree mass, volume or surface area is therefore not a reliable index of absolute population over an extended period of years. For short periods, however, and especially during the violent population changes of an epidemic, the errors involved will not be large until repeated defoliation starts to kill the trees. Then branch surface will decline very rapidly and population per surface area of the host plant will have little meaning. Insect numbers in a given forest area may be estimated using the general means for branch surface per hectare. In the 80-year age class the low population is 404 larvae per hectare while an epidemic population of 250 larvae per m² represents 2 million larvae per hectare. In dense 55-year stands the corresponding estimates are 910 and 4.5 million. If 90 of the 250 larvae survive to the pupal stage, and the mean pupal weight is roughly 100 mg, this would represent nearly 16.5 kg of pupae per hectare in the 55-year stand.

Much of the sampling research done in forest entomology is designed primarily for use in extensive surveys and control operations. KNIGHT (1967) has reviewed the subject of forest insect evaluations. Sequential sampling systems have been developed for a number of insects: the lodgepole needle miner (STARK, 1952), STEVENS and STARK (1962); the larch sawfly (IVES and PRENTICE, 1958); the red pine sawfly (CONNOLA, WATERS and NASON, 1959); white grubs (IVES and WARREN, 1965); and the winter moth (REEKS, 1956). KULMAN and HODSON (1962) used the terminal cluster of shoots to sample populations of jack pine budworm. MARTIN (1966) sampled the crown fauna by spraying with pyrethins and capturing the falling arthropods in cloth funnels suspended beneath trees. MORRIS and BENNETT (1967) used one annual census of nests of fall webworm from a moving vehicle for comparing changes in population and natural control factors over diverse climatic areas. LYONS (1964) determined that although sawfly cocoon density could be estimated per unit area of forest, whole trees were the only practical sampling unit for egg and larval estimates. Aerial survey procedures have been developed for many forest insects and these are an absolute necessity in economical determination of extent and intensity of insect infestations. Some of the procedures are: HELLER et al. (1955), a sketch mapping technique for southern pine beetle; HELLER and SCHMIEGE (1962), a strip viewer operation recorder technique for spruce budworm. WEAR et al. (1964) used aerial photos to estimate Douglas-fir beetle infestations. COLWELL (1964) pointed out the uses and possibilities of remote sensing through various types of spectral analysis especially infrared photography.

Management Programs

Economic Damage in the U. S.

In commercial forests, insect damage may be measured in loss of yield, loss of quality, and increased costs of forestry operations. Timber mortality for one year in the U. S. is shown in Table 4. Growth loss caused by insect feeding per year at least

Table 4. *Annual insect caused mortality of growing stock and of sawtimber on commercial forest land in the United States, 1962*[a]

	Softwoods	Hardwoods	Total
Growing stock (1000 cubic feet)	1,096,488	87,619	1,184,107
Sawtimber (1000 board feet)	5,289,801	126,385	5,416,186

[a] Timber Trends in the United States. Forest Resource Report No. 17. Forest Service, U. S. Dept. Agriculture

equals and probably exceeds the total volume of mortality. Other effects of timber mortality on the forest ecosystem are poorly understood. Changes in timber stand composition or destruction of timber stands may have profound influence upon the wildlife populations which were present prior to the timber mortality.

Nutrient losses from the forest ecosystem following insect-caused mortality most certainly must be significant. BORMANN et al. (1968) studied the nutrient cycles in clear-cut and undisturbed forest ecosystems. Their results suggest that clear-cutting tends to deplete the nutrients of a forest ecosystem by (1) reducing transpiration and so increasing the amount of water passing through the system; (2) simultaneously reducing root surfaces able to remove nutrients from the leaching waters; (3) removal of nutrients in forest products; (4) adding to the organic substrate available for immediate mineralization; and (5) in some instances, producing a microclimate more favorable to rapid mineralization. Insect caused mortality undoubtedly has much the same effect as clear-cutting.

Control Programs — Ecosystem Level

Forest canopy insects have been controlled with dramatic success by aerial spraying with insecticides. Ecosystem contaminations and injury to nontarget organisms has drawn criticism for the usual methods of application. Recently HIMEL and MOORE (1967) have shown, through a technique of using insoluble micron size fluorescent particles in the spray solution, that small spray droplets are responsible for insect kill. Spray droplets 50μ or smaller accounted for 98% of the mortality of spruce budworm larvae. Droplets apparently reach the larvae through atmospheric transport and diffusion. This was less than 5% of the total volume of spray. Thus 95% of the total volume of conventional sprays, most droplets larger than 50μ, contribute chiefly to environmental contamination. The economic and ecological implications, arising from the high order of magnitude of potential reduction in insecticide requirements, indicate a need for redesign of spray delivery systems and for a considerable increase in knowledge of atmospheric transport and diffusion of fine droplets in the forest.

Ecological control of insects employs diverse techniques to manipulate three main environmental factors: food, microclimate, and organisms antagonistic to the pests. Food and microclimate factors may be altered by cutting practices, age distribution, vigor, density, and spatial distribution of trees in a stand. Good silvicultural practices often have entomological value. Economic necessities of forest management, however, are sometimes incompatible with ecological control of insects. Suitable compromise of various control methods usually can and should be developed.

Summary

Although economic pest species often have been studied in detail, entomologists have given little attention to the role of the entire insect population in determining the composition, structure, and succession of vegetation. The forest is always in a state of change — often orderly, sometimes cataclysmic. One thing seems evident: the interactions between plants and consumer insects in forest ecosystems are complex and involve many factors, both physical and biotic. The multitude of variables involved create complexities that challenge and often defy our capacity to collect, analyze, and synthesize. The problems of forest insect control seem to illustrate particularly well the diversity of the disciplines involved. At one time or another populations of forest insects may be affected significantly by any of the components of the ecosystem. Analyses of the functional role of insects in forested ecosystems must integrate contributions from many disciplines, especially physiology and forest biology.

References

ANDERSON, R. F.: Forest and shade tree entomology. New York: John Wiley & Sons, Inc. 1964.

BECKER, C. C., and D. M. BENJAMIN: The biology of *Neodiprion nigroscutum* (Hymenoptera: Diprionidae) in Wisconsin. Can. Entomol. **99**, 146—159 (1967).

BLAIS, J. R.: Spruce budworm outbreaks in the past three centuries in the Laurentide Park, Quebec. Forest Sci. **11**, 130—138 (1965).

BORMANN, F. H., G. E. LIKENS, D. W. FISHER, and R. S. PIERCE: Nutrient loss accelerated by clear-cutting of a forest ecosystem. Science **159**, 882—884 (1968).

BRAY, J. R.: Primary consumption in three forest canopies. Ecology **45**, 165—167 (1964).

BROWN, C. E.: Mass transport of forest tent caterpillar moths, *Malacosma disstria* Hubner, by a cold front. Can. Entomol. **97**, 1073—1075 (1965).

COLWELL, R. N.: Aerial photography — a valuable sensor for the scientist. Amer. Sci. **52**, 17—49 (1964).

CONNOLA, D. P., W. E. WATERS, and E. R. NASON: A sequential sampling plan for red-pine sawfly, *Neodiprion nanulus* Schell. J. Econ. Entomol. **52**, 600—602 (1959).

COULSON, R. N., and R. T. FRANKLIN: Frequence of occurrence of cone and seed insects on shortleaf pine in the Georgia Piedmont. J. Econ. Entomol. **61**, 1026—1027 (1968).

CROSSLEY, D. A.: Radioisotope measurement of food consumption by a leaf beetle species, *Chrysomela knabi* Brown. Ecology **47**, 1—8 (1966).

DROOZ, A. T.: Some relationships between host, egg potential, and pupal weight of the elm spanworm, *Ennomos subsignarius* (Lepidoptera: Geometridae). Ann. Entomol. Soc. Amer. **58**, 243—245 (1965).

DUNCAN, D. P., and A. C. HODSON: Influence of the forest tent caterpillar upon the aspen forests of Minnesota. Forest Sci. **4**, 71—93 (1958).

EMBREE, D. G.: Effects of the winter moth on growth and mortality of red oak in Nova Scotia. Forest Sci. **13**, 295—299 (1967).

FINNEGAN, R. J.: The pine needle miner, *Exoteleia pinifoliella* (Chamb.) (Lepidoptera: Gelechiidae), in Quebec. Can. Entomol. **97**, 744—750 (1965).

GIESE, R. L., and D. M. BENJAMIN: The insect complex associated with maple blight. Part II. Studies of maple blight. Univ. Wisconsin Res. Bull. 250. June, 1964.

GRAHAM, S. A., and F. B. KNIGHT: Principles of forest entomology, 4th Ed. London-New York: McGraw-Hill Book Comp. 1965.

HEDLIN, A. F.: Cone insects of grand fir, *Abies grandis* (Douglas) Lindley, in British Columbia. J. Entomol. Soc. Brit. Columbia **64**, 40—44 (1967).

HELLER, R. C., J. F. COYNE, and J. L. BEAN: Airplanes increase the effectiveness of southern pine beetle surveys. J. Forestry **53**, 483—487 (1955).

—, and D. C. SCHMIEGE: Aerial survey techniques for the spruce budworm in the Lake States. J. Forestry **60**, 525 —532(1962).

HIMEL, C. M., and A. D. MOORE: Spruce budworm mortality as a function of aerial spray droplet size. Science **156**, 1250—1251 (1967).

IVES, W. G. H., and L. D. NAIRN: Effects of defoliation on young upland tamarack in Manitoba. Forest. Chron. **42**, 137—142 (1966).

—, and R. M. A. PRENTICE: Sequential sampling techniques for surveys of the larch sawfly. Can. Entomol. **90**, 331—338 (1958).

—, and G. L. WARREN: Sequential sampling for white grubs. Can. Entomol. **97**,596—604 (1965).

KNIGHT, F. B.: Evaluation of forest insect infestations. Ann. Rev. Entomol. **12**, 207—228 (1967).

KULMAN, H. M.: Effects of artificial defoliation of pine on subsequent shoot and needle growth. Forest Sci. **11**, 90—98 (1965a).

— Effects of disbudding on the shoot mortality, growth, and bud production in red and sugar maples. J. Econ. Entomol. **58**, 23—26 (1965b).

LAMBERT, H. L., and R. T. FRANKLIN: Tanglefoot traps for detection of the balsam woolly aphid. J. Econ. Entomol. **60**, 1525—1529 (1967).

LINZON, S. N.: The effect of artificial defoliation on various ages of leaves upon white pine growth. Forest. Chron. **34**, 51—56 (1958).

LYONS, L. A.: The spatial distribution of two pine sawflies and methods of sampling for the study of population dynamics. Can. Entomol. **96**, 1373—1407 (1964).

McCAMBRIDGE, W. F.: Nature of induced attacks by the black hills beetle, *Dendroctonus ponderosae* (Coleoptera: Scolytidae). Ann. Entomol. Soc. Amer. **60**, 920—928 (1967).

MARTIN, J. LYNTON: The insect ecology of red pine plantations in central Ontario IV. The crown fauna. Can. Entomol. **98**, 10—27 (1966).

MILLER, C. A.: The black-headed budworm in Eastern Canada. Can. Entomol. **98**, 592—613 (1966).

MORRIS, R. F.: Influence of parental food quality on the survical of *Hyphantria cunea*. Can. Entomol. **99**, 24—33 (1967).

— The dynamics of epidemic spruce budworm populations. Mem. Entomol. Soc. Can. No. 31. 1963.

— The development of sampling techniques for forest insect defoliators, with particular reference to the spruce budworm. Can. J. Zool. **33**, 225—294 (1955).

—, and C. W. BENNETT: Seasonal population trends and extensive census methods for *Hyphantria cunea*. Can. Entomol. **99**, 9—17 (1967).

O'NEIL, L. C.: Some effects of artificial defoliation on the growth of jack pine (*Pinus banksiana* Lamb.). Can. J. Bot. **40**, 273—280 (1962).

REEKS, W. A.: Sequential sampling for larvae of the winter moth, *Operophtera brumata* (Linn.). Can. Entomol. **88**, 241—246 (1956).

REICHLE, D. E., and D. A. CROSSLEY: Investigation on heterotrophic productivity in forest insect communities, pp. 563—587. In: Secondary productivity of terrestrial ecosystems, (K. PETRUSEWICZ, Ed.), Panstwowe Wydawnictwo Naukowe, Warsaw-Krakow 1967.

ROSS, D. A.: Bionomics of the maple leaf cutter, *Paraclemensia acerifoliella* (Fitch), (Lepidoptera: Incurvariidae). Can. Entomol. **94**, 1053—1063 (1962).

SMEREKA, E. P.: The life history and habits of *Chrysomela crotchi* Brown (Coleoptera: Chrysomelidae) in Northwestern Ontario. Can. Entomol. **97**, 541—549 (1965).

STARK, R. W.: Sequential sampling of the lodgepole needle miner. Forest. Chron. **28**, 57—60 (1952).

STEVENS, R. E., and R. W. STARK: Sequential sampling for the lodgepole needle miner, *Evagora milleri*. J. Econ. Entomol. **55**, 491—494 (1962).

TAUBER, M. J., and C. A. TAUBER: Biology and leaf-mining behaviour of *Phytomyza lanati* (Diptera: Agromyzidae). Can. Entomol. **100**, 341—349 (1968).

WATERS, W. E.: The ecological significance of aggregation in forest insects. Proc. 11th Intern. Congr. Entomol. **2**, 205—210 (1962).

WEAR, J. F., R. B. POPE, and P. G. LAUTERBACH: Estimating beetle-killed Douglas-fir by aerial photo and field plots. J. Forestry **62**, 309—315 (1964).

WELLINGTON, W. G.: Some maternal influences on progeny quality in the western tent caterpillar, *Malacosoma pluviale* (Dyar). Can. Entomol. **97**, 1—14 (1965).

WICKMAN, BOYD E.: Mortality and growth reduction of white fir following defoliation by the Douglas-fir tussock moth. Pacific SW. Forest & Range Expt. Sta. 1963, PSW-7, 15 pp.

WILKINSON, R. C., G. C. BECKER, and D. M. BENJAMIN: The biology of *Neodiprion rugifrons* (Hymenoptera: Diprionidae) a sawfly infesting jack pine in Wisconsin. Ann. Entomol. Soc. Amer. **59**, 786—792 (1966).

WILSON, L. F.: Effects of different population levels of the European pine sawfly on young Scotch pine trees. J. Econ. Entomol. **59**, 1043—1049 (1966).

WOOD, D. L., and J. P. VITÉ: Studies on the host selection behavior of *Ips confusus* (Le Conte) (Coleoptera: Scolytidae) attacking *Pinus ponderosa*. Contrib. Boyce Thompson Inst. **21**, 79—96 (1961).

9

Estimation of the Effects of Phytophagous Insects on Forest Production

P. M. RAFES

Insects may damage woody plants in a variety of different ways and, consequently, their potential to reduce the photosynthetic capacity or net primary production of forests is considerable:

1. Foliage consumption deteriorates the physiological condition of the tree, decreases the photosynthetic biomass, and indirectly affects the wood increment and the formation of reproductive organs.

2. Sap consumption also deteriorates the physiological condition of the tree and reduces foliage and wood production, but not to the extent as does the consumption of photosynthetic tissues by phytophages.

3. Consumption of woody tissues decreases wood production; it also can disrupt sap flow and, therefore, deteriorates the overall condition of the tree. However, wood consumption generally affects weakened plants or their parts, and damage to the entire forest stand caused by xylophages is not as great as with foliage consumers. In fact, dead wood consumption may be beneficial to plants, since it thins the nonproductive, but light-intercepting parts of trees.

4. Insect damage to reproductive organs (at all stages of their development) may decrease fruit production and reduce the reproductive potential, but normally it does not influence the foliage and wood increment of growing trees.

The production and turnover of foliage occurs over a short seasonal cycle and may show rapid recovery from insect attack. Wood production, however, is cumulative through time and the effects of reduced increments incurred during previous years are lasting. Therefore, the influence of foliage consumers on the wood increment of forests has received foremost attention.

In the USSR there have been many recent studies on the wood increment losses of forests resulting from insect damage to the photosynthetic structure. These investigations were made by A. I. VORONTSOV with his collaborators. They proposed an approach based on analysis of a series of publications (GOLOSOVA, 1963; IERUSALIMOV, 1965; MOZOLEVSKAYA and TUDOR, 1967; TURCHINSKAYA, 1963), which indicated that during periods of high population density of herbivorous insects not only did the quantity of foliage consumed increase but the wood increment also decreased accordingly.

The Effect of Foliage Consumption on the Wood Increment of Forests

Foliage Consumption

The amount of foliage consumed by an insect population can be calculated from the equation (RUDNEV, 1935):

$$X = \frac{N \cdot F \cdot 100}{M}$$

where X is the estimated percentage foliage loss, N is the number of feeding individuals in the crown, F is the average total food requirement of an individual during its lifetime, and M is the biomass of foliage in the crown. ILINSKII (1952) and ILINSKII and TROPIN (1964) published tables in which the estimated foliage consumption by a number of different insect species was calculated from this relationship. In such calculations, however, density patterns of the population must be considered. Larvae are seldom uniformly distributed in a crown and theoretical levels of foliage consumption may never occur, since competition between consumer species may lead to localized starvation in the midst of an overabundance of food (RAFES, 1967, 1968).

The appraisal of increment loss of foliage due to insect damage can be made only if the normal expected foliage production of the forest is known. Therefore, it is necessary to determine the increment production of foliage and other tree components which would be formed in the absence of insect infestation.

Increment Production

VORONTSOV's (1963) review shows that the increment production of damaged trees may be compared with "normal" production rates calculated in one of three ways: 1) the expected increment calculated from established precipitation relations; 2) the increment rate of uninfested similar forest area; 3) the mean increment of the forest for some years preceding insect attack. In each case, however, there is no assurance that the increment accurately characterizes that which would have been formed in the given forest if the trees had not been damaged.

It is recognized widely that precipitation greatly influences the increment rate of wood formation, e.g., ILINSKII and KOBOZEV (1939) found a positive correlation between the amount of precipitation and the wood increment of trees. However, correlations with precipitation alone neglect the influence of other environmental variables. The second alternative using values from adjacent stands tends to remove environmental factors as variables, but it is applicable only if there is a previously demonstrated similarity in increment production by the forests being compared. The third method considers the variation in increment production as being solely dependent upon foliage consumption.

Soil condition is another obvious factor affecting wood production. The decreased increment rate during attack by the gypsy moth (*Porthetria dispar* L.) on an oak forest in the Voronezh region was greater for stands on better soils than for stands on solonetz and solonetz-like loam, but the increment rate following termination of the outbreak was greater (ILINSKII and KOBOZEV, 1939). This site correlation with soil conditions is illustrated in Tables 1 and 2 (VORONTSOV, 1967).

Increment growth relations with precipitation might be expected as a result of the high moisture requirements of growing tree parts: foliage, one-year-old shoots, and inner bark of stem and branches. Mentioning the importance of water relations, SMIRNOV (1964) noted that the rate of shoot and foliage increments increased with solar radiation, air and soil temperature, and atmospheric humidity deficit; there was no such relation between diameter growth and these weather conditions. Seasonally, maximum annual production is obtained first by foliage, then first-year shoots, still later the stem increment, and finally the roots.

Table 1. *Mean percentage increment losses in the different types of forests damaged by leaf-eating lepidopteran larvae* (Vorontsov, 1967)

Types of forests[a]	Habitat conditions[a]	Stand (site) quality[b] understory, grass cover	Phenological groups of insects					
			spring		spring-summer		summer-autumn	
			Duration of foliage damage					
			one year	two years	one year	two years	one year	two years
Steppe Zone								
Dry and fresh maple-shrub oak forests	Dry and very dry (rarely fresh), leached chernozems, dark loams, skeletal-carbonate soils; on plateaus, river and gully slopes	Oak of stand quality IV-V (rarely III); Norway maple in 2nd layer or only light understory (spindle tree, *Acer tataricum*, hawthorn)	30	65	25	55	0	20
Humid and damp lily-of-the-valley and dewberry oak forests in bottomlands	Alluvial loamy soils, degraded chernozems; in rarely-flooded bottomlands, above flood-plain terrace	Oak of quality II-III; Ulmaceae, maple, lime; understory of average age density (*Acer tataricum*, buckthorn and others)	25	70	25	60	0	25
Forest-steppe Zone								
Fresh and dry maple-lime, sedge and goutweed oak forests	Dark grey and grey forest soils, at south leached chernozems; on plateaus, elevations, and dried-up gully and ravine slopes	Oak of quality II is predominant; maples, lime, ash; honeysuckle, spindle tree, hazel and others in understory	20	55	30	75	0	30
Fresh maple-lime-oak forests with herb layer	Grey soddy podzolic loamy sand and sandy or meadow boggy soils; gentle slopes along left banks of rivers	Oak if quality II-III with birch, aspen; Norway maple, lime in 2nd layer; light understory (spindle tree, *Sorbus*, buckthorn)	25	60	25	60	0	25
Forest Zone (Moscow Region)								
Fresh and humid sedge weaselsnout goutweed oak forests with lime and hazel	Slightly and moderately podzolized loams, slightly gleied; undulating microrelief, gentle slopes with southern exposure	Oak of quality II-III, coppice forest: lime, aspen, spruce; understory: hazel, honeysuckle, spindle tree and others	10	22	0	25	0	0
Mixed pine and spruce forests with oak	Loamy sands and loams, podzolized to various degrees; undulating microrelief, locally covered with mounds	Oak of quality III-IV coppice forest with conifers; understory: hazel, buckthorn, honeysuckle	10	20	0	20	0	0

[a] "Fresh" types or stands refer to (at least intermittendly) moist but (mostly) well aerated substrates.   [b] Or Bonität.

Table 2. *Percentage tree mortality in forests damaged by leaf-eating Lepidoptera larvae* (after Vorontsov, 1967)

Forest Type	Phenological groups of insects					
	spring		spring-summer		summer-autumn	
	Duration of foliage damage					
	2 years	3 years	2 years	3 years	2 years	3 years
Steppe Zone						
Dry and fresh maple-shrub oak forests	0	2.0	3.0	9.0	2	—
Humid and damp bottomland oak forests	4.0	8.5	5.0	10.0	2.0	—
Forest-Steppe Zone						
Fresh and dry maple-lime-oak forests	2.0	4.0	4.0	12.0	0	—
Fresh-maple-lime-oak forests (herbaceous layer present)	4.0	6.0	5.0	10.0	—	—
Forest Zone (Moscow Region)						
Fresh and humid oak forests with lime and hazel	1.5	3.0	1.5	—	—	—
Mixed pine-spruce forests with oak	1.5	3.0	1.5	—	—	—

Table 3. *Relation of diameter increment loss to volume loss of growing trees in stands attacked by leaf-eating insects* (after Vorontsov et al., 1966)

Age of stand	Ratio of percent volume loss: percent diameter increment loss		
(years)	minimum	mean	maximum
20	1.01	1.18	1.48
25—39	0.52	0.73	0.86
40—59	0.59	0.83	0.96
60—79	0.67	0.86	0.97
80—100	0.59	0.79	0.90
all age classes	0.52	0.88	1.48

Since light intensity is typically the limiting physical variable on tree production in forests, the correlation between precipitation and increment growth often is obscured because of the inverse relationship between radiation flux and precipitation (Ilinskii and Kobozev, 1939). It is necessary to define "increase" in this context as the change in variables within the limits of natural environmetal conditions, which affect the potential increment. Increased intensity of solar radiation favors foliage production and photosynthesis; at constant radiation intensity, production increases with increasing temperature. The greater the humidity deficit of the air the more transparent to radiation is the atmosphere. Increased soil moisture (within the limits of wilting point and saturation) accelerates evapotranspiration and nutrient transport.

Effect of Foliage Consumption on the Wood Increment

Investigations of the effects of foliage consumption on the production and wood increment of forests in the USSR have concentrated on oak (*Quercus robur* L.) forests, since this species is economically very important (Vorontsov et al., 1966; Vorontsov, 1967). Foliage consumption during various periods of the growing season influences the increment formation of wood to different degrees. For this reason, leaf-eating insects were divided into phenological groups characterizing the period of foliage-feeding activity: 1) spring insects (Tortricidae; *Operophthera brumata* L.*, *Phigalia pedaria* F., *Biston hispidaria* Schiff., and *Porthetria dispar* L.), 2) spring-summer insects (*Notodonta anceps* Goeze), and 3) summer-autumn insects (*Phalera bucephala* L. and *Dasychira pudibunda* L.). These data are summarized in Table 1 using mean percentage loss in diameter increment, although repeated insect attacks and defoliation may affect height growth. These damages apply to situations where grazing was intense — percentages of foliage loss in excess of 75%. When foliage damage is less severe (under 50% consumption), increment losses are negligible.

Increment losses vary between stands. On better sites, where the growing season of oak is longer, the greatest losses are caused by insects (e.g., *P. dispar* and *N. anceps*) which feed during early and midsummer. On poorer sites, where the period of growth is shorter, the greatest increment losses are caused by early vernal pest species (loopers, tortricids). Such seasonality occurs with two oak forms; in the early one (var. *praecox*) the greatest increment losses are caused by the spring group of insects, but in the late form (var. *tardiflora*) they are caused by summer insects (Vorontsov, 1967).

Since diameter increments at the beginning of the growing season before leaf development is completed are supported by reserve food materials, spring foliage consumption has little effect upon this initial increment. Increments produced later in the season are much more dependent upon photosynthetic biomass and the activity of phytophagous insects. Regeneration of foliage from dormant buds during the summer will increase the photosynthetic biomass, but usually will not affect the increment. In estimating increment losses, such behavioral peculiarities as the grazing habits of larvae—either proceeding from stem to crown periphery or from periphery to stem—may be important. In the first case "shade" leaves and in the second case "sun" leaves will be eaten, and these will obviously affect the photosynthetic capacity of the tree differently.

Each tree in a forest has individual characteristics, and responses of different trees to infestation will not necessarily be similar. Moreover, it is evident that quantitative changes of foliage biomass not only affect increment growth but also may result in changes in vegetational composition, i.e., tree mortality (Table 2). Tables 1 and 2 were constructed for middle-aged coppice stands; the forest types have been grouped for convenient analysis. The values given, therefore, may vary considerably even within the same climatic zone.

Using the data of stand increment, mean diameter, and standing crop for each age class, increment losses in volume can be estimated (Table 3). From Tables 1—3 it is apparent that data on foliage consumption (past and present) permit estimation of the

---

* Lepidoptera: Geometridae, *O. brumata*, *P. bucephala*, *B. hispidaria*, *P. pedaria*; Notodontidae, *N. anceps*; Liparidae, *P. dispar*, *D. pudibunda*.

reduction in stand performance. Accurately predicting future increment losses and tree mortality from existing levels of insect infestation enables decisions to be made with respect to the need for preventive control practices. Vorontsov's method (Vorontsov et al., 1966; Vorontsov, 1967) is potentially valuable, and for widespread application comparable data need to be developed for other tree species.

Insect-Plant Relationships

Questions underlying the functional nature of the insect-host plant relationship in forest ecosystems still remain unanswered. It is noteworthy that increment losses are insignificant until defoliation exceeds limits of about 75% of the photosynthetic biomass (Vorontsov, 1967). Researchers commonly have drawn several conslusions from this phenomenon: a) that if phytophagous insect outbreaks do not occur, foliage and incremental losses due to foliage consumption are negligible; b) that foliage loss affects only the increment, but the ecosystem remains unaltered. Both of these assumptions are erroneous.

The standing crop of consumers typically exists at some "stable" threshold. These populations normally consume a rather small fraction of the foliage biomass, and the host plants are adapted to these losses. Therefore, the consequences of severe population fluxes of foliage-feeders should be compared with the effects of normal population levels. Rafes (1964, 1968) calculated from the data of Ilinskii (1959) the quantities of foliage consumption during outbreaks of the gypsy moth over a seven-year period. During the first two years, foliage losses did not amount to more than a few percent of the host-plants' production. However, during the third year the population had increased so greatly that larvae consumed the entire spring production of foliage (which partly regenerated during the summer), and the increment losses became very significant.

Concurrent with tree increment inhibition, some other phenomena also take place in the ecosystem. The quantity and quality of foliage consumed by insects, in part, determines their gross production and thereby their impact on the energy flow through forest food chains. For this reason one also must consider: a) the effects on those phytophagous insect populations developing later in the growing season, whose food resource may be limited by previous foliage damage; b) the increase of saprophagous populations as a consequence of the increased frass input to litter; and c) the increases of predator populations which prey upon both these phytophagous and saprophagous populations. Qualitative and quantitative changes in plant and animal remains in litter (especially in the ratios of carbon, nitrogen, and phosphorus) can influence the soil flora, rates of decomposition, and future primary production through increased nutrient availability. Ierusalimov (1965, 1967) and Vorontsov et al. (1967), working in oak forests, showed that extraordinary foliage consumption increases the light intensity reaching lower strata of the forest, changes the microclimate, alters the growth rate of various strata differently, accelerates the mortality of repressed trees, and may affect changes in the species composition of tree and herbaceous strata. Many of the potential changes in forest conditions resulting from insect damage of foliage have been reviewed by Sukachev and Dylis (1964) and Rafes (1968).

Summary

Even the limited examples considered here illustrate that the increment losses resulting from foliage consumption by insects have economic consequences, but increment losses alone may be an inadequate measure of the overall response of the forest ecosystem. It is difficult to categorize all the effects of insect activity as either detrimental or beneficial to the ecosystem, but attempts should be made to interpret their significance in relation to other components and processes of the system. In order to describe, and especially predict, insect-plant-ecosystem interactions, attention must be given to the population dynamics of the consumer — life tables, predation, and parasitism, and the seasonal aspects of population growth.

References

Golosova, M. A.: Biology and economic values of loopers *Biston* and *Phigalia* in oak woods of steppe zones. Doctoral Dissertation 22p. Moscow Forest Engineering Institute (1963).

Ierusalimov, E. N.: Change in accretion in mixed oak forests following defoliation by leaf-eating insects. Izv. Vyssh. Ucheb. Zaved., Lesn. Kh. **6**, 52—55 (1965).

— Influence of canopy defoliation by leaf-eating insects on certain elements of the forest biocenosis. Doctoral Dissertation, 18 p. Moscow Forest Engineering Institute (1967).

Ilinskii, A. I. (Ed.): Control of leaf-eating pests in forests and prediction of their outbreaks. State Pulp and Paper Pub. House, 138 p (1952).

Ilinskii, A. I.: The Gipsy moth and its control measures. State Pulp and Paper Pub. House (1959).

—, and I. A. Kobozev: Gipsy moth invasions in the Tellerman State Forestry Farm and their influence on accretion of oak. Nauchn. Zap. Voronezh Institute of Forest Economy **5**, 11—28 (1939).

—, and I. V. Tropin (Eds.): Survey and recording of leafeating pests and prediction of their outbreaks. Lesn. Prom-St' (Forest Industry), 525 p. (1964).

Mozolevskaya, E. G., and I. Tudor: Effect of *Tortrix viridana* L. on the state and accretion of oak forests, pp. 6—14. In: Problems in protection of forests, 15. Moscow: Forest Engineering Institute 1967.

Rafes, P. M.: Insect pest outbreaks as a special case in the material and energy turnover in a forest biogeocenosis, pp. 3—57. In: Forest Protection from Pests. Moscow: Science Pub. House 1964.

— The food relations of forest insects. Trans. 17th and 18th Memorial Sessions of Kholodkovskii. Moscow: Science Pub. House 1967.

— The role and significance of phytophagous insects in forests, 234 p. Moscow: Science Pub. House 1968.

Rudnev, D. F.: Methods of surveying and recording infestations of *Panolis flammea* Schiff. In: Forests, Proc. Inst. Zool. and Biol. Ukrainian Acad. Sci., Kiev (1935).

Smirnov, V. V.: Seasonal growth of main tree species. Moscow: Science Pub. House 1964.

Sukachev, V. N., and N. V. Dylis: Fundamentals of forest biogeocenology, 574 p. Moscow Science Pub. House 1964. (English translation published by Oliver and Boyd, 1968).

Turchinskaya, I. A.: Effect of defoliation by the Gipsy moth and other leaf-eating pests on the growth of oak. Zool. Zh. **42** (1963).

Vorontsov, A. I.: The biological bases of forest protection, 348 p. Moscow: Vyssh. Shkola 1963.

— Criteria for chemical control of pests in deciduous plantings, pp. 19—29. In: Problems in protection of forests, 15. Moscow: Forest Engineering Institute (1967).

—, M. A. Golosova, and E. G. Mozolevskaya: Criteria for the basis of chemical control of leaf-eating insects, pp. 3—10. In: Problems of forest protection. Moscow: Central Sci. Res. Institute 1966.

—, E. N. Ierusalimov, and E. G. Mozolevskaya: The role of leaf-eating insects in forest biogeocenoses. Zh. Ob. Biol. **28**, 172—187 (1967).

10

Secondary Production of Birds and Mammals

DALE R. McCULLOUGH

Few studies of birds and mammals have attempted to place these animals in an ecosystem context. In most terrestrial ecosystems, and especially forest ecosystems, the proportion of the total energy and nutrients cycling through birds and mammals is relatively small. However, the importance of birds and mammals cannot be evaluated only in these terms since they are a primary source of animal proteins in human diets.

The emphasis on calories as a measure of nutrition on a world-wide basis has not obscured the fact that the supply of quality protein is a more critical consideration. In addition, birds and mammals perform numerous functions in ecosystems due to activities other than feeding (i.e., mechanical disturbance, selective feeding, seed dispersal, etc.). Because of their mobility, they play an important role in transport of materials within and between both terrestrial and aquatic communities.

Methods of Determining Standing Crop Biomass

No attempt will be made here to cover in detail the myriad of techniques which have been developed for the study of birds and mammals (e.g., MOSBY, 1963; PETRU-SEWICZ, 1967a; GOLLEY, 1968) and especially large herbivores (GOLLEY and BUECH-NER, 1968). Usually, legally harvested animals can be checked as a prime source of information, and scientific collecting permits can be obtained for closed season periods or where the legal kill is inaccessible. Live-trapping devices (TABER and COWAN, 1963), although more expensive to obtain and operate, have been developed for animals ranging in size from song birds and bats (mist nets) to elk (box and corral traps). The recent development of drugs and projectile syringes makes it possible to capture most large mammals.

Determination of the standing crop of biomass (weight) of birds and mammals at any instant in time involves the estimation of three basic parameters: 1) the mean biomass (weight) of individuals by sex and age class, 2) the composition of sex and age classes in the population, and 3) the total number of animals in the population. Standing crop of energy, water, Ca, P, etc. can be determined by the same general procedures.

Among both mammals and birds the weight of an individual can vary greatly depending upon its age, sex, and general health (Tables 1—3). Age is of particular importance in the larger-sized species where growth is usually prolonged and may extend over several years (Table 1). Similarly, sexual dimorphism is common among birds and mammals.

Birds and mammals in temperate regions show seasonal fluctuations in weight with the low during the winter. Game birds appear to lose weight normally over the

Table 1. *Weights (kg) by age and sex for some representative large mammals*

Age class (years)	Elephant[a] (*Loxodonta africanus*) both sexes	Lechwe[b] (*Kobus leche*)		Black-tailed deer[c] (*Odocoileus hemionus columbianus*)	
		♂	♀	♂	♀
0		4.8	4.5	3.1	3.4
1	90.0	54.6	60.0	36.4	35.4
2	204.5	67.1	72.7	54.9	49.6
3	318.1	85.5	78.1	71.8	56.7
4	454.5	103.2	80.0	74.5	63.5
5	613.6	104.5	75.4	87.3	41.4
6	795.4	105.9	83.2	96.4	42.7
7	1000.0	100.9	78.1		
8	1204.5	101.8	80.9		
9	1409.1				
10	1613.6				
.					
.					
.					
25	4090.9				

[a] Petrides and Swank (1966).
[b] Robinette and Child (1964).
[c] Brown (1961) except *0* age class Golley (1957).

winter (Keith, 1962) and may lose from 40 to 50 percent of body weight before death by starvation (Keith, 1962; Jordan, 1953; Tester and Olson, 1959). White-tailed deer males typically lose weight during the rutting season (White, 1968) and both sexes may lose up to 30 percent of their fall weight during the winter under poor food conditions (Davenport, 1939). Other mammals show a similar weight loss during seasonal stresses such as cold or drought. Migratory birds typically build fat reserves in the fall and spring prior to migration (e.g., King, Barker and Farner, 1963). These fluctuations in weight can have a considerable impact upon the standing crop biomass, for under poor conditions production drops to a low level, and body tissues are metabolized with resultant weight loss.

Determination of Sex and Age

The sex of all mammals and most birds can be determined in the hand, and many have secondary sexual characteristics which can be observed in the field on free-roaming individuals. Most modern "field guides" include such information. Age criteria are now available for most groups of mammals and birds (Madsen, 1957).

In birds, cloacal examination and/or juvenile plumage characteristics will separate young from adults (Taber, 1963), but this usually requires that the birds be in hand for careful examination. There are at present no satisfactory methods of determining age of adult long-lived birds such as gulls and birds of prey, other than records of birds banded as juveniles (e.g., Kadlec and Drury, 1968). However, many birds show deterministic growth (although sexual dimorphism in size is common) and, for purposes of deriving biomass estimates, the various adult age classes can be combined for each sex.

In mammals, juveniles can be separated from adults by the eye lens weight (FRIEND, 1967), epiphiseal closure, and milk tooth replacement techniques (TABER, 1963). None of these characteristics will adequately separate year classes of older animals. For many years, the degree of wear on the permanent teeth was the principle method used to assign ages to adult mammals, but this is somewhat variable due to the variation in abrasiveness of the diet in time and space. Annual horn rings in some bovids allows accurate age determination (COWAN, 1940; BRANDBORG, 1955), but for most mammals, no such method is available. The development of the cementum ring technique (LAWS, 1952), which has been applied to many mammals has resulted in accurate age determinations of older animals.

Table 2. *Weights (g) by age and sex for some representative small mammals*

Age (wks)	House mouse[a] (*Mus musculus*) ♂	♀	Norway rat[a] (*Rattus norvegicus*) ♂	♀	Cottontail rabbit[b] (*Sylvilagus floridanus*) both sexes
0	1.25	1.26	6.08	5.75	25.6
1	4.03	4.04	17.5	16.2	58.3
2	6.54	6.52	36.9	33.5	68.5
3	8.76	8.72	59.6	53	92.4
4	12.13	11.94	92.9	79.5	134.5
5	14.94	14.73	138	113	234.5
6	17.60	16.77	188	147	274.5
7	20.06	18.42	233	172	374.5
8	21.59	19.81	274	196	435.6
9					414.0
10			339	227	497.4
11					517.5
12	24.78	22.99	393	251	659.7
13					765.3
14					785.4
15			440	274	945.0
16	27.76	26.11			975.2
19					981.0
20	29.91	29.03	490	303	
21					1106.0
24	31.07	31.00			
28	33.07	33.00			
30				335	
32	34.07	34.52			
36	35.49	36.38			
40				358	

[a] ALTMAN and DITTMARS (1964).
[b] LORD (1963).

Population Composition

A major problem in determining composition is obtaining a representative sample, since many sources of data are biased. Sport hunting kill, for example, is biased towards trophy animals, brightly colored males among birds, etc. Conversely, certain sex and age classes may be inherently susceptible to hunting mortality because of

Table 3. *Weights (g) by age and sex of some representative birds*

Age (days)	House Sparrow[a] (*Passer domesticus*) both sexes		Ruffed Grouse[b] (*Bonasa umbellus*) ♂	♀	Wild Turkey[c] (*Meleagris gallopavo*) ♂	♀
0			12	11	40	40
1	2					
4	10					
7			21	20		
8	18					
10	20					
14			39	32	80	80
15	25					
28			95	81	280	250
42			180	63	560	480
56			295	260	850	690
70			400	352	1220	960
84			479	428	1640	1240
98					2100	1580
112			550	498	2600	1980
126					3320	2520
140			579	511	4050	3000
154					4620	3320
168					5100	3480
182					5500	3620
196					5780	3710
210					5950	3770
252					6260	3910
280					6350	3960
Adult	♂30.5	♀30.0	651	587		

[a] Summers-Smith (1963).
[b] Bump et al. (1947).
[c] Altman and Dittmar (1964).

behavorial traits (see, for example, Anderson, 1953; Imber, 1958; Eberhardt and Blouch, 1955). Composition based upon hunter-killed or even "randomly" collected animals should be cross-checked against composition obtained from another method. Similar difficulties pertain to live-captured animals because of variation in trap susceptibility (or avoidance) by sex, age, and season (Young et al., 1952; Geis, 1955; Bailey, 1969).

Field observations must also be treated with caution. In large ungulates, adult males are often separated socially and geographically for much of the year (Darling, 1937; Murie, 1951; Welles and Welles, 1961; Cowan, 1956; Robinette and Child, 1955). In territorial species (e.g., passerine birds) the pairs may be particularly obivous, but a "floating population" of excess birds (Stewart and Aldrich, 1951; Hensley and Cope, 1951) may be extremely difficult to sample. Very young animals are often difficult and sometimes impossible to account for in composition counts.

However, with adequate precaution to obtain a representative sample, population composition can be determined with reasonable satisfaction, with the exception of very small young. An intimate knowledge of the life history and behavior of the species under study is indispensible in avoiding pitfalls due to biases in data.

Population Census

Methods of determining total animals in a population are far from satisfactory. Large animals in openlands can be counted successfully from the air (e.g., SINIFF and SKOOG, 1964; TALBOT and STEWART, 1964), but little success has been obtained in habitats with an appreciable amount of cover such as the deciduous forest. There is no dearth of methods, of which only the more satisfactory ones will be given here.

Aerial counts are ineffective in deciduous forests because of the lack of contrast between the animal and its background. The recent use of aerial infrared scanners to detect deer by heat rather than visible light (CROON et al., 1968; McCULLOUGH, OLSON and QUEAL, in press) shows promise of giving accurate counts under the specific conditions to which it can be applied. It requires a snow background, which limits the technique to the temperate and more polar regions and to the winter season. It will be effective in detecting only large animals about the size of a deer or larger. It will not penetrate a green canopy, so it will not be effective in dense conifers or in a deciduous forest in foliage. Thus, this technique is quite restricted in applicability, but it promises to give good counts where it will work.

Strip counts (HAYNE, 1949a; ROBINETTE et al., 1954; HIRST, 1969) are unreliable because of the difficulty in determining the effective width of the strip, overlooking of concealed animals and the escape of animals unobserved ahead of the observer. The behavior of animals varies daily because of weather conditions (HAHN, 1949; KOZICKY, 1952) and seasonally with the stage of the life cycle, which further complicates strip counts.

Drive censuses, where a line of men pass through an area, were at one time considered to give accurate counts. A good example of its application is the George Reserve, Michigan deer herd. This herd is contained in a two-square-mile (5.2 km²) enclosure through which a line of about 100 drivers passes systematically with a spacing of about 20 to 30 m between drivers. Incredible as it may seem, the count of deer has been found to be about 20 percent low (JENKINS, 1964). Subsequently, procedures have been improved to assure that failure of the count is not due to gaps in the line. Apparently some of the deer are showing the well-known behavior of seeking out heavy cover to lie down and drivers pass by without seeing the animals.

Mark and recapture (or resight) methods (see TEPPER, 1967 for a bibliography), often referred to as the Lincoln Index when applied to terrestrial populations, require immense effort to obtain representative samples both in the marking and recapture phases of the census. Particularly discouraging is the report by STRANDGAARD (1967) that more than two-thirds of a roe deer population in Denmark had to be marked in order to obtain a reliable estimate of the total population. Nevertheless, this method is commonly used to estimate populations of birds and small mammals, and it will probably continue to be one of the more widely used methods.

The change in age or sex ratios approach (PAULIK and ROBSON, 1969) requires a determination of the population sex and age composition before and after a harvest of known numbers and composition. It assumes that the observed changes in the remaining population were caused by the harvest. While this method is logically sound, the assumption that the harvest is responsible for the observed changes is greatly complicated by natural mortality as well as differential reporting of the kill. The latter situation is a very severe problem where certain classes of animals are illegal. For example, many states in the U. S. restrict the kill to males; however, an appreciable number of females may be killed by design or accident (LEEDY and HICKS, 1945; DAHLBERG and GUETTINGER, 1956) and these go unreported. But perhaps the

Table 4. *Comparison of some census methods for estimating deer populations*

Method	Mule deer (Utah)[a]	Black-tailed deer (Calif.)[b]	White-tailed deer (George-reserve)[c]
Drive census	a) 93 (5 men, 282 meters apart) b) 143 (35 men, 40 meters apart)		80
Infrared scanning			98
Reconstructed population[d]			101
Winter reconnaissance	c) 64 (two counters on horseback)		
Strip census	d) 128 (1940) e) 143 (1941)		
Lincoln index		109 (June-Nov. 1952) 86 (Nov. '52-March '53)	
Sample area count		100 (June-Nov. 1952) 83 (Nov. '52-March '53)	
Pellet group count		93 (June-Nov. 1952) 120 (Nov. '52-March '53)	
Total deer count		103 (June-Nov. 1952) 94 (Nov. '52-March '53)	
Actual	a) 143 b) 141 c) 147 d) 434 e) 172		

[a] RASMUSSEN and DOMAN (1943).
[b] DASMANN and TABER (1955).
[c] Author's data.
[d] Nearly every animal in the population is eventually harvested. By aging, the minimum number alive in a given year can be determined (JENKINS, 1964).

Table 5. *Comparison of some census methods for estimating small rodent populations*[a]

	House mouse *Mus musculus* (1964) Adult		Oldfield mouse *Peromyscus polionatus* (1964) Juv.		Adult		Oldfield mouse *P. polionatus* (1965) Juv.		Adult	
	♂	♀	♂	♀	♂	♀	♂	♀	♂	♀
Lincoln index	27.9	13.9	4.0	2.0	19.8	3.0	0	0	16.2	8.0
Hayne method[b]	18.2	12.2	3.7	1.9	13.7	2.8	10.0	6.0	14.0	17.0
Actual no./ha[c]	53.8	51.1	45.9	48.9	38.0	41.0	25.7	15.3	38.3	43.5

[a] SMITH (1968).
[b] HAYNE (1949b).
[c] Number from trapping and digging from burrows (attempted total removal).

greatest drawback to the method is that it requires three estimates, each with a more or less large error factor, which when combined in one computation give a dubious result at best.

The rate of capture method (ZIPPIN, 1958; DAVIS, 1963; EBERHARDT, 1969) relies upon a catch per unit effort which changes as the population fluctuates in number. In the case of small mammals, the catch per unit effort declines as the population is reduced by trapping, and the plotting of animals caught against effort can be projected to the point where catch reaches zero to indicate the total population. Unfortunately, catch per unit effort fluctuates because of factors other than population density. Reductions of populations by trapping usually result in substantial immigration of individuals from neighboring areas.

Pellet group (feces) counts (MCCAIN, 1948) have been used in many studies to estimate the number of large herbivores. The method assumes a constant defecation rate on a given area over a known length of time. Defecation rates have been established for some species for the winter period. However, defecation rates seem to fluctuate seasonally, and perhaps also within season depending upon the nature of the food. Perhaps of greater importance is the observer error in that many pellet groups are overlooked in the typical survey (see NEFF, 1968 for a review of the technique). There are many similar census methods which rely upon calls of birds, numbers of burrows, beds, cast antlers, etc.

While this overview of census methods must be cursory, it does serve to illustrate the generally unsatisfactory status of population enumeration of wild birds and mammals. In most cases, the margin of error is great (Tables 4 and 5). Given this discouraging state of affairs, most workers have used more than one method, hoping to get some sort of agreement, and very cautiously interpreting their results within the limitations of the techniques used.

The prognosis for the development of satisfactory census methods is not good. Perhaps the development of sophisticated instrumentation is the most hopeful sign for future advances. However, even that approach is not likely to overcome one of the most basic difficulties with animal censuses — the problem of sampling. The discontinuous distribution and mobility of animal populations, particularly in deciduous forests which are mosaics of types, sites, microhabitats, results in a situation where neither adequate stratification nor adequate sample size can be achieved without a prohibitive amount of manpower and time.

Methods of estimating individual animal parameters (e.g., sex, age, weight) are generally satisfactory. Methods of determining population composition are satisfactory if done with due precaution. Methods of total population enumeration are largely unsatisfactory. Unfortunately, biomass and production figures which are derived from computations involving all of these values cannot transcend the poorest estimate in reliability. Obviously, development of better methods of population estimation are a prerequisite for valid biomass studies.

Flux of Energy and Materials

An estimate of standing crop biomass at any instant in time is a useful bit of information even though it is valid for only that particular time. Many of the biomass estimates of birds and mammals in the literature are of this nature (Table 6). These

Table 6. *Some representative examples of standing crop biomass of herbivores for various geographic and vegetational regions*

Area	Vegetation	Herbivore	kg/ha	
Uganda	grass-brush	elephant, hippo, buffalo	175.0	(Petrides and Swank, 1966)
Kenya Masailand	acacia-savanna	wild ungulates	122.5—175.0	(Talbot and Talbot, 1963)
Kenya Masailand	acacia-savanna	cattle, sheep, goats	19.25—28.0	(Talbot and Talbot, 1963)
Kenya-Tanganyika	acacia-commiphoro brushland	wild ungulates	52.5	(Talbot and Talbot, 1963)
Kenya-Tanganyika	acacia-commiphoro brushland	sheep and goats	3.5— 14.0	(Talbot and Talbot, 1963)
Northern Rhodesia	marsh-grassland	wild ungulates and cattle	96.3	(Robinette and Child, 1964)
Uganda	grassland	Uganda kob	21.7	(Buechner and Golley, 1967)
U. S. Midwest	tall grass prairie	cattle	49.0	(Watts et al., 1936)
California	managed chaparral	black-tailed deer	18.6	(Taber and Dasmann, 1958)
Russian Steppe	poor grasses	saiga antelope	0.5	(Bourlière, 1963)
Isle Royale, Michigan	mixed forest	moose	3.8	(Mech, 1966)
Michigan Upper Peninsula	mixed forest	white-tailed deer	3.5[a]	
Czechoslovakia	beech-maple-fir forest	herbivores and diversivores	6.0	(Turček, 1969)
Czechoslovakia	hardwood forest	herbivores and diversivores	4.5	(Turček, 1969)
Scotland	mixed forest	red deer	1.3	(Bourlière, 1963)
Ontario	coniferous forest	woodland caribou	0.02	(Simkin, 1965)
Canadian Arctic	tundra	barrenground caribou	0.79	(Banfield, 1954; Kelsall, 1957)

[a] From estimates made by Michigan Conservation Department, several publications.

estimates usually involve animals of economic importance to man, and commonly are made at the instant in time appropriate for exploitation. Such estimates do not indicate seasonal fluctuations in biomass, or the period of time over which the biomass has been accumulating.

Study of the dynamic changes in biomass can best be made with reference to the flux of energy and materials between birds and mammals and other components of the ecosystem. While the examples given here are concerned mainly with energy, with suitable modification (basically deletion of the respiration component) they can be applied to fluxes of nitrogen, minerals, etc.

Notation used in conceptualizing movement of energy and materials has not been uniform in work on primary and secondary production. An attempt will be made here to integrate the two kinds of production into a common notation where possible. The scheme is based upon papers designed to give uniformity of method to IBP productivity projects: Newbould (1967) for primary production and Petrusewicz (1967b) for secondary production.

Energy Budget for an Individual Animal or a Species Population

Fig. 1 is a schematic diagram of energy passing through an individual animal or population. Material removed (MR) includes all material functionally killed by the activites of an animal whether or not it is consumed (C). Thus, the browsing activities

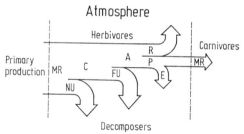

Fig. 1. Schematic diagram of energy transfer in an individual animal or a species population. Notation follows Petrusewicz (1967b). *MR*, material removed; *NU*, material removed but not consumed; C, consumption; *A*, assimilation; *FU*, rejecta; *R*, respiration; *P*, production; *E*, elimination. See text for further explanation

of a herbivous mammal may so weaken a plant that it subsequently dies. This material (NU) lost from the living plant biomass will pass more or less quickly into the decomposer cycle. The total impact of a herbivore upon its food base cannot be evaluated without taking into account this kind of influence. Other examples include trampling, materials removed for nest construction and animals killed by predators but not consumed.

The distinction between consumed (C) and not used (NU) is basically ingestion verses other uses and waste. Nesting materials are considered NU but certainly are important in the biology of some species. In human exploitation of ecosystems, the material removed (MR) may be consumed (C) as food or used for other purpose (e.g., lumber, medicine, etc.). The latter would be in the NU category.

Of the material consumed (C) part is digested and assimilated (A) while the remainder passes through the digestive tract and is lost as feces (FU = rejecta).

8*

Losses in the form of urine, secretions, metabolic gases, molting of hair and feathers, casting of antlers, etc., present particular problems in the scheme. PETRUSE-WICZ (1967b) considers these as rejecta, but physiologically they are assimilated and may be lost at a considerable time after assimilation. It seems best to erect a separate category for these products eliminated (E). See below for a discussion of Petrusewicz's E category.

Assimilated materials (A) pass into production (P) or are lost as respiration (R) or elimination (E). Production is the elaboration of new tissue which is added to the standing crop biomass (B). Respiration is the metabolic cost of living and includes the basal metabolic rate plus energy expended in digestion and movements.

At present, there are no adequate methods of determining respiration (R) of animals in the natural state. Basal metabolic rate is closely correlated with body size in birds (LASIEWSKI and DAWSON, 1967) and mammals (KLEIBER, 1947). However, the cost of energy for movements, behavioral interactions and heat production fluctuates greatly. Laboratory determinations have been made in some species, but it is not valid to extend these results to the wild state where activity, heat production and food consumption vary enormously. Radionuclides (ODUM and GOLLEY, 1963; GOLLEY, 1967) have been used to study metabolism in the field. The excretion rate of ^{65}Zn may be correlated with metabolism but the relationship is not clear. Similarly the $D_2{}^{18}O$ method of LIFSON, GORDON and McCLINTOCK (1955) and LeFEBVRE (1964) involves many variables. Perhaps the use of radio telemetry to monitor respiration rates or the use of portable infrared radiometers (GATES, in press) to study heat budgets may result in a breakthrough in this area. In the meantime, however, only the most crude estimates of respiration are possible (e.g., PEARSON, 1948; DAVIS and GOLLEY, 1963; LAMPREY, 1964; BUECHNER and GOLLEY, 1967).

Energy Flux and Biomass Distribution between Trophic Levels

The relationship of species populations at various trophic levels in this conceptual scheme is shown in Fig. 2. Newbould's symbol G for grazing has been changed to MR for material removed to be consistent throughout the trophic levels. Also, the concept of food availability is added to indicate that, particularly in forests, only a small proportion of the total plant biomass is of sufficient nutritional quality to be potential food for herbivores. Potential food typically comes from the latest growth (B_{2N}, apparent growth increment of NEWBOULD) and includes meristematic tissue in twigs and stems, leaves, and reproductive parts including seeds and fruits. All or part of the available food may be removed depending upon the state of the herbivore population, as will be discussed further below. Designation of availability in the consumer trophic levels is equivalent to the standing crop biomass.

Production removed by subsequent animals in the food chain is the material removed (MR) from this particular animal or species population. The distinction between material removed (MR) and elimination (E) is made on the nature of the process. Under this definition, elimination involves losses of energy or materials through physiological processes which the organism has evolved, and which presumably have survival value. By contrast, material removed is usually to the detriment of the organism, and involves the activities of other organisms which are beyond its control.

Note that this concept differs from that of Petrusewicz (1967b) who uses elimination for removal by other organisms, and includes losses of gases, secretions, etc. in rejecta (FU). The reason for not considering other waste products in rejecta are given above. Petrusewicz's category of elimination is here included as material removed so that there is a consistent category for transfer between trophic levels (Fig. 2).

Methods of determining material removed are reasonably well developed (Golley, 1967). They include measurements of vegetative parts removed, stomach contents

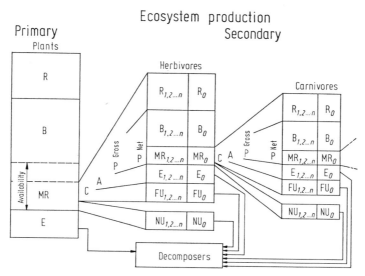

Fig. 2. Schematic diagram of energy transfer in an ecosystem following Petrusewicz (1967b) and Newbould (1967). Notation as in Fig. 1 except B — standing crop biomass. Subscript numbers are age classes of the animals involved and are distinguished by italics from the time interval subscripts of Petrusewicz (1967b). See text for further explanation

analysis, direct observation and penned feeding trials. Usually a suitable set of methods can be found to estimate consumption by a particular organism or from a particular food source. Species utilizing a narrow food base such as in obligate or specialized feeders obviously are simpler to work with and require less effort than those possessing diverse food habits.

Turček (1969) has emphasized that considerable plant material is destroyed by the activities of herbivores and omnivores (= diversivores) in addition to that consumed. These materials should be included in the NU portion which passes to the decomposer level.

Production of birds and mammals differs importantly from primary production in plants and secondary production in fish (Chapman, 1968) since the former tend towards deterministic growth. While woody plants and fish tend to grow over long periods of time, most of the increment of biomass in birds and mammals occurs while the animals are young.

This problem is handled in Fig. 2 by partitioning the population according to age classes into an adult component (with *n* age classes) which shows little growth, and a juvenile component (*o* subscript) with a relatively high growth rate. If more than one age class produces significant amounts of growth, then an additional column should be added for each such age class. Similarly, if there are sexual differences in growth, these should be partitioned also. Although the use of mean growth rates (or mean weights) makes good sense in organisms with fairly continuous growth (e.g., trees and fish), such an approach with birds and mammals results in a loss in power of the analysis, particularly in populations heavily weighted towards adults. Birds and mammals have such a skewed distribution of growth on age that the mode, rather than the mean, is more meaningful in estimating production.

Change of Biomass with Time

The biomass of birds and mammals can, and usually does, fluctuate markedly through the season and often from year to year. As a general rule, the interval that defines age classes is an interval which represents a comparable point in the reproductive cycle, most conveniently the interval from just before one reproductive

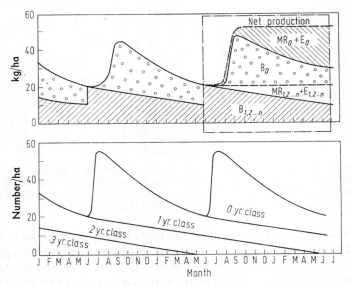

Fig. 3. Fluctuation of numbers and biomass of a hypothetical population in time. Assumptions are: a life span of three years, one litter or clutch per year, maturation in one year, completion of growth of young in four months, and a stable population and reproductive rate from year to year. The inset shows partitioning of biomass for net production per year. The net production exceeds the biomass peak because of the production of animals dying prior to the time of biomass peak. See text for further explanation

period to just before the next (Fig. 3). Sampling at this interval will give a bare minimum of "benchmarks" of biomass. The census results can suggest a stable population from year to year; but the bulk of the population may be young indicating high turnover, or old indicating low turnover with low survival of young. Classification into age classes will give some idea of population turnover. A population which fluctuates markedly between years is usually characterized by a preponderance or scarcity of animals in one or more age classes. This kind of data may suggest what sort of fluctuation in biomass has occurred during the year, but directly estimates only the standing crop biomass at the particular point in time at which the sample was taken.

A full accounting of the fluctuations in biomass, such as that in Fig. 3, requires a complete accounting of births, deaths, and growth rates throughout the season. In practice this is difficult or even impossible to accomplish. The period of early life of the young is particularly difficult. Reasonable approximations must be sought.

Because animal respiration is presently unmeasurable in the wild, gross production must be based upon crude estimates. The best estimate of net production is the peak in standing crop biomass (Fig. 3). This peak occurs at or near the time when juvenile animals reach adult size. Net production determined by this method will be an underestimate of the actual figure, since early mortality and predation losses will not be included (as shown in the inset in Fig. 3). The magnitude of error will be greatest in species that are slow to mature and where mortality of young extends well over the period of maturation. Mortality of young in the wild tends to occur during a short period after birth when immature animals constitute a small proportion of the total biomass.

A second shortcut to determine the seasonal amplitude of biomass is based upon the difference in mortality rates between adult and juvenile segments of the population. Mortality of young tends to be considerably higher than that of adults. If a total population count and composition can be determined at the beginning and end of the reproductive interval, then estimates of adult to young composition ratios can indicate changes occurring in the juvenile component, assuming the general shape of the adult survival curve is known or can be estimated.

Biomass Versus Production

Biomass and production, terms which are so frequently misunderstood, should be carefully distinquished because they are primarily antithetical concepts. Because resources are finite, the relationship between rate of production and standing crop biomass of most bird and mammal populations operates in a density-dependent manner. At low population densities there is a low standing crop biomass but a high rate of production. This, in turn, results in a buildup of standing crop biomass and eventually a decrease in the rate of production as the environmental capacity to support the population is approached.

The relationship between cumulative biomass and rate of production also varies due to inherent genetic differences between species (Table 7). The elephant, because of its longevity, large body size, great reserve capacity, and low maintenance metabolism in relation to weight, has a tremendous capacity to accumulate biomass. The elephant's biology is not geared for production; its rate of biomass production is low. Thus, management favoring elephants in equatorial Africa would not be a good solution to human protein dificiencies, since sustained yields are based upon production. To prevent population extinction and exhaustion of the protein supply, the long-term harvest rate cannot exceed the population's production rate.

The meadow mouse (Table 7) represents the other extreme; it is an organism with a tremendous capacity to produce new tissue but a low capacity to accumulate biomass due to a high metabolic maintenance cost, little reserve for periods of food shortage, and susceptibility to a wide range of predators. Consequently, the potential yield is low (although higher than in elephants) because the inherent production capacity is counteracted by inability of the species to maintain population densities sufficiently high to establish large standing crop biomasses.

The white-tailed deer represents an intermediate situation where capacity to produce and capacity to accumulate biomass achieve a reasonable balance. The

Table 7. *Standing crop (high density) and growth of three representative mammals in kilocalories per square meter per year* (modified from Petrides and Swank, 1966)

	Meadow mouse (*Microtus penn-sylvanicus*)	White-tailed deer (*Odocoileus virginianus*)	Elephant (*Loxodonta africanus*)
Standing crop	0.2	1.3	7.1
Growth per unit standing crop	2.5	0.5	0.048
Net production	0.50	0.65	0.34

potential yield from animals with this kind of balance is typically greater than that of animals at either extreme. Generally among mammals this balance is best in species with about 40 to 500 kg adult weight. It is perhaps not accidental that the most important domestic mammals raised for meat fall within this general size range. Among birds, most of which are restricted in size by constraints imposed by flight, the greatest yield comes from species with a combination of large body size and large clutch size, such as ducks, geese, grouse, pheasants and turkeys.

The Systems Approach

The field of population dynamics has developed along rather restricted lines, despite the epithet "dynamics". Results have been deficient in that they usually represent the operation of a population at one state of a highly variable system. Such results are poorly suited to predicting the outcome of changes in the state of the system or extrapolating to adjacent areas and populations which, by chance, are unlikely to be in the same state. To do this we must know the dynamics of the dynamics. In other words, our concept must encompass a higher level of integration than it has previously. The capacity for integration is the greatest advantage of the systems approach, with due respect to the mathematical methods and facilitation of computer modeling which are obvious strengths.

The inadequacy of traditional population dynamics and need for the systems approach can be traced to three interrelated problems. The first is that nearly all population parameters are variables, not constants. The establishment of a value under one set of conditions is not necessarily appropriate for other sets. For example, the rate of growth of an animal will depend upon the quantity of food, quality of food, preference for one species of food over another, mix of palatable and unpalatable species, predator population, parasites and diseases, and seasonal fluctuations in all of these factors. Changes in growth rates in turn will influence survival, age at maturity, mature weight and reproductive capacity.

The second reason, the interaction between trophic levels, should be obvious from the list of factors already given as influencing population parameters. At present we are able to predict only at the grossest level the outcome of 1) increasing primary production on herbivore production and subsequently carnivore production, 2) removal of predators on the herbivore populations, or 3) the impact of herbivore harvest by man on the other trophic levels.

The third reason is that we live in a world where exploitation by man has accentuated inherent fluctuations in ecosystems. Steady states, as are normally associated with climax* situations are found in only a few instances. Except for the arctic, severe desert, and some rapidly dwindling areas of lowland tropical forest, man has altered the vegetation of the earth to obtain high rates of harvestable production normally associated with early successional stages. In deciduous forests there are pulpwood cycles or short-term sawlog cycles, and climax is reached in only a few areas protected for esthetic or scientific reasons. The remainder are in a constant state of fluctuation introduced by managment — or mismanagement.

This is not to say that ecosystems exist in total chaos; but rather that parts of the system are so integrated that no component functions in isolation, and changing any component may affect other components and the new system state. It is this very integration, however, which establishes limitations (as will be discussed below) on the various components and promises to give predictability once functional relationships between components are understood.

The systems approach, unfortunately, will not make it possible to get by without a good set of data; neither will it make results easier to obtain. However, it will suggest the kinds of data which must be obtained and identify components and rates which are most important, the estimation of which will give the greatest return for the research investment.

Ecosystem Function — An Hypothesis

The ecosystem model which will be given here is in skeleton form, since it is the concept I wish to convey rather than specific examples. The ideas were originally derived from research on large herbivorous mammals, but I believe that they may apply generally to ecosystems.

Ecologists speak of "energy flow" through ecosystems in contrast to the cycling and reutilization of nutrients. In reality, energy does not "flow" in ecosystems; in the form of food it is located, captured and digested by organisms at the expense of a considerable performance of work. Far from flowing, energy (i.e., food) is dragged forcefully from one trophic level to the next.

Each functional unit in a model (Fig. 4) consists of an input rate, storage, and an output rate. The behavior of the compartment is such that its standing crop biomass tends to accumulate until the subsistence level is reached. With change in biomass, there is a corresponding shift in the utilization of energy. At low biomass, most energy is used for growth (net production) while at high biomass, most incoming energy is used to maintain the existing biomass. Because maintenance of existing biomass has priority in energy utilization, the inherent tendency of a compartment

---

* The terms "climax" and "succession" are used here only in the sense of relative rates of change.

will be to accumulate a high biomass with low maintenance metabolism rather than a low biomass with rapid growth and turnover. In animals showing deterministic growth, the population will consist primarily of adults because they are stronger competitors for food than juveniles; failure to accomplish adequate growth results in death of young animals. In organisms showing continuous growth, young individuals may survive in stunted condition. In either case, as food becomes scarce growth ceases and the standing crop biomass exists on a maintenance basis.

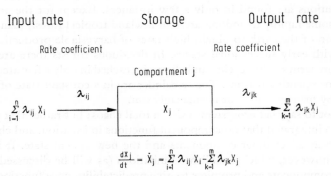

Fig. 4. Diagram of a functional unit of a food chain model, consisting of a compartment (j) with standing crop (X_j), an input rate $\left(\sum\limits_{i=1}^{n} \lambda_{ij} X_i\right)$ and an output rate $\left(\sum\limits_{k=1}^{m} \lambda_{jk} X_j\right)$; the λ's are factional transfer coefficients. The input rate and output rate are comparable to the a_i's and z_i's of SMITH (Chapter 2). The donor compartment in the food chain is X_i (e.g., vegetation), and the recipient compartment of X_j (herbivore) is compartment X_k (predator). The rates are interpreted as fluxes or units of material transfer per unit time

Obviously, inherent characteristics of organisms will influence the observed behavior of the compartment. The characteristics of elephants which give them an enormous capacity to store energy and those of meadow mice which are poor for storage already have been discussed. Similarly, a forest is capable of storing far more energy than a grassland. In effect, each compartment in an ecosystem tends to maximize its input rate and minimize its output rate. This is accomplished by selective pressures on predators for more efficient mechanisms of locating, capturing and assimilating food and on prey for mechanisms of avoidance or inhibition of consumption by enemies. The latter include such things as distastful or poisonous substances and spines in plants, and predator avoidance mechanisms in animals.

Turning to compartment interactions, one can consider the simplest case of plant, herbivore and predator (Fig. 4). We might have added additional dashed arrows going from recipient to donor compartments to indicate influence (feedback) of one compartment on another. Most of the influence in this example would occur as consumption (i.e., the material or energy flux arrows already in the diagram) but other influences such as die-back of clipped shoots, trampling and incomplete consumption of carcasses by predators also occur. Conceptually, it is more lucid to indicate forces in both directions. The impact of feedback influences of a compartment on a previous compartment (i.e., predator on herbivore; herbivore on plant) is to decrease standing crop biomass, and hence increase its production. In effect, the consumer compartment increases in size while inducing a high rate of turnover in the donor (food) compartment, a self-reinforcing process within limits.

As a point of debate, it can be estimated that a recipient compartment always has an influence on donor compartments in the system, but influence varies considerably depending upon the organisms involved. For example, the influence of seed eaters on the plant component is indirect by affecting the future number of individuals composing the plant compartment. The effects of defoliators are direct and more obvious, particularly in the case of large herbivores and insect outbreaks. Similarly, kills by predators reduce the standing crop biomass of herbivores. Litter decomposers, by hastening the release and recycling of nutrients, maintain plant production and, indirectly insure subsequent litter availability. Inherent characteristics of the interaction between consumer and consumed usually set limitations (usually referred to as feedback mechanisms) on the interactions. Rarely is material removed to the point of disappearance of the consumed. While the quantitative aspects of feedback mechanisms are frequently mentioned, there are a corresponding set of qualitative mechanisms which are less well known, but which may be of even greater importance.

It has already been pointed out that, of the total standing crop of green plants, only a part is of high enough quality to be food (availability). In general, quality is highest in meristematic tissues and grades off in structural tissues (e.g., ALDOUS, 1945; CAMPBELL and CASSADY, 1954; TABER and DASMANN, 1958; KRUEGER, 1967). Even in ruminant animals, which use microorganisms to digest cellulose, limitations are set by the high energy required to masticate structural components and the slow rate at which digestion occurs (SHORT, 1963). Also, protein and other nutrients must be present in high enough quantities to sustain the microorganisms and, consequently, the host. Availability is a variable; a hectare of brushland may yield only 300 kg of food with an average of 13 percent protein, but 500 kg of food with an average of 8 percent protein. Since the highest quality food is selected (e.g., HAGEN, 1953; SWIFT, 1948; MILLER, 1968), a low population consumes the high quality forage and can sustain production. At high population density, more food with a lower average quality is consumed up to the point at which the nutrient level will no longer sustain growth but only maintain the existing biomass. Furthermore, seasonal fluctuations in food quality (e.g., GORDON and SAMPSON, 1939) reinforce this limitation upon animals which do not have dormant stages.

Quality of food varies not only with the anatomical part of the plant, but among plant species. For each species of herbivore, some plants are highly palatable while others are not consumed under any circumstances. (e.g., MARTIN, ZIM and NELSON, 1951; McMAHAN, 1964; STEVENS 1966; TALBOT, 1962). At low densities, most food is obtained from favored species. However, at high densities, the less favored species are consumed as the more palatable ones are exhausted.

Similar constraints pertain to the interaction between prey and predators. Successful killing by predators reduces prey populations and increases subsequent time required to locate and capture prey. The model of HOLLING (1965) emphasizes the time element in the activity of predators. The work of ERRINGTON (1946) illustrates the importance of prey population density in relation to environmental resources on the success of predators in capturing prey. With lowered density, animals in the prey population show a greater individual size and are less subject to attack if the predator is of similar size. Better health results in greater alertness and less susceptibility to predisposing factors for predation such as malnutrition, parasitism and disease. Loss of alertness under inadequate nutrition such as was reported in starving pheasants by TESTER and OLSON (1959) is well known to field workers. Undoubtedly this leads to a greater susceptibility to predation, although direct evidence is somewhat meager. The indirect evidence of MURIE (1944), CRISTLER (1956), BORG (1962), MECH (1966) and others suggests that weaker animals (if they are present in the population) and the more vulnerable very young and very old are selectively removed by predation.

Control of compartment size can come from either the energy-nutrient pathway, the influence-material removed pathway, or from the physical state of the system (most commonly weather). In most cases, the outcome is the result of varying degrees of impact of all three factors. While one factor may predominate under a given set of circumstances, at a different state of the system the predominate impact may come from another factor. Similarly the nature of the organism will greatly modify the relative impact of each of the pathways. Inimical weather, which results in great mortality in an insect population, may have no measurable effect upon a mammal population. Control pathways tend to operate in a compensatory fashion; a great impact by one of the three factors usually results in a minor impact of the other two. If weather severely reduces a population, neither predators nor food limitations will be very important. If favorable weather prevails, then predators or food (usually both) will limit the population. Food limitation is the ultimate population control mechanism if all other factors are optimal.

A "healthy" system is one in which all of the compartments are functioning within the limits of the compensatory mechanisms of the system. This encompasses a wide range of states, from energy passing quickly through a series of rapidly cycling trophic levels to states where a few trophic levels show significant accumulations of biomass while others cycle rapidly. In terms of yield, the rapidly cycling systems are most productive, and the usual strategy of management is to maintain rapid cycling as typified by successional states.

Any system functioning within its resiliency (i.e., the system's capacity to adjust to modification) may be said to be in a state of stability. Ecosystem stability by this definition should be clearly distinguished from "equilibrium" where net change in state of the system has halted (i.e., climax). The continuing change during community succession does not necessarily, or even usually, involve instability. This change may be orderly and well within the system's regulatory capacity, so that successional cycles may be repeated indefinitely without harm, e.g., in seres repeatedly set back by fire. Similarly, many management schemes induce continuing fluctuations in states of the ecosystem with no loss of stability (i.e., controlled harvest of animals and timber); others require periodic replacement of nutrients to compensate for those removed by harvest (most agricultural crops involving monoculture). Thus, it is of paramount importance to recognize that species composition and population sizes within an ecosystem can undergo changes without necessarily disrupting the stability of the system as a whole.

The biomass of animal populations in a stable system also can vary considerably. The concept of carrying capacity (a term with many different definitions) too often has resulted in thinking in terms of a single level, while in actuality a whole range of levels are within the capacity of the system. These range from the subsistence level at the upper extreme and all intermediate levels down to that level at which animals are being removed by man or predators at a rate equal to the maximum replacement capacity of the population. This latter is the maximum sustained yield definition of carrying capacity used by range managers.

Disruption of ecosystem stability can occur via either the energy or influence pathway. Such disruption doubtlessly occurred naturally throughout geologic time, particularly during periods of upheaval with associated changes in the environment and mixing of biota. However, on a human time-scale such natural disruptions appear to be relatively rare in occurrence compared to the major disruption of the landscape caused by man. The emergence of twentieth-century technology has accelerated the rate of such disturbance.

Eutrophication is a disruption of the energy pathway where overenrichment of aquatic ecosystems with nutrients results in an abnormal accumulation of algae. Conversely, over-exploitation of herbivore populations (e.g., whales, SIMON, 1965), while not necessarily exceeding the regulatory mechanisms of the system, often has impaired the efficiency of the system. Disruption by removal of predators has occurred repeatedly. The irruption of deer, elk, and moose populations in North America (e.g., LEOPOLD et al., 1943; CLIFF, 1939; ALDOUS and KREFTING, 1946) involved both an influence pathway (control of large predators) and an energy pathway (widespread creation of successional stages of vegetation which favor these species). However, stability has been maintained in systems where hunting by man has replaced other forms of predation.

This hypothesis, although differing in many respects, is more closely allied to the thinking of HAIRSTON et al., (1960); SLOBODKIN et al., (1967) than that of their critics (MURDOCK, 1966; EHRLICH and BIRCH, 1967). It affirms the existence of order in nature — a balance which functions mainly in setting limits on deviations. It does not deny the role of catastrophy in regulating certain populations at certain times, but certainly not all populations at all times. The significant point about a catastrophy is not the fact that it is independent of prior events, but rather that subsequent events are not independent of it.

However, I take exception to the conclusion of SLOBODKIN, SMITH and HAIRSTON (1967) that the herbivore trophic level is not food-limited. Their observation that most of the quantity of vegetation is not consumed is certainly true, but I propose that the reason lies with limitations in quality. Even if predators were entirely removed from a system, herbivores could not be expected to consume anywhere near the total plant biomass. Too much of it is cellulose and other refractory complex carbohydrates. If the average quality of all plant tissues was as great as that of seeds, nectar and pollen, then high levels of consumption of plant tissue would be the usual case.

But more basic, I see little reason to consider a trophic level to be "food-limited" or "predator-limited", when its regulation is the outcome of several controlling factors acting in concert. Indeed, if the hypothesis that the inherent tendency of an ecosystem is towards equalizing control among adjacent trophic levels is valid, overwhelming control by a single factor should occur only when compensation of the system is taxed at an extreme.

Finally, I would modify SLOBODKIN, SMITH and HAIRSTON's partly stated and partly implied contention that the quantity (presumably biomass) or density of organisms in a trophic level will tend to be the same despite changes in species composition. This is somewhat a side issue to their major point that there are not abundant empty spaces in nature (which with I agree). It is probably the usual case that a displaced organism is replaced by a similar kind of organism (in which case biomass and density frequently would tend to be quite similar), but exceptions do occur. The mouse versus elephant example illustrates the changes in biomass and density which are possible with species changes within a trophic level. While this is a greatly exaggerated case, replacement of grass by brush and small trees or large herbivores by smaller ones, for example, are common under human influence. Neither does energy consumption by herbivores remain the same under a shift in species composition. A large animal has a lower maintenance requirement per unit biomass than a smaller one, and it can survive on consumption of plant biomass with a lower average caloric and nutrient content. The contents of a trophic level cannot be combined irrespective of the characteristics of the various species in that level.

Acknowledgements

For discussion and criticism of this manuscript I wish to thank my colleagues, CHARLES F. COOPER, FRANK F. HOOPER, JOHN A. KADLEC, and FRED E. SMITH. However, I am solely responsible for the ideas and opinions expressed herein, and all errors are my own. Concepts presented herein were developed from studies supported in part by National Science Foundation Grant GB-6171.

References

ALDOUS, C. M.: A winter study of mule deer in Nevada. J. Wildl. Mgmt. 9, 145—151 (1945).
ALDOUS, S. E., and L. W. KREFTING: The present status of mouse on Isle Royale. Trans. N. Amer. Wildl. Conf. 11, 296—306 (1946).
ALTMAN, P. L., and D. S. DITTMAR: Biology data book. Washington D. C.: Federation of Amer. Soc. Exp. Biol. 1964.
ANDERSON, J.: Analysis of a Danish roe-deer population. Danish Rev. Game Biol. 2, 127—155 (1953).
BAILEY, J. A.: Trap responses of wild cottontails. J. Wildl. Mgmt. 33, 48—58 (1969).
BANFIELD, A. W. F.: Preliminary investigation of the barren ground caribou. Can. Wildl. Serv., Wildl. Mgmt. Bull. Ser. 1, No. 10A, 1—79 and 10B, 1—112 (1954).
BORG, K.: Predation on roe deer in Sweden. J. Wildl. Mgmt. 26, 133—136 (1962).
BOURLIÈRE, F.: The wild ungulates of Africa: ecological characters and economic implications. Publ. IUCN NS 1, 102—105 (1963).
BRANDBORG, S. M.: Life history and management of the mountain goat in Idaho. Idaho Dept. Fish Game, Game Bull. No. 2. 142 (1955).
BROWN, E. R.: The black-tailed deer of western Washington. Wash. State Game Dept. Biol. Bull. No. 13, 124 (1961).
BUECHNER, H. K., and F. B. GOLLEY: Preliminary estimation of energy flow in Uganda kob (*Adenota kob thomasi* Neumann), pp. 243—254. In: Secondary productivity of terrestrial ecosystems (K. PETRUSEWICZ, Ed.). Warsaw: Panstwowe Wydawnictwo Naukowe 1967.
BUMP, G., R. W. DARROW, F. C. EDMINSTER, and W. F. CRISSEY: The ruffed grouse: life history — propagation — management. Albany 1947: New York Conservation Dept.
CAMPBELL, R. S., and J. T. CASSADY: Moisture and protein in forage on Louisiana forest ranges. J. Range Mgmt. 7, 41—42 (1954).
CHAPMAN, D. W.: Production, pp. 182—196. In: Methods for assessment of fish production in fresh waters (W. E. RICKER, Ed.). IBP Handbook No. 3. Oxford-Edinburgh: Blackwell Sci. Publ. 1968.
CLIFF, E. P.: Relationship between elk and mule deer in the Blue Mountains of Oregon. Trans. N. Amer. Wildl. Conf. 4, 560—569 (1939).

Cowan, I. M.: Distribution and variation in the native sheep of North America. Amer. Midland Nat. **24**, 505—580 (1940).

— Life and times of the coast black-tailed deer, pp. 523—617. In: The deer of North America (W. P. Taylor, Ed.). Harrisburg: Stackpole Co.; Washington, D. C.: Wildl. Mgmt. Inst. 1956.

Crisler, Lois: Observations of wolves hunting caribou. J. Mammol. **37**, 337—346 (1956).

Croon, G. W., D. R. McCullough, C. E. Olson Jr., and L. M. Queal: Infrared scanning techniques for big game censusing. J. Wildl. Mgmt. **32**, 751—759 (1968).

Dahlberg, B. L., and R. C. Guettinger: The white-tailed deer in Wisconsin. Wisconsin Cons. Dept., Tech. Wildl. Bull. No. 14. (1956).

Darling, F. F.: A herd of red deer. London: Oxford Univ. Press 1937.

Dasmann, R. F., and R. D. Taber: A comparison of four deer census methods. Calif. Fish and Game. **41**, 225—228 (1955).

Davenport, L. A.: Results of deer feeding experiments at Cusino, Michigan. Trans. N. Amer. Wildl. Conf. **4**, 268—274 (1939)

Davis, D. E.: Estimating the numbers of game populations, pp. 89—118. In: Wildlife investigational techniques (H. S. Mosby, Ed.). Washington, D. C.: Wildlife Soc. 1963.

—, and F. B. Golley: Principles in mammalogy. New York: Reinhold Publ. 1963.

Eberhardt, L. L.: Population estimates from recapture frequencies. J. Wildl. Mgmt. **33**, 28—39 (1969).

—, and R. I. Blouch: Analysis of pheasant age ratios. Trans. N. Amer. Wildl. Conf. **20**, 357—367 (1955).

Ehrlich, P. R., and L. C. Birch: The "balance of nature" and "population control". Amer. Naturalist. **101**, 97—108 (1967).

Errington, P. L.: Predation and vertebrate populations. Quart. Rev. Biol. **21**, 144—177, 221—245 (1946).

Friend, M.: A review of research concerning eye-lens weight as a criterion of age in animals. New York Fish Game J. **14**, 152—165 (1967).

Gates, D. M.: Infrared measurement of plant and animal surface temperature and their interpretation. In: Remote sensing in ecology (P. L. Johnson, Ed.). Athens: Univ. Georgia Press (in press).

Geis, A. D.: Trap response of the cottontail rabbit and its effect on censusing. J. Wildl. Mgmt. **19**, 466—472 (1955).

Golley, F. B.: Gestation period, breeding and fawning behavior of Columbian black-tailed deer. J. Mammol. **38**, 116—120 (1957).

—, Methods of measuring secondary productivity in terrestrial vertebrate populations, pp. 99—124. In: Secondary productivity of terrestrial ecosystems (K. Petrusewicz, Ed.). Warsaw: Panstwowe Wydawnictwo Naukowe 1967.

— Secondary productivity in terrestrial communities. Amer. Zool. **8**, 53—59 (1968).

—, and H. K. Buechner: A practical guide to the study of the productivity of large herbivores. IBP Handbook No. 7. Oxford and Edinburgh: Blackwell Scientific Publ. 1968.

Gordon, A., and A. W. Sampson: Composition of common California foothills plants as a factor in range management. Univ. Calif. Agr. Exp. Sta. Bull. 627 (1939).

Hagen, H. L.: Nutritive value for deer of some forage plants in the Sierra Nevada. Calif. Fish Game **39**, 163—175 (1953).

Hahn, H. C. Jr.: A method of censusing deer and its application in the Edwards Plateau of Texas. Texas Game, Fish Oyster Comm., Fed. Aid. Proj. **25-R**. 24 (1949).

Hairston, N. G., F. E. Smith, and L. B. Slobodkin: Community structure, population control, and competition. Amer. Naturalist **94**, 421—425 (1960).

Hayne, D. W.: An examination of the strip census method for estimating animal populations. J. Wildl. Mgmt. **13**, 145—157 (1949a).

— Two methods for estimating population from trapping records. J. Mammol. **30**, 399—411 (1949b).

Hensley, M. M., and J. B. Cope: Further data on removal and repopulation of the breeding birds in a spruce-fir forest community. Auk **68**, 483—493 (1951).

Hirst, S. M.: Road-strip census techniques for wild ungulates in African woodland. J. Wildl. Mgmt. **33**, 40—48 (1969).

Holling, C. S.: The functional response of predators to prey density and its role in mimicry and population regulation. Mem. Entomol. Soc. Canada, No. 45. 60 (1965).

Imber, M. J.: Sex ratios in Canada goose populations. J. Wild. Mgmt. 32, 905—920 (1968).

Jenkins, D. H.: The productivity and management of deer on the Edwin S. George Reserve, Michigan. Unpubl. Ph. D. Thesis. Ann Arbor: Univ. Michigan 1964.

Jordan, J. S.: Effects of starvation on wild mallards. J. Wildl. Mgmt. 17, 304—311 (1953).

Kadlec, J. A., and W. H. Drury: Structure of the New England herring gull population. Ecology 49, 644—676 (1968).

Keith, L. B.: Fall and winter weights of Hungarian partridges and sharp-tailed grouse from Alberta. J. Wildl. Mgmt. 26, 336—337 (1962).

Kelsall, J.: Continued barren-ground caribou studies. Canadian Wildl. Serv., Wildl. Mgmt. Bull. Ser. 1, No. 12, 1—148 (1957).

King, J. R., S. Barker, and D. S. Farner: A comparison of energy reserves during the autumnal and vernal migratory periods in the white-crowned sparrow *Zonotrichia leucophrys gambelii*. Ecology 44, 513—521 (1963).

Kleiber, M.: Body size and metabolic rate. Physiol. Rev. 27, 511—541 (1947).

Kozicky, E. L.: Variations in two spring indices of male ring-necked pheasant populations. J. Wildl. Mgmt. 16, 429—437 (1952).

Krueger, K. W.: Nitrogen, phosphorus and carbohydrate in expanding and year-old Douglas-fir shoots. For. Sci. 13, 352—356 (1967).

Lamphery, H. F.: Estimation of the large mammal densities, biomass and energy exchange in the Tarangire Game Reserve and the Masai Steppe in Tanganyika. E. Afr. Wildl. J. 2, 1—46 (1964).

Lasiewski, R. C., and W. R. Dawson: A re-examination of the relation between standard metabolic rate and body weight in birds. Condor 69, 13—23 (1967).

Laws, R. M.: A new method of age determination for mammals. Nature 169, 972—973 (1952).

Leedy, D. L., and L. E. Hicks: The pheasants in Ohio, pp. 56—130. In: The ring-necked pheasant (W. L. McAtee, Ed.). Washington, D. C.: Amer. Wildl. Inst. 1945.

le Febvre, E. A.: The use of $D_2{}^{18}O$ for measuring energy metabolism in *Columba livia* at rest and in flight. Auk 81, 403—416 (1964).

Leopold, A., E. F. Bean, and N. C. Fassett: Deer irruptions. Trans. Wisc. Acad. Sci. Arts Letters 35, 351—366 (1943).

Lifson, N., G. B. Gordon, and R. McClintock: Measurement of total carbon dioxide production by means of $D_2{}^{18}O$. J. Appl. Physiol. 7, 704—710 (1955).

Lord, R. D. Jr.: The cottontail rabbit in Illinois. Carbondale: S. Ill. Univ. Press 1963.

Madsen, R. M.: Age determination of wildlife: a bibliography. Biblio. No. 2. Washington, D. C.: U. S. Dept. Int. 1967.

Martin, A. C., H. S. Zim, and A. L. Nelson: American wildlife and plants: a guide to wildlife food habits. New York: McGraw-Hill Book Co., Inc. 1951.

McCain, R.: A method for measuring deer range use. Trans. N. Amer. Wildl. Conf. 13, 431—441 (1948).

McCullough, D. R., C. E. Olson Jr., and L. M. Queal: Large animal census by thermal mapping — a progress report. In: Remote sensing in ecology (P. L. Johnson, Ed.). Athens: Univ. Georgia Press (in press).

McMahan, C. A.: Comparative food habits of deer and three classes of livestock. J. Wildl. Mgmt. 28, 798—808 (1964).

Mech, L. D.: The wolves of Isle Royale. National Parks Fauna Series, No. 7, 210 pp. (1966).

Miller, G. R.: Evidence for selective feeding on fertilized plots by red grouse, hares, and rabbits. J. Wildl. Mgmt. 32, 849—853 (1968).

Mosby, H. S. (Ed.): Wildlife investigational techniques. Washington, D. C.: Wildlife Soc. (1963).

Murdock, W. W.: Community structure, population control, and competition — a critique. Amer. Naturalist 100, 219—226 (1966).

Murie, A.: The wolves of Mount McKinley. National Parks Fauna Series, No. 5, (1944).

Murie, O. J.: The elk of North America. Harrisburg: Stackpole Co.: and Washington, D. C.: Wildl. Mgmt. Inst. 1951.

NEFF, D. J.: The pellet-group count technique for big game trend, census, and distribution: a review. J. Wildl. Mgmt. 32, 597—614 (1968).

NEWBOULD, P. J.: Methods for estimating the primary production of forests. IBP Handbook No. 2. Oxford-Edinburgh: Blackwell Sci. Publ. 1967.

ODUM, E. P., and F. B. GOLLEY: Radioactive tracers as an aid to the measurement of energy flow at the population level in nature, pp. 403—410. In: Radioecology (V. SCHULTZ, and A. W. KLEMENT, JR. Eds.). New York: Reinhold Publ. Corp. 1963.

PAULIK, G. J., and D. S. ROBSON: Statistical calculations for change-in-ratio estimators of population parameters. J. Wildl. Mgmt. 33, 1—27 (1969).

PEARSON, O. P.: Metabolism and energetics. Sci. Monthly 66, 131—134 (1948).

PETRIDES, G. A., and W. G. SWANK: Estimating the productivity and energy relations of an African elephant population. Proc. Internat. Grassland Cong., Sao Paulo, Brazil (1965). 9, 831—942 (1966).

PETRUSEWICZ, K. (Ed.): Secondary productivity of terrestrial ecosystems. Warsaw: Panstwowe Wydawnictwo Naukowe 1967a.

— Concepts in studies on the secondary productivity of terrestrial ecosystems, pp. 17—49. In: Secondary productivity of terrestrial ecosystems (K. PETRUSEWICZ, Ed.). Warsaw: Panstwowe Wydawnictow Naukowe 1967b.

RASMUSSEN, D. I., and E. R. DOMAN: Census methods and their application in the management of mule deer. Trans. N. Amer. Wildl. Conf. 8, 369—379 (1943).

ROBINETTE, W. L., and G. F. T. CHILD: Notes on biology of the lechwe (Kobus leche). Puku 2, 84—117 (1964).

—, D. A. JONES, J. S. GASHWILER, and C. M. ALDOUS: Methods for censusing winter-lost deer. Trans. N. Amer. Wildl. Conf. 19, 511—515 (1954).

SHORT, H. L.: Rumen fermentations and energy relationships in white-tailed deer. J. Wildl. Mgmt. 27, 184—195 (1963).

SIMON, N.: Of whales and whaling. Science 149, 943—946 (1965).

SIMKIN, D. W.: A preliminary report of the woodland caribou study in Ontario. Ontario Dept. Lands and For. Sect. Rept. (Wildl.) No. 59, 1—76 (1965).

SINIFF, D. B., and R. O. SKOOG: Aerial censusing of caribou using stratified random sampling. J. Wildl. Mgmt. 28, 391—401 (1964).

SLOBODKIN, L. B., F. E. SMITH, and N. C. HAIRSTON: Regulation in terrestrial ecosystems, and the implied balance of nature. Amer. Naturalist 101, 109—124 (1967).

SMITH, M. H.: A comparison of different methods of capturing and estimating numbers of mice. J. Mammol. 49, 455—462 (1968).

STEVENS, D. R.: Range relationships of elk and livestock, Crow Creek Drainage, Montana. J. Wildl. Mgmt. 30, 349—363 (1966).

STEWART, R. E., and J. W. ALDRICH: Removal and repopulation of breeding birds in a spruce-fir forest community. Auk 68, 471—482 (1951).

STRANDGAARD, H.: Reliability of the Peterson method tested on a roe-deer population. J. Wildl. Mgmt. 31, 643—651 (1967).

SUMMERS-SMITH, D.: The house sparrow, London: Collins 1963.

SWIFT, R. W.: Deer select most nutritious forages. J. Wildl. Mgmt. 12, 109—110 (1948).

TABER, R. D.: Criteria of sex and age, pp. 119—189. In: Wildlife investigational techniques. (H. S. MOSBY, Ed.): Washington, D. C.: Wildl. Soc. 1963.

—, and I. M. COWAN: Capturing and marking wild animals, pp. 250—283. In: Wildlife investigational techniques. (H. S. MOSBY, Ed.): Washington, D. C.: Wildl. Soc. 1963.

—, R. F. DASMANN: The black-tailed deer of the chaparral. Calif. Fish. and Game, Game Bull. No. 8. 163 pp. (1958).

TALBOT, L. M.: Food preference of some East African wild ungulates. E. Afr. Agric. For. J. 27, 131—138 (1962).

—, and D. R. M. STEWART: First wildlife census of the entire Serengeti-Mara region, East Africa. J. Wildl. Mgmt. 28, 815—827 (1964).

—, and M. H. TALBOT: The high biomass of wild ungulates on East African savanna. Trans. N. Amer. Wildl. Conf. 28, 465—476 (1963).

TEPPER, E. E.: Statistical methods in using mark-recapture data for population estimation. Biblio. No. 4. Washington, D. C.: U. S. Dept. Interior 1967.

TESTER, J. R., and L. OLSON: Experimental starvation of pheasants. J. Wildl. Mgmt. 23, 304—309 (1959).

TURČEK, F. J.: Large mammal secondary production in European broad leaved and mixed forests: some results and methods of recent research. Biologia (Bratislava) 24, 173—181 (1969).

WATTS, L. R., G. STEWART, C. CONNAUGHTON, L. J. PALMER, and M. W. TALBOT: The management of range lands. U. S. Senate Doc. 199, 501—522 (1936).

WELLES, R. E., and F. B. WELLES: The bighorn of Death Valley. National Parks Fauna Series, No. 6 (1961).

WHITE, D. L.: Condition and productivity of New Hampshire deer, pp. 69—113. In: The white-tailed deer of New Hampshire (H. R. SIEGLER, Ed.). New Hampshire Fish and Game Dept. Survey Rept. No. 10 (1968).

YOUNG, H., J. NEESS, and J. T. EMLEN, JR.: Heterogeneity of trap response in a population of house mice. J. Wildl. Mgmt. 16, 169—180 (1952).

ZIPPEN, C.: The removal method of population estimation. J. Wildl. Mgmt. 22, 82—90 (1958).

11

Soil Flora:
Studies of the Number and Activity of Microorganisms in Woodland Soils

L. Steubing

Litter fall is of great importance in the nutrient cycle of forest ecosystems. Microbes attack and transform this organic material which is accumulated on and in the soil. There have been a number of investigations on the decomposition of litter in woodlands. It is obvious that the destruction of organic substances depends upon the number, species and activity of soil microflora and fauna. In investigating the role of soil microflora two approaches have been used: counting and isolating soil microorganisms or determining microbial induced biochemical effects in the soil.

It will become apparent that there are many techniques to count microorganisms using direct or indirect methods. Direct methods are used to yield information about the number of germs and interrelationships of soil microorganisms. Rossi and Riccardo (1927) and Cholodny (1930) developed the buried slide technique which was later modified by some authors. Waid and Woodman (1957) replaced the glass slides by nylon gauze, and Kerr (1958) and Gams (1959) used cellophane strips. It is possible to estimate populations of soil microorganisms after the microscopical examination of buried slides, gauzes or strips. Soil suspensions also can be directly examined microscopically by the techniques of Strugger (1949), Jones and Mollison (1948), Tchan (1952) and Alexander and Jackson (1954). The quantitative evaluation of these direct methods is difficult — one can not distinguish between living and dead cells, between bacteria and spores of actinomycetes. This distinction can be made indirectly through plate count methods. These have been used for counting and isolation of pure cultures of microorganisms or of special physiological groups.

The estimate of the total population of bacteria, actinomycetes, algae and fungi refers to the number of viable cells or mycelial fragments in the soil sample capable of growing on the agar medium employed in the test. Composition of the agar medium can be varied to favor or select for certain elements of the flora. Every special agar or nutrient solution has a selective effect on the development and growth of specific microorganisms. Therefore, the results are relative numbers of soil germs; these are comparable only with other results obtained by the same technique. The lack of a general standard technique illustrates why the results of the different plate counting methods are not comparable.

I do not want to attempt to include all methods of analysis of soil flora that might be suitable. They are referred to in special papers, e.g., Glathe (1966) or in manuals, e.g., Johnson et al. (1960), Janke (1946), Pochon and Tardieux (1962), and Steubing (1965). It is obvious that the methods here discussed are not completely satisfactory. Eventually, it may be possible obtain information about the number of some groups of

9*

microbes by chemical analyses of special cell constituents. The data presented in this chapter illustrate the type of information needed for analysis of the functional relationships of the soil flora. Results from both deciduous and coniferous woodlands compare the contrasting constituents of the soil flora of these two forest types.

Origin of the Soil Samples

In the Federal Republic of Germany, the research contribution to the International Biological Program is concentrated on a group of sample areas established in the Solling. This is a centrally-located mountainous terrain with a maximum height of a little over 500 m. The dominant plant community is the red beech forest with an associated acid humus (Luzulo-Fagetum). Adjacent to the beech stands are forests of spruce in man-made plantations. It advantageous for our measurements that the beech and spruce forests grow on comparable habitats. The geological substratum is generally of a uniform structure. It consists of Triassic sandstone covered by a thin sheet of loess-loam and belongs to a type of oligotrophic brown earth. Our soil investigations in the Solling beech and spruce forests are carried out under both field and laboratory conditions.

Algae and Chlorophyll Content of the Soil
Methods

Relative numbers of soil algae were determined by plate counting methods (STEUBING, 1965).

Extraction and Determination of Chlorophyll: 100 ml of 80% acetone were added to 50 ml air-dried and pulverized soil in an erlenmeyer flask. The flask was stoppered and placed in the dark on a shaker for one hour. The soil suspension was then centrifuged, the overstanding solution filtered (glass fritte G 4) and the filtrate measured with a Beckman spectrophotometer at 663 nm*. Because of the innate color of the soil, another part of the soil filtrate was shaken first with 10 ml and once more with 5 ml ether and then the two solutions were separated; the chlorophylls were further partitioned in the ether while colored soil substances remained in the acetone. To this acetone phase, 80% acetone was added up to 10 ml and then measured with the spectrophotometer. This soil sample blank was subtracted from the chlorophyll determinations.

It is evident in Fig. 1 that the chlorophyll extracts of fresh green leaves (a), beech soil (b) and spruce soil (c) do not differ in absorption spectra.

Results and Discussion

In contrast to other soil microorganisms, the autotrophic algae contain chlorophyll. In a study of plant pigments in woodland soils, GORHAM and SANGER (1967) found low concentrations of carotinoids and chlorophyll derivatives and a rapid decline in the concentration of each between the L and F layers of the soil. They concluded that the pigments are decomposed at a rate greater than for soil organic matter as a whole. While leaves are turning brown on the trees, chlorophylls are degraded much faster than carotinoids. Chlorophyll also will be destroyed immediately by acids. The beech and spruce soils examined in these experiments had pH values between 3.5

---

* nm $= 10^{-9}$ meters $=$ millimicrons.

and 4.0. Therefore, we assumed that the chlorophyll of the litter was decomposed in these soils and that measured chlorophyll had its origin in the soil algae.

First it was necessary to examine the assumption that chlorophylls are destroyed in the acid Solling soils. In the laboratory, fresh green beech leaves were incubated at 24° C in soil samples for 3 weeks. After this time the chlorophyll extracts of these

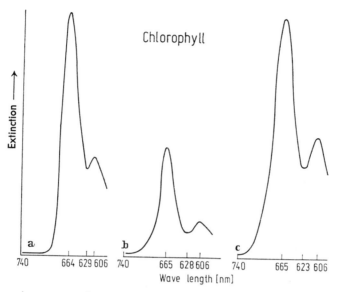

Fig. 1. Absorption spectra of acetone extracts of chlorophyll: a green beech leaves, b beech soil, c spruce soil

samples and untreated controls were analyzed. There was negligible difference in the chlorophyll content between tests (Table 1). This difference certainly would be smaller if natural litter with nearly completely degraded chlorophyll was buried.

Table 1. *Chlorophyll content of soil samples from the H layer of beech soil (pH 3.7) stored for 3 weeks at 60% water capacity at 24° C*

Treatment	Chlorophyll content in 100 g air-dried soil
untreated control	29.5 mg
5% green beech leaves added	33.0 mg

Other data indicate that the concentration of chlorophyll is higher in the spruce than in the beech soils (Table 2, Fig. 2). The F layers had the highest chlorophyll content which declined especially between the H and A horizons of both soils. The number of algae (plate count method) seems to be directly related to the chlorophyll concentration of these acid soil layers. Further tests on other forest and soil types are necessary to substantiate the generality of these results.

Table 2. *Chlorophyll content and number of algae from different layers of beech and spruce soils*

Horizon		Chlorophyll (mg/100 ml soil)	Number of algae/1 ml soil
beech	F	41	15,000
	H	32	2,500
	A₁	2.9	600
spruce	F	65	25,000
	H	51	4,000
	A₁	4.7	740

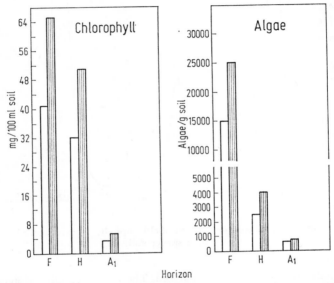

Fig. 2. Number of algae and chlorophyll content in samples of beech soil (light column) and spruce soil (shaded column)

The soil algae can be found among the Cyanophyceae or blue green algae, Chlorophyceae or green algae, Bacillariophyceae or diatoms and Xanthophyceae or yellowgreens. Only the first three groups are numerous. In our acid woodland soils of the Solling the Chlorophyceae are predominant with *Chlamydomonas, Chlorococcum, Chlorella, Chladophora, Cosmarium, Dactylococcus, Dictyosphaerium, Euglena, Gloeocystis, Hormidium, Muriella, Protococcus,* and *Scenedesmus.* Besides these, often there are encountered in other soils species of *Ankistrodesmus, Borodinella, Coccomyxa, Cylyndroclastis, Interfilum, Mesotaenium, Pleurastum, Stichococcus, Ulothrix, Vaucheria,* and *Zygonium.*

The Cyanophyceae are represented only by a sparse abundance of *Nostoc, Anabaena, Phormidium, Schizothrix* and *Lyngbya.* This class is sensitive to the hydrogen ion concentration. Therefore many soil genera are missing as, e.g., *Cylindrospermum, Nodularia, Plectonema, Scytonema* or *Tolypothrix.*

Diatoms are likewise less frequent in acid soils, while in neutral or alkaline environment can be found *Achnanthes, Actinoptychus, Cymbella, Eunotia, Fragilaria, Hantzschia, Navicula, Nitzchia, Pinnularia, Surirella,* and *Synedra.*

Amino Acids in the Soil

Methods

Determination of Free Amino Acids: Following the technique of GILBERT and ALTMAN (1966) two-dimensional paper chromatography was employed for identification of the free amino acids extracted from soils. The extracts were assayed on an amino acid analyzer.

Determination of α-ε-Diaminopimelic Acid (DAP): Extraction of DAP: 25 g air-dried and pulverized soil was shaken with 100 ml 0.1 N HCl and centrifuged for 40 min. The decantate was discarded, the residue treated with acetone and also centrifuged for 20 min. The supernatant liquid was then removed, and the remaining residue dried at 105° C. The dried residue was pulverized and hydrolyzed in a kjeldahl flask in 100 ml of 6 N HCl for 24 h at 105° C. The hydrolysate was hot filtered, and the residue washed with boiling 0.1 N HCl (150 ml). The filtrate was concentrated to about 10 ml under vacuum.

DAP was separated using Dowex 50 W resin. The extract was introduced on the top of a 100 ml column and eluted with 300 ml citrate buffer (pH 3.8). The eluate was dried under vacuum. The residue was leached with 5 ml CO_2-free water; 0.1 ml 3.1 N KCl was added and heated to 40° C. By shaking the pH was adjusted to 8.9 with 1 N NaOH. After adding 0.5 ml dinitrofluorbencol, the liquid was placed in the dark (control to pH 8.9 also!). After 50 min, the excess dinitrofluorbencol was removed by extracting twice with 5 ml ether. Addition of 0.5 ml conc. HCl and shaking the solution 5 times with 4-ml portions of ether dissolves most of the amino acids in the upper ether fraction; DAP remains in the lower water-soluble fraction.

Paper Chromatography: 1 ml extract was punctiformly coated on the paper. The developing solvent was isopropanol-formic acid-water (80—4—20 by vol.). After 24 hr the chromatogram was dried, the fluorescent blue-black spot cut out and then leached in a tube with 0.5 ml 0.1 N HCl. To 0.5 ml of the eluate was added 0.5 ml ninhydrin + 0.5 ml acetic acid and the mixture placed in boiling water for exactly 5 min. The cooled solvent was assayed on a Beckman spectrophotometer at 465 nm. The standard curve was made for 0.5—10 γ DAP. Two blanks were necessary to consider (1) the color of the ninhydrin reagent and (2) the color of the analyzed extract (soil extract + acetic acid).

Results and Discussion

Another way to obtain information about the number of microorganisms and their activity was used by MISCHUSTIN and PETROWA (1963). These authors fastened strips of linen upon solid plates and buried these vertically in the soil; after one week they removed the linen strips from the soil. The strips were dried and sprinkled with ninhydrin, so the zones were visible where amino acids (protein) had contact with the strips. The authors assumed that the ninhydrin-positive zones were related to biological activity in the soil.

It is known that free amino acids are important for microbial nutrition. Therefore, it may be possible to relate the amount of amino acids and the number of microorganisms in the soil (PAUL and TU, 1965). The amino acid content of soils declined rapidly with depth in both beech and spruce forests (Table 3). In general, the amount of amino acids was not high, about 60 ppm in the F layer, about 30 to 10 ppm in the H layer and only traces in the mineral horizon (A_1).

Table 3. *Amount of amino acids in different horizons of beech and spruce forest soils*

Horizon	Depth	Relative concentration of amino acids
F	0— 3 cm	100
H	3— 5 cm	50
A_1	15— 20 cm	20
A_1	45— 50 cm	5
A_1	95—100 cm	(<1)

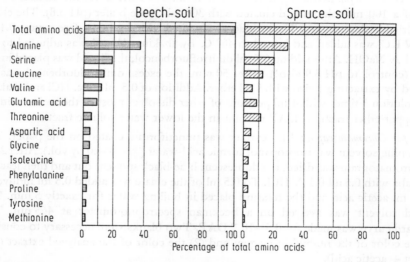

Fig. 3. Qualitative analyses of free amino acids in the beech and the spruce soil of the Solling

In Fig. 3 the identified amino acids in the test soils are shown as the percentage of the total amount present. It can be seen that there is no qualitative difference between the beech and the spruce soil; the quantity however is not similar (Table 4). Most of the amino acid concentrations are higher in the spruce soil. Only the content of valine was lower in the spruce soil than in the beech soil. Extensive plate counts on microbial populations were conducted, and it was obvious that the total number of microorganisms was equally higher in the spruce soil than in those samples taken from under beech.

Besides the well-identified amino acids, some ninhydrin positive substances were found. The concentration of these tentatively determined substances was registered as if they were real amino acids. Their identities could be:

(1 + 2) methionine sulfoxide, cysteine,
(3 + 4) amino adipic acid,
(5 + 6) diaminopimelic acid,
(7 + 8) tryptophan,
(13 + 14) γ-aminobutyric acid.

Table 4. *Column-chromatographical analysis of the amount of amino acids in the F and H layers of beech and spruce soils*

Amino acids (abbr.)	Beech soil		Spruce soil	
	F(μmol)	H(μmol)	F(μmol)	H(μmol)
* ? (1)	(0.14)	—	(0.22)	—
Asp	1.01	0.38	1.65	0.37
? (2)	(0.31)	(0.27)	(0.30)	—
Thr	1.22	0.65	3.75	0.24
Ser	5.04	1.02	7.25	0.58
Glu	2.31	1.55	2.93	0.93
Pro	0.47	0.35	1.07	0.15
Gly	1.01	0.54	1.26	0.18
Ala	8.52	1.89	10.37	0.70
? (3)	(0.18)	(0.06)	(0.12)	—
? (4)	(0.31)	(0.20)	(0.23)	(0.14)
Val	2.81	0.98	1.71	0.37
? (5)	(0.11)	(0.10)	(0.10)	(0.09)
? (6)	(0.19)	(0.10)	(0.10)	(0.05)
Met	0.26	0.16	0.35	0.10
? (7)	(0.12)	(0.19)	(0.13)	(0.11)
? (8)	(0.31)	(0.12)	(0.21)	(0.06)
Ileu	0.88	0.43	1.01	0.17
Leu	3.52	2.96	4.44	0.53
? (9)	(0.17)	(0.08)	(0.26)	(0.06)
? (10)	(0.18)	(0.05)	—	(0.01)
? (11)	(0.21)	(0.01)	(0.20)	—
? (12)	(0.11)	—	(0.09)	—
Tyr	0.27	0.08	0.42	0.03
Phe	0.63	0.29	0.80	0.08
? (13)	(0.15)	—	(0.09)	—
? (14)	(0.25)	(0.10)	(0.27)	—

* Ninhydrin positive substances; discussion in text

Among these amino acids, diaminopimelic acid (DAP) has become of special interest. DAP is not found in higher plants. It is a constituent of the cell wall of bacteria, especially of spore-forming strains (WORK, 1951, 1957; MEISTER, 1965). SHAZLY and HUNGATE (1966) present results from the application of DAP determinations to measurements of rumen bacterial growth. Contrary to most other amino acids, the amount of DAP was higher in the beech soils. The relationship for DAP in the H-layer was beech: spruce = 1 : 0.6. It was interesting that the relation of the number of bacteria in the two soils was similar, beech: spruce = 1 : 0.58. The effect of acidity on the decline of DAP in the soils corresponded with that for bacteria. The beech soil had lower acidity than the spruce soil. Therefore, there were more bacteria and less fungi present in the beech soil.

If further investigations should confirm the outcome of these tests, then it may be that chemical determinations will be advantageous in that results are more reproducible than direct counting of individuals. Statements about specific systematic groups, however, are not yet possible with chemical techniques.

Cellulose Content in the Soil and Cellulose Decomposition
Methods

The content of cellulose in soil samples was determined by the method of BLACK (1950). The carbohydrates in soils were measured with anthron (BRINK et al., 1960). Measurements on cellulose decomposition under environmental conditions followed the method of UNGER (1960). Nylon net bags containing cotton were placed in different layers of the soil. After 12 weeks the loss of weight was determined.

Results and Discussion

Instead of counting the number of cellulolytic microorganisms most investigators prefer to determine the cellulose-degrading, activity of these microbes (SIU and REESE,

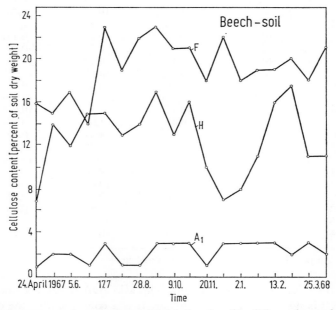

Fig. 4. Cellulose content (percent dry weight) in beech soil in different layers during the year

1953; IMSCHENEZKI, 1959). Cellulose is one of the most important substances annually entering the soil with litter. The amount of cellulose would increase, if it were not for the microbial attack on this carbohydrate (TÖRNE, 1967). The rate at which cellulose is metabolized depends upon a number on environmental influences such as temperature, moisture, aeration and some chemical factors. It is known that the

decomposition of cellulose is affected by the presence of other carbohydrates. Therefore, not only should the disappearance of cellulosic matter (buried cotton) in the
field be studied but also the natural cellulose content of soil samples.

Figs. 4 and 5 show the seasonal variation in cellulose content of forest soils under
beech and spruce stands. The highest amount of cellulose was found in the F layer
with a slight decline in the H layer. Differences between these two layers remained

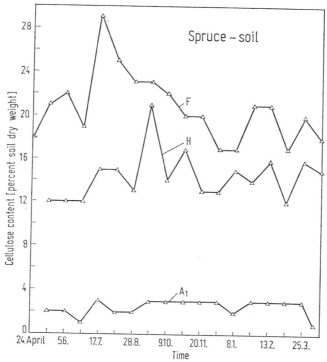

Fig. 5. Cellulose content (percent dry weight) in spruce soil in different layers during the year

small compared with the cellulose content in the A_1 horizon. Variations in the content
of cellulosic substances during the year were greatest in the two organic layers (F and
H) and more constant in the mineral horizon (A_1). Cellulose values were higher in
samples of spruce soils than in those of beech soils (Table 5).

Table 5. *Average annual value for cellulose content in soil samples under beech and spruce stands of the
Solling* (values expressed as per cent of soil dry weight)

Layer	Beech soil	Spruce soil
F	19	21
H	13	14
A_1	2	2

Fig. 6 indicates that the weight loss, i.e., the degradation of cotton, under environmental conditions in most cases was higher in the beech soil than in the spruce soil.
The decomposition was better in the F and H layer than in the mineral horizon (A_1).

The stronger microbial attack on cellulose in the beech soil samples appeared to be a result of the amount of nitrogen and total carbohydrate in the test soils. Table 6 shows that the beech soil had higher nitrogen and smaller carbohydrate content than did the spruce soil. Nitrogen enhances cellulose breakdown in the soil, while high amounts of carbohydrates have a retarding effect (ALEXANDER, 1967).

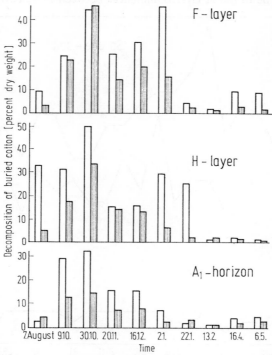

Fig. 6. Decomposition of cotton in different forest floor layers: beech soil (light column), spruce soil (black column)

Table 6. *Content of nitrogen and total carbohydrates in samples of beech and spruce soil* (values expressed as percent of dry weight)

Layer	Beech soil		Spruce soil	
	Nitrogen (%)	Hexose (%)	Nitrogen (%)	Hexose (%)
H	0.65	19.5	0.5	63.8 mg
A_1	0.12	2.4	0.1	3.0 mg

Lignin Content in the Soil and the Conversion of Lignin Fractions
Methods

The method of MAEDER (1960) was used for the determination of lignin.
Decomposition of Vanillin in Soil Samples: To 5 g soil in 100 ml dist. H_2O was added 0.01% vanillin. The flask with this suspension was stoppered and stored for 48 hr at

25° C. After this a 10 ml-suspension was extracted with ether, the water phase was removed and the ether fraction dried. The residue was leached in methanol and used for paper chromatography (KUNZE, 1968). Vanillin can be seen by blue fluorescence in UV-light. It was eluted from the chromatograms and measured spectophotometrically at 308 nm. The vanillin decomposition in the same liquid, but without soil extract was used as blank (KUNZE, in press).

Results and Discussion

Lignin belongs to the most abundant constituents of plant organic matter. Lignin disappears very slowly in soils, and little is known of the environmental variables governing the microbiological decomposition of this substance (HENDERSON, 1960; HAIDER and SCHETTERS, 1967). It is very difficult to obtain purified lignin fractions

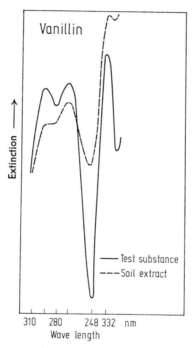

Fig. 7. Absorption spectra of pure vanillin and soil extract in methanol solutions

suitable for quantitative studies. Therefore, vanillin, one of the degradation products of lignin (HAIDER et al., 1964; FLAIG, 1966), was used as a substitute for lignin in decomposition studies.

Fig. 7 shows the absorption spectrum of vanillin from soil extracts and for the pure substance. Simultanous with analyses of the lignin content of soil samples, in the laboratory determinations were made of decomposition of vanillin in the presence of extracts from the different soil horizons (Fig. 8).

The highest lignin contents were found in the F and H layers and only negligible amounts in the mineral horizons (A_1). Samples of spruce soils contained more lignin than those of beech soils. The rate of decomposition of vanillin was similarly affected by the depth of the soil sample; the most rapid breakdown of vanillin occured in samples of the upper soil layers. Slow metabolizing processes developed in the lower A_1 horizon of the two soils.

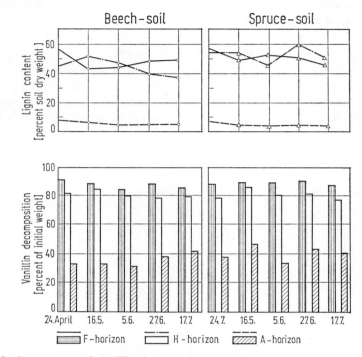

Fig. 8. Lignin content and vanillin decomposition in different layers of beech and spruce soils

Ecology of the Soil Microflora

The soil includes five major groups of microorganisms, the bacteria, actinomycetes, algae, fungi and protozoa. Soil algae, divided into Cyanophyceae, Chlorophyceae, Bacillariophyceae and Xanthophyceae, are never as numerous as the other groups. As photoautotrophic organisms they normally require sunlight. Therefore, they are abundant upon, directly below and a few centimeters under the surface (Fig. 2). On the other hand, algae are obtained from soil samples of 50 to 100 cm depth, where no light penetrates. These algae can live only as heterotrophs. It seems to be possible that they are carried downward in the dark by the seepage of rainwater or the movement of the soil fauna — especially the earthworms. The subterranean algae have to compete with the dominant heterotrophs, the bacteria and fungi, for the supply of available organic matter.

Bacteria and fungi dominate in the soils. The number of bacteria is great, but the single organism is very small; the fungi have more mass as result of their extensive filamentous mycelium. It is possible to differentiate the bacteria physiologically as aerobes, anaerobes and facultative anaerobes or as biochemical groups, e.g., cellulolytic, proteolytic, ammonifying or nitrifying bacteria. Taxonomically, the most frequent bacteria in the soils belong to the orders of Pseudomonadales (*Pseudomonas*) Eubacteriales (*Achromobacter, Arthrobacter, Bacillus, Chromobacterium, Clostridium, Flavobacterium, Sarcina*) and Actinomycetales (*Mycobacterium, Nocardia, Streptomyces, Streptosporangium*) (BECK, 1968).

Especially in forest soils, the fungi play an important role in the decomposition of organic matter. Although several hundred species of Basidiomycetes are present in soils, their isolation is so difficult that our knowledge of the hyphal growth and distribution in different kinds of habitats is sparse. The other classes of fungi are easier to isolate. Common genera of Fungi imperfecti and Ascomycetes are *Aspergillus, Alternaria, Botrytis, Chaetomium, Cladosporium, Cylindrocarpon, Gliocladium, Fusarium, Monilia, Penicillium, Trichoderma* or *Verticillium*. Less abundant but numerous are certain genera (*Mucorales, Saprolegniales*) in the Phycomycetes (GILMAN, 1957).

Commonly, the microorganisms are concentrated in the surface layers of the soil and their mass declines with depth. Estimates of algae, bacteria and fungi vary according to the technique of determination. Algae, which were emphasized in this paper, were often enumerated by plate counts or because of their chlorophyll content by direct fluorescent microscopy. Values from about 500 in the lower horizons to a maximum of 650,000 algae per gram near the surface were found in our beech and spruce soils. All these numbers are of limited value because they change with the method and they do not indicate the significance of algae in the soil; some species are unicellular and therefore one finds a high value while others are filamentous and give only low counts.

When it is possible to measure primary production by the chlorophyll content of the flora then it also will be possible to estimate the significance of autotrophic algae in soils by the quantitative analysis of their pigments. This assumes that chlorophyll in the litter does not confuse the results. The tests demonstrate that in our Solling soils the chlorophyll of incubated leaves is destroyed by the high hydrogen ion concentration. Therefore, the detected chlorophyll must have its origin in the living soil algae and should be related to algal biomass.

The same problem in counting procedure applies to the bacteria. Looking for a chemical method to estimate the amount of bacteria in the soils, it seems to be reasonable to determine a special bacteriological compound. In the extracts of free amino acids in the samples of beech and spruce soil (Table 4) there can be found such a substance — diaminopimelic acid (DAP). The relation of the number of bacteria in the two soils was similar to the amount of DAP.

On the whole, the annual litter fall takes from 3 to 5 years to decompose (OVINGTON, 1962). The rate of leaf breakdown depends on the soil population and its activity, but also on the tree species and the chemical composition of the leaves (SHANKS and OLSON, 1961). Table 7 demonstrates that the oligotrophic brown earth

Table 7. *Ratio of numbers of different groups of microorganisms between beech and spruce soil* (plate count method, Oct. 1967)

Microorganisms	Beech soil pH (KCl) 3.1	Spruce soil pH (KCl) 2.7
Bacteria	1	0.62
Fungi	1	2.3
Algae	1	2.0

under spruce and beech stands in the Solling does not have the same microbial composition; in the beech soil there are more bacteria but less fungi and algae than in the spruce soil.

Soil moisture and aeration, temperature, acidity, organic matter and inorganic nutrient supply influence the microflora. The major microbial groups find optimal conditions at 50 to 80 percent of the water holding capacity of the soil. Therefore, microbial density declines rapidly when soils become desiccated. Bacteria and actinomycetes are more sensitive to periods of drought than fungi. Chlorophyceae and Cyanophyceae exhibit more resistance to desiccation than diatoms. High temperatures favor the biochemical processes taking place in the soil. Most microorganisms are mesophiles with the highest activity between 25 to 35° C, but there are also a few thermophilic and psychrophilic strains. Decomposers of cellulose and lignin are able to metabolize substrates between 5 and 65° C. Often, there is a gradient of species within the soil profile related to the optimal temperature range of specific strains.

The hydrogen ion concentration is another important determinant. Commonly, bacteria and actinomycetes have optimal development in neutral or alkaline conditions, while fungi can grow over a wide pH range. Therefore, the fungi dominate in acid soils (Table 7). The algal distribution is also influenced by acidity of the substrate. Cyanophyceae and diatoms are less abundant in acid soils in contrast to the Chlorophyceae; the latter species are not appreciably limited by the soil reaction.

The content of available organic matter is related to the development of the soil heterotrophs. Humus-rich localities have the largest microbial mass. Only the autotrophic algae do not need organic carbon. In contrast, they create organic matter from inorganic substance. Only the heterotrophic variants participate in the many biochemical transformations of the organic material. Constituents of the soil include a variety of organic components which can be separated into 6 categories (Alexander, 1967):

1. cellulose, varying from 15 to 60 percent of the dry weight,
2. hemicelluloses, making up 10 to 30 percent,
3. lignin, 5 to 30 percent of the dry weight,
4. the water soluble fraction contains amino acids, aliphatic acids, simple sugars, tannins,
5. ether and alcohol-soluble fraction, containing pigments, resins, waxes, fats and oils,
6. proteins which contain most of the nitrogen and sulfur.

Usually the concentration of these fractions declines with depth of the soil, i.e., cellulose (Table 5), lignin (Fig. 8), amino acids (Table 3) and chlorophyll (Table 2).

The rate at which cellulose and lignin are metabolized in the soil is influenced by the previously-discussed environmental factors. Besides these, nitrogen content, the relation between cellulose and lignin and the presence of other carbohydrates are important as well (Table 6). The high amount of carbohydrates and small content of nitrogen in spruce soils has a retarding effect on the decomposition of cellulose compared with conditions in the beech soils. The fungal attack on the lignin leads to aromatic compounds of low molecular weight such as vanillin. Various soil fungi are able to oxidize vanillin to vanillic acid. These substances could be extracted from our soils; therefore, vanillin was used as substrate to study the conversion of a special lignin fraction. Fig. 8 demonstrates that the microbial decay of vanillin was more rapid in the F than in the A_1 horizons of our soils.

In a homeostatic forest soil, neither lignin nor cellulose accumulate from year to year. An equilibrium exists between litter fall with its high content of cellulose and lignin and the rate of microbial decomposition. Therefore, relatively constant values of these substances are found in the mineral horizons; some variations in the amount of cellulose and lignin occur in the organic horizons where the influences of environmental factors (including season) become more important.

Acknowledgements

I am grateful to Dr. Söchtig, Braunschweig, for the quantitative determination of free amino acids, to my assistant Dr. Kunze for the lignin analyses and to Miss Gramlich for technical help. This investigation was supported by the Deutsche Forschungsgemeinschaft.

References

Alexander, M.: Introduction to soil microbiology. New York-London-Sidney: Wiley & Sons 1967.

Alexander, F. E., and R. M. Jackson: Examination of soil microorganisms in their natural environment. Nature **174**, 750—751 (1954).

Beck, Th.: Mikrobiologie des Bodens. München-Basel-Wien: Bayrischer Landw. Verlg. 1968.

Black, W. A. P.: The seasonal variation in the cellulose content of the common scottish Laminariaceae and Fucaceae. J. Mar. Biol. Assoc. U. Kingd. **29**, 379—387 (1950).

Brink, R. H., P. Dubach, and D. L. Lynch: Measurement of carbohydrates in soil hydrolyzates with anthrone. Soil Sci. **89**, 157 (1960).

Cholodny, N.: Eine neue Methode zur Untersuchung der Bodenmikroflora. Arch. Mikrobiol. **1**, 620—652 (1930).

Flaig, W.: The chemistry of humic substances. In: The use of isotopic in soil organic matter studies. Report of the FAO/LAEA technical meeting Brunswick-Völkenrode Sept. 1963. New York: Pergamon Press 1966.

Gams, W.: Isolierung von Hyphen aus dem Boden. Sydowia. Ann. Mycol. II, **13**, 87—94 (1959).

Gilbert, R. G., and J. Altman: Ethanol extraction of free amino acids from soil. Plant and Soil **24**, 229—238 (1966).

Gilman, J. C.: A manual of soil fungi. Ames: Iowa State Coll. Press 1957.

Glathe, H.: Die Mikroorganismen des Bodens und ihre Bedeutung, pp. 551—649. In: Handb. d. Pflanzenern. u. Düng. II, Teil 1 (Linser, H., Ed.). New York-Berlin: Springer 1966.

Gorham, E., and J. Sanger: Plant pigments in woodland soils. Ecology **48**, 306—308 (1967).

Haider, K., S. Lim u. W. Flaig: Experimente und Theorien über den Ligninabbau bei der Weißfäule des Holzes und bei der Verrottung pflanzlicher Substanz im Boden. Holzforschung **18**, 81—88 (1964).

—, and C. Schetters: Mesophile ligninabbauende Pilze in Ackerböden und ihr Einfluß auf die Bildung von Humusstoffen, pp. 425—439. In: Progress in soil biology (Graff, O., and J. E. Satchell, Eds.) Braunschweig: Vieweg & Sohn 1967.

Henderson, M. E. K.: Studies on the physiology of Lignin decomposition by soil fungi, pp. 286—296. In: The ecology of soil fungi (Parkinson, D., and J. S. Waid, Eds.). Liverpool University Press 1960.

Imschenezki, A. A.: Mikrobiologie der Cellulose. Berlin: Akademie Verlag 1959.

Janke, A.: Arbeitsmethoden der Mikrobiologie. Dresden-Leipzig: Steinkopf 1946.

Johnson, L. F., E. A. Curl, J. H. Bond, and H. A. Fribourg: Methods for studying soil microflora-plant disease relationships. Minneapolis: Burgess Publ. Co. 1960.

Jones, P. C. T., and J. E. Mollison: A technique for the quantitative estimation of soil microorganisms. J. Gen. Microbiol. **2**, 54—69 (1948).

Kerr, A.: The use of cellophane in growth studies on soil fungi. Trans. Brit. Mycol. Soc. **41**, 14—16 (1958).

Kunze, C.: Die Identifizierung von Ligninspaltstücken anhand von Papier- und Dünnschichtchromatographie. Experientia **24**, 844—845 (1968).

Kunze, C.: Die biologische Aktivität von Bodenproben, gemessen an der CO_2-Abgabe und dem Vanillin-Abbau (in press).

Maeder, H.: Chemische und pflanzenphysiologische Untersuchungen in Rottestroh. Diss., Univ. Gießen, 1960.

Meister, A.: Biochemistry of the amino acids. New York-London: Academic Press 1965.

Mischustin, E., and A. Petrowa: Bestimmung der biologischen Aktivität der Böden. Mikrobiol. **32**, 479 (1963).

Ovington, J. D.: Quantitative ecology and the woodland ecosystem concept, pp. 103—192. In: Advances in ecological research. (J. B. Cragg, Ed.): London-New York: Academic Press 1962.

Paul, E. A., and C. M. Tu: Alteration of microbial activities, mineral nitrogen and free amino acid constituents of soils by physical treatment. Plant Soil **22**, 207—219 (1965).

Pochon, J., and P. Tardieux: Techniques d'analyse en microbiologie du sol. Collection "Techniques de base" St. Mandé (Seine): Tourelle 1962.

Rossi, G., and S. Riccardo: L'esame microscopico e bacteriologici diretto del terreno agrario. Nuovi Ann. Minist. Agric. **7**, 457—470 (1927).

Shanks, R. E., and J. S. Olson: First year breakdown of leaf litter in Southern Appalachian Forests. Science **134**, 194—195 (1961).

Shazly el, K., and R. E. Hungate: Method for measuring diaminopimelic acid in total rumen contents and its application to the estimation of bacterial growth. Appl. Microbiology **14**, 27—30 (1966).

Siu, R. G., and E. R. Reese: Decomposition of cellulose by microorganisms. Bot. Rev. **19**, 377—416 (1953).

Steubing, L.: Pflanzenökologisches Praktikum. Berlin-Hamburg: Parey 1965.

Strugger, E.: Fluoreszenzmikroskopie und Mikrobiologie. Hannover: Schaper 1949.

Tchan, Y. T.: Counting soil algae by direct fluorescence microscopy. Nature **170**, 328—329 (1952).

Törne, E. V.: Summarized remarks on the rate of wet decomposition of cellulose under biologically different experimental conditions, pp. 473—474. In: Progress in soil biology (Graff, O., and J. E. Satchel, Eds.). Braunschweig: Vieweg & Sohn 1967.

Unger, H.: Der Zellulosetest, eine Methode zur Ermittlung der zellulolytischen Aktivität des Bodens in Freilandversuchen. Z. Pflanzenernähr. Düng. Bodenk. **91**, 44—52 (1960).

Waid, J. S., and M. J. Woodman: Symposium, Methodes d'études microbiologiques du sol. June 1957, Louvain, Belgium (1957).

Wood, E.: A new naturally occurring amino acid. Nature **165**, 74—75 (1950).

Work, E.: The isolation of α, ε-diaminopimelic acid from Corynebacterium diphtheriae and Mycobacterium tuberculosis. Biochem. J. **49**, 17—23 (1951).

— Reaction of ninhydrin in acid solution with straight-chain amino acids containing two amino groups and its application to the estimation of α, ε-diaminopimelic acid. Biochem. J. **69**, 416—423 (1957).

12

The Role of Soil Invertebrates
in Turnover of Organic Matter and Nutrients

C. A. Edwards, D. E. Reichle, and D. A. Crossley, jr.

Substantial proportions of organic substances which originate in plants or animals eventually reach the soil, where they may remain for only a few hours or days if they are readily decomposable (Heath et al., 1966), or as long as several decades if they decay slowly (Kendrick, 1959). The plant material ranges from bacteria, fungal hyphae and soft leaf tissue to tough woody substances. Animal materials range from single-celled protozoa, nematodes, soft-bodied worms and insect larvae through arthropods with tough exoskeletons to large vertebrate animals and excreta.

Plants and small animals with soft tissues usually are decomposed by soil microflora alone, but tougher and more chemically stable tissues usually are broken down by a complex series of events involving both soil fauna and microflora. Some tissues decompose at the soil surface and their products leach into the soil, or raw organic matter may be incorporated directly into the soil and decay there.

Soil animals contribute to the breakdown of organic matter in several ways:

1. they disintegrate plant and animal tissues and make them more easily invaded by microorganisms;
2. they selectively decompose and chemically change parts of organic residues;
3. they transform plant residues into humic substances;
4. they increase the surface area available for bacterial and fungal action;
5. they form complex aggregates of organic matter with the mineral part of soil; and
6. they mix the organic matter thoroughly into the upper layers of soil.

The Turnover of Plant Organic Matter

The form and amount of incorporation of plant organic matter into the soil is one way of classifying soil types. Plant remains break down least rapidly in an acid raw humus, where there are plant residues with well-preserved cellular structure, few animals and numerous fungi. A less acid raw humus or mor soil contains more animals, but these are mainly collembolans and oribatid mites. In a moder soil most of the plant remains have been converted into animal feces or other residues, although some plant tissues and cellular structures may still be discerned. Animals present are mostly mites and collembolans, and the production of humic substances has not proceeded very far. Mineral matter, animal feces and organic residues form a fairly loose mixture.

As a moder becomes transformed to a mull humus form, the larger arthropods and invertebrates increase in numbers. These ingest mineral substances with their food, and soil aggregates begin to be formed (although much of the upper soil is still only a mechanical mixture). In a true mull humus where aggregation is complete, some soil aggregates are formed during passage through the intestines of animals, the humus and clay fractions have become inseparable, all plant structures have been destroyed, and there is a numerous and varied fauna.

Litter Breakdown

The rate of breakdown of litter varies considerably between seasons and sites. In temperate mull soils, oak litter disappears in eight to fifteen months, whereas in tropical regions a leaf may decompose within weeks after falling to the forest floor.

Bocock and Gilbert (1957) in northern England found the following percentage losses of dry leaf weight during six months (January to June): birch *(Betula)*, 82.9%; lime *(Tilia)*, 55.6%; and oak *(Quercus)*, 17.4 to 26.2%. Edwards and Heath (1963) reported a weight loss of 90% for oak and 70% for beech *(Fagus)* leaves between July and April in southern England. Witkamp (1966) followed the weight loss of four leaf species in Tennessee, USA, from November to November: they were mulberry *(Morus rubra)*, 90%, redbud *(Cercis canadensis)*, 70%; white oak *(Quercus alba)*, 55%; and pine *(Pinus)*, 40%. In a similar study (Heath et al., 1966), beech leaves lost 64% of their weight, oak lost 89% and elm *(Ulmus)*, birch and ash *(Fraxinus)* leaves broke down completely after one year. The leaf species greatly influences the animal and microflora populations which feed on them, particularly because of such physical and chemical properties of the leaves as the carbon/nitrogen ratio. Nutritious and succulent leaves often have small C/N ratios, whereas tough leaves containing resistant constituents such as cellulose and lignin, which are unpalatable to litter fauna, often have large C/N ratios (Witkamp, 1966).

The sequence of events after fresh plant material reaches a mull soil varies widely and is not yet fully understood. The general pattern of breakdown and incorporation into soil can, however, be outlined.

For several weeks after fresh deciduous litter reaches the forest floor, it is not eaten by the litter fauna, although it is invaded by some microorganisms, especially fungi. During this period the litter darkens becomes weathered, and water-soluble substances, mainly sugars, organic acids and polyphenols are leached out. Edwards and Heath (1963) noted that beech and oak leaves which did not darken appreciably after reaching the soil, began to be fragmented by litter animals much sooner than leaves which darkened rapidly. The darker leaves were "sun" leaves (Heath and Arnold, 1966) and had a much greater percentage of both polyphenols and sugars than did "shade" leaves (Heath and King, 1964). As the water soluble polyphenols were removed by weathering the litter became more palatable to arthropods (Heath et al., in press). Edwards and Heath (1963) suggested that the increase in polyphenolic materials precipitated protein complexes in leaves and made them less digestible by the microfauna. King and Heath (1967) confirmed that the amount of polyphenol in beech litter was inversely proportional to the rate at which the litter was consumed. Feeny and Bostock (1968) also found that most feeding by phytophagous insects (when leaves were still on the tree) occurred when total tannin content was minimal and condensed tannins were practically absent. Satchell and Lowe (1967), who studied the palatability of many kinds of leaf litter to the earthworm *Lumbricus terrestris*, reached similar conclusions. They tested the influence of cuticle thickness, water-holding capacity, nitrogen content, carbohydrate and polyphenol content on palata-

bility. Although some of these factors correlated with palatability, only the polyphenol content could be experimentally demonstrated to affect palatability.

Water-soluble polyphenols and other substances are leached from litter into the
soil. Various anions and cations are released which percolate into deeper soil horizons.
During the early stages of litter breakdown there may be a large but inactive microbial
population. EDWARDS and HEATH (1963) showed that leaves buried in soil, in nylon
bags with mesh of 0.003 mm opening that excluded invertebrate animals, changed
little in physical appearance and weight. In later experiments (unpublished data)
leaves remained superficially intact even after three years. Apparently, without
preliminary fragmentation by soil animals, the omnipresent microorganisms cannot
decompose the constituents of many leaf species.

In temperate regions, litter appears fragmented mainly by earthworms (particularly *Lumbricus terrestris*), enchytraeid worms, diplopods, isopods, dipteran larvae,
collembolans and oribatid mites. If this fragmentation is retarded experimentally, the
whole process of decomposition is slowed. For example, KURCHEVA (1960) used
napthalene to exclude some invertebrates from litter. She found that one tenth of oak
litter was decomposed in 140 days when both soil invertebrates and microorganisms
were present; however, less than one twentieth disappeared when most invertebrates
were excluded. WITKAMP and CROSSLEY (1966) also excluded arthropods from litter
with napthalene and found that 60% of the oak litter in untreated soil broke down in
one year compared to only 45% of the litter treated with napthalene. Other chemicals
(insedicides have even greater influence on decomposition. In aldrin-treated soils with
few Collembola, oak litter disappears more slowly than in untreated soil, whereas in
DDT-treated soil with larger Collembola populations (because their mesostigmatid
mite predators are eliminated) breakdown is accelerated, when earthworms are excluded from the litter (EDWARDS, 1965).

During the initial fragmentation of litter the main contribution of the fauna is to
provide a more suitable physical substrate for microfloral growth, because there is
little chemical difference between the food and feces of the microfauna. Estimates of
annual amounts of leaf litter converted to animal feces range from 20% to 100% (BOR-
NEBUSCH, 1930; NICHOLSON et al., 1966). A succession of species of microflora has
been identified on fecal pellets (VAN DER DRIFT and WITKAMP, 1960). DUNGER (1958)
found only slight chemical changes occurring after passage of litter through the alimentary tracts of diplopods, isopods and other small soil arthropods. The increase in
surface area resulting from comminution of litter by microarthropods has been considered the most important contribution to its ultimate breakdown (KEVAN, 1962).

There are, however, certain chemical changes which occur during fragmentation,
because some invertebrates hydrolyze not only sugars, proteins and fats but also stable
substances such as cellulose, keratin and chitin. Cellulases have been reported in some
invertebrates (TRACEY, 1951; GASDORF and GOODNIGHT, 1963). Invertebrates that
break down cellulose usually do so with cellulases and other appropriate enzymes
from their symbiotic gut flora (GHILAROV, 1962), although some animals can synthesize humic substances in their digestive tract (FRANZ and LEITENBERGER, 1948).

GASDORF and GOODNIGHT (1963), however, demonstrated proportionate increases
in lignin and decreases in cellulose in feces of oribatid mites *(Peloribates* and *Herman-
niella)* feeding on oak litter. BOCOCK (1963) found increases in lignin and soluble
nitrogen in the feces of *Glomeris marginata*. Nevertheless, most of the evidence indi-

cates that the chemical processes of humification are caused more by the microflora than by the fauna. Knowledge of the mechanisms of the humification process is inadequate, with few advances since Waksman's account in 1936.

At abscission, leaves already have an extensive internal and external microflora, and in the litter layer fungi invade the leaf tissues. Fragmentation by the fauna accelerates microbial invasion; tissue breakdown initiated by the microflora in turn favors further attack by other soil animals. Collembola can be differentiated into the surface-dwelling species, which feed on decaying leaves, and deeper-living species, which graze on microorganisms (Zachariae, 1962).

The activities of the fauna and microflora are complementary and intricately interrelated, and where soil animals are very numerous, microorganisms (especially bacteria) are also abundant. The distribution of soil invertebrates and microorganisms in different soil layers of the same soil is often very similar (Ghilarov, 1947). Hartenstein (1962) demonstrated that oribatid mites would not feed on wood or leaves which did not contain fungi. Engelmann (1961) concluded that fungi were the food source of oribatid mites although oribatids do ingest litter (Hartenstein, 1962). Kollmansperger (1956) estimated that soils worked by earthworms and millipedes contained about two-thirds more bacteria than unworked soils. Went (1963) reported a similar increase in soils with earthworms and concluded that the mixing of earthworm excrements with upper soil layers increased the numbers of microbes. Bacteria may survive unfavorable climatic conditions in the intestinal tract of arthropods, but there is evidence that large increases in numbers of fungi in soil animals are harmful. Kühnelt (1963) found that some fungi inhibited increases in numbers of amoebae, rotifers and Collembola whereas other fungi encouraged them. Parasitiform predatory mites may be injured by heavy growths of fungi, and the alternating dominance of acarid and oribatid mites in soil may reflect such changing microbiological processes (Karg, 1963).

Thus, the gradual decomposition of organic matter and its incorporation into soil depends on (1) a sequence of feeding by soil animals alternating with growth of microorganisms, combined with (2) animal activity in the soil which promotes the formation of soil aggregates. As the decomposition process continues, humus substances increase in the organic matter fraction with less cellulose, sugars and proteins remaining. During decomposition, the C/N ratio of the litter progressively decreases from 30:1 to about 10:1. Most of the sugars and starches are transformed to carbon dioxide and water, and the proteins to carbon dioxide, water and ammonia (which is rapidly converted to nitrates). Mineral elements also disappear (Attiwill, 1968). Sodium is leached rapidly from the litter horizons. Magnesium, phosphorus and, to a lesser extent, potassium are rapidly lost from fresh litter but are held in the lower litter layers. In a typical mull soil less than one-fourth of the fresh organic matter reaching the soil is converted into humus; the remainder disappears by oxidation, leaching and erosion. The humus has little structure but is a dark amorphous colloid, usually intimately mixed with the mineral fraction of soil. The composition of humus is unknown, but it probably consists of complex polymers of phenolic materials (the main source of phenols being the breakdown products of lignin by microorganisms, especially basidiomycetes and some ascomycetes).

When humification is advanced, the soil invertebrates, especially earthworms, millipedes and dipteran larvae, are still important in mixing the humified material through the vertical soil strata. In soils with a rich fauna, the upper soil horizons become ill-defined and have a uniform structure.

Breakdown of Woody Materials

Much of the biomass of organic material that reaches the forest floor is wood which ranges in size from small twigs to whole trees and which also is attacked and decomposed by a succession of animals and microbes. In some regions termites are the most important animals that decompose wood, but many species of Diptera and Coleoptera also commonly attack wood at various stages of decay (BRAUN, 1954). Decomposition can follow different pathways; the symbiotic gut flora of many insects can digest cellulose, but lignin is only partly digested and fragmented and remains for attack by other microorganisms (see section on litter breakdown).

Fresh wood can be attacked by scolytid, curculionid, buprestid and cerambycid beetles or sawfly larvae and later by larvae of Diptera such as Tipulidae, Mycetophilidae, Cecidomyidae and Sciaridae. Some beetles carry ambrosia fungi which proliferate in the tunnels and provide food for their larvae. Moisture is important and soaking for long periods makes the woody tissues more susceptible to attack by microorganisms. Fungal mycelia ramify through cracks and soften tissues, and then many insects and larvae, especially Scarabaeidae, millipedes and isopods begin to invade the moist tissues. The feces of these animals are a fertile substrate for microorganisms. Often, mosses may begin to grow on the outside of the wood and provide a moist habitat for invasion by arthropods. Similarly the holes that occur in tree stumps fill with water and provide a suitable habitat for diverse invertebrate animals (PARK and AUERBACH, 1954).

Eventually the wood becomes soft, fragmented, gradually incorporated into the humus layer, and finally thoroughly distributed through the upper layers of soil. The whole process of breakdown of woody tissues is slow and may take several decades, and the detailed succession of flora and fauna is still poorly known.

Root Breakdown

The primary production of the underground parts of plants in forests is difficult to measure, and existing records of the live weight of roots are both scarce and subject to large errors (OVINGTON, 1962). The contribution of roots to the total organic-matter biomass in soil is not easily separated from that of the organic matter contributed by above-ground plant parts. The feeding of underground phytophagous invertebrates undoubtedly reduces the amount of live roots in the soil, but the rates and relative proportions of such losses caused by invertebrates feeding below the surface remain largely unknown. A wide variety of species of root-feeding invertebrates, especially insects and nematodes, have been studied in cultivated areas. Many references in RITCHER's (1958) review of scarabaeid beetle biology concern damage done by larvae feeding on the roots of grasses or crop plants, but less is known about the feeding by beetle larvae on roots of forest trees (NAKAMURA, 1965). Some soil-inhabiting species of beetles and other invertebrates which are normally saprophagous may occasionally feed on roots (EDWARDS and GUNN, 1961; PARIS and SIKORA, 1965; RITCHER, 1958). Herbivores which normally feed on above-ground vegetation may attack roots of trees during certain seasons (GHILAROV, 1964). Only fragmentary information is available on root-feeding invertebrates of forest soils, but their abundance and importance in cultivated areas suggests that root feeders may considerably reduce underground primary production.

Nymphs of the periodical cicada, which suck juices from the roots of deciduous forest trees, may be responsible for considerable loss of primary production (Lloyd and Dybas, 1966) as well as soil turnover (Beck, 1968). Numbers of nymphs quoted by Lloyd and Dybas (1966) and Dybas and Davis (1962) ranged up to 311 ± 41 per square yard, and commonly averaged about 10 to 25 per square yard. Dybas and Davis estimated live weights of mature periodical cicada nymphs at between 230 and 428 kg/ha in an upland forest stand, and between 1913 and 3685 kg/ha in a floodplain area in Illinois (USA). In the years immediately preceding emergence, nymph biomass is probably large enough for feeding to reduce productivity measurably. In apple orchards in Connecticut (USA), periodical cicadas have extensively damaged rootlets and decreased the yield of fruit (Hamilton, 1961).

Methods of Studying Litter Breakdown by Soil Animals

Populations

There is now considerable evidence that litter decomposes at a rate directly related to the number of invertebrate animals in the litter and underlying soil. Populations of animals in litter are so large that they can be estimated only by taking cores or quadrats of litter and soil of a known area or volume. The dimensions of these samples depend on the size and distribution of the various animals to be sampled; the less uniformly distributed an animal is, the larger the samples must be. There are many methods of extracting invertebrates from soil and litter samples: these rely either on the animals migrating from the samples (dynamic methods), or on the relative density or the lipophilic properties of the arthropod cuticle (mechanical methods). Edwards and Fletcher (in press) reviewed the eleven most common methods and showed that, for the majority of invertebrates, the most efficient method was a high gradient funnel of the type described by Macfadyen (1962). Extraction of some invertebrates requires specific methods, e.g., the use of dilute formalin to bring earthworms to the soil surface (Raw, 1959). Detailed information on these techniques can be obtained from Murphy (1962) and Edwards and Fletcher (in press).

The invertebrates in soil range in size from a few microns (Protozoa) to many centimeters in length, so their relative importance in energy turnover cannot be expressed in terms of mere numbers. A more meaningful estimate is to express the populations in terms of live or dry weight or biomass. Individual weights can be calculated from regressions on linear dimensions of preserved specimens, and total biomass calculated from the numbers per unit area (Edwards, 1967). Populations can also be expressed as biovolume but, since the specific gravity of most litter invertebrates ranges between 1.0 and 1.1, estimates of biovolume differ little from those of biomass. A better criterion for many soil invertebrates might be the external surface area of the organisms or "biosurface", because this would take into account the faster metabolism of small organisms (Murphy, 1953) (see following section, page 158, "Respiration of Organisms").

Loss of Weight, Nutrients and Energy by Litter

Calculations of the amount of organic matter intake needed to sustain animal populations can give indirect estimates of the turnover of organic matter, but there are experimental techniques that measure litter decay and the role of different faunal groups in this process more directly and precisely. Nylon litter bags and screened frames, or similar devices, can be used to exclude particular animals and allow litter samples to be handled and analyzed repeatedly.

CROSSLEY and HOGLUND (1962) adapted the nylon bag method to examine the succession of microarthropods in leaf litter. Different sizes of mesh have been used to decide which animals have access to the litter, for example CROSSLEY and WITKAMP (1964) used 1 mm mesh to exclude large invertebrates, and to allow the influence of microarthropods on the breakdown of litter from various leaf species to be assessed. EDWARDS and HEATH (1963) used different mesh sizes (0.003, 0.5, and 7.0 mm) in a similar way, and later (HEATH et al., 1966) compared the decay of seven leaf species exposed to these different size classes of invertebrates. NICHOLSON et al. (1966) used such bags to study coprophagy by the millipede *Glomeris marginata*.

Screened frames (and nylon net bags) were used (EDWARDS and HEATH, 1963) to contrast the physical appearance of litter attacked by microarthropods and earthworm populations. RAW (1962), KURCHEVA (1964) and SATCHELL and LOWE (1967) also put litter in frames in studies of arthropods of the forest floor, and of leaf palatability and selection by *Lumbricus terrestris*. Alternatively, screen mesh buried in soil prevents animals entering the overlying litter (HEATH and ARNOLD, 1966). CROSSLEY and WITKAMP (1964) suggested the use of screen soil boxes to study how different soil types influenced litter breakdown. Other investigators have strung leaves on threads through the petioles to allow them to be collected later for analysis.

These techniques, although useful, do not maintain leaves in completely natural conditions (WITKAMP and OLSON, 1963). Leaves in bags often become compacted and moist. Animals such as earthworms which physically remove litter from the soil surface may penetrate bags with intermediate sizes of mesh but still be prevented from removing leaves. Large mesh sizes also increase experimental error by loss of small fragments of litter during handling. Tullgren funnel extraction of microarthropods from litter bags can destroy litter flora and, when bags are replaced, can obliterate synergistic effects of microbial and arthropod populations. Thorough consideration of the specific variables of each experiment is needed to select the appropriate technique.

Tagging organic detritus with radioisotopes is a convenient way to measure loss of elements during decomposition. This technique has several advantages including nondestructive analysis and sampling, ease of analysis, and allowance for contamination by elements in rainfall and soil. Disadvantages include the difficulties of obtaining isotopes with suitable radiological characteristics and of obtaining detritus or organisms tagged with the isotope incorporated naturally into tissues. Many isotope studies have been concerned with the cycling of different elements in natural ecosystems: e.g., cycling of ^{106}Ru, ^{60}Co and ^{85}Sr in *Pinus*, *Quercus* and *Cornus* leaf litter (OLSON and CROSSLEY, 1963); ^{137}Cs in lichens, mosses and needle litter in a *Pinus* forest (WITKAMP and FRANK, 1967); ^{45}Ca in decomposing *Cornus* litter (THOMAS, 1969), and ^{14}C in water-soluble organic substances which form humic soil compounds (CHEKALOV and ILLYUVIYEVA, 1962). Other studies have been more specifically concerned with determining the role of invertebrates in the breakdown and release of nutrients from litter (Crossley and Witkamp, 1964; Witkamp and Crossley, 1966). Engelmann (1961) used ^{14}C-glycine in yeast to measure the intake and assimilation of the isotope by oribatid mites. Reichle and Crossley (1965, 1967) used tracer ^{137}Cs to delineate litter food chains and measure food consumption. Hubbell et al. (1965) used [85]Sr and Paris and Sikora (1967) used ^{32}P to measure the rate of intake by isopods and uptake into their predators. Reichle (1967) followed the turnover of ^{137}Cs and calculated energy flow in isopod populations. Reichle and Crossley (1969) and Reichle et al. (1969) have listed the concentrations of ^{137}Cs, [24]Na, [42]K and [45]Ca in invertebrates of different trophic levels and used radiocesium turnover to predict the potassium content of animals in higher trophic levels. Isotope turnover and systems analysis techniques have been used to study organic and mineral fluxes through the biota of soil and litter systems of varying degrees of complexity (Patten and Witkamp, 1967). Nitrogen-15, although not a radioisotope, can be traced in decomposing litter by mass spectrometry.

Such techniques permit studies of trophic relationships and the role of various invertebrates in soil processes under natural field conditions. To study individually numbers of species of soil invertebrates is a formidable task because of their diversity and minute size. We can, however, assess the function of major ecological groups, e.g., microarthropods, earthworms or soil flora, by creating artificially controlled populations. It has been difficult to manipulate populations of specific animals without incidentally affecting micro- and mycoflora of the soil. Although insecticides are less specific than is required, they have been used effectively (Edwards, 1963, 1965; Kurcheva, 1960; Crossley and Witkamp, 1964). The toxicity of fungicides to insects was reported by Sander and Cochy (1967). Chlordane (Edwards, 1965) and phorate (Edwards et al., 1968) readily kill earthworms, whereas aldrin, DDT and lindane do not. Davey (1963) reviewed the effects of pesticides on earthworms. A fumigant such as DD (dichloropropane-dichloropropene) kills almost all soil invertebrates, and recolonization is primarily aerial (Edwards and Buahin, 1964). Aldrin (or dieldrin) only slightly reduces numbers of predatory mites, but DDT is very toxic to these mites. Of the organophosphorus insecticides, menazon is very toxic to aphids but to no other soil animals (Raw, 1964). Most saprophagous species are less susceptible than predaceous species and often they increase in numbers after an insecticidal treatment (Grigor'eva, 1952; Sheals, 1956; Edwards and Dennis, 1960; Edwards et al., 1967). Edwards et al. (1969) described a technique which uses a broadly toxic soil fumigant (DD), an insecticide (aldrin), and formalin as an earthworm extractant (Raw, 1959). DD fumigation eliminated earthworms and killed 98.3% of the microarthropods. Periodic applications of aldrin to the fumigated soil kept recolonization by microarthropods to 6.7% of that in control plots after eight months. In other plots repeated extraction of earthworms with formalin decreased numbers by 80% after six months, but there was no significant effect on numbers of soil microarthropods. Invertebrates could be added to plots treated in these ways so that litter breakdown caused by such groups as earthworms and microarthropods could be assessed.

Physical Disappearance of Litter Leaf Area

Methods of assessing litter decomposition based on weighings were described in a previous section; these estimate total breakdown but do not distinguish the type of consumer. A better estimate of the invertebrate contribution to leaf breakdown can be made by determining the total numbers of leaves or leaf area that has disappeared (EDWARDS and HEATH, 1963, HEATH et al., 1964). Disappearance can be measured visually, photometrically, graphically, or by tracing the original outline of each leaf and the part that has disappeared and weighing both pieces of paper. A more precise method is to use disks of leaf tissue of standard initial area with an area photometer of the type described by HEATH et al. (1964).

Effects of Climatic Conditions on Organic Matter Turnover

The turnover of organic matter in temperate woodlands is reasonably well documented in papers by KENDRICK and BURGES (1962), EDWARDS and HEATH (1963), SAITO (1967), OLSON and CROSSLEY (1963), VAN DER DRIFT (1963), DUNGER (1958), GERE (1963), WITTICH (1943), and HANDLEY (1954), but much less is known about the cycle in tropical woodlands where primary productivity is much greater. OLSON (1963) presents mathematical models for litter production and decomposition in evergreen and deciduous forests.

Litter production has been estimated as varying from as little as 0.5 metric tons/ha/yr in alpine and arctic forests through about 2.5—3.5 metric tons/ha/yr for a stable temperate forest to 5.5—15 metric tons/ha/yr tropical forests (BRAY and GOR-HAM, 1964). Litter fall in temperate woodlands is markedly seasonal, whereas in the tropics it is much more continuous although more leaves fall during the wet season.

Although much more leaf litter is deposited in tropical forests, decomposition by the abundant microflora and fauna is rapid during the wet season and litter does not accumulate. The alternating periods of heavy rainfall and hot sunshine greatly favor multiplication of microorganisms and provide a suitable environment for very rapid desomposition by microbes. MADGE (1965, 1966) estimated that litter breaks down six to ten times as fast in a tropical forest as in a temperate forest.

Populations of soil and litter animals, although usually greater in tropical than in temperate woodlands, do not differ sufficiently in number to account for the greater turnover of organic matter in the tropics. Seasonal differences in numbers of soil animals are greater in tropical than in temperate woodlands, the peak numbers coinciding with the periods of highest rainfall. Earthworms seem much less important in organic matter turnover in the tropical woodlands; much of their function in litter and soil is replaced by the numerous termite populations. The state of equilibrium in organic matter turnover in a tropical forest is such that there is slightly more humus than in a temperate woodland; although humus accumulates faster, it also is broken down more rapidly.

Temperature greatly influences the rate of turnover of organic matter, but its overall effect is complicated by the interaction between temperature and moisture. There is no doubt that high temperatures and adequate moisture favor decomposition. In temperate woodlands it is usually too cold and dry for rapid decomposition except during spring and autumn, whereas in a tropical forest the wet season coincides with higher temperatures. MADGE (1966) determined the optimum temperature for humus accumulation to be 20—25° C, and the most favorable temperature for humus disappearance to be 30—35° C with adequate moisture. Between 25 and 35° C humus forms and decomposes at about equal rates, and therefore does not accumulate.

Either moisture or temperature can be limiting factors in the rate of litter decomposition. Temperature is a more important factor in a relatively moist climate such as Tennessee, USA (Witkamp, 1966). Madge (1966) reported no relationship between organic matter turnover and temperature in tropical soils with the same rainfall but different temperatures, and thought moisture was more important there.

Energy Flow and Nutrient Cycles Through Soil Populations
Rates of Exchange Along Food Chains

Although some soil organisms clearly belong to a specific trophic level in the detritus food chain, the more detailed structure of the food web is still uncertain and complicated for any given soil community (Fig. 1). As more refined mathematical

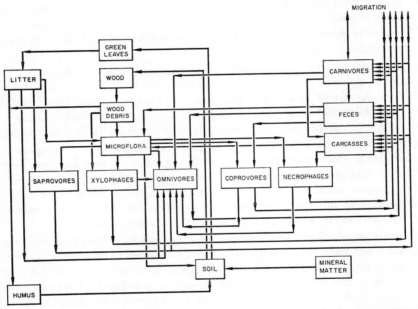

Fig. 1. A trophic model of the decomposer community of forest floor and soil. This diagrammatic description of the detritus food web illustrates the innate complexity of the trophic relationships. The trophic level as a distinct entity is not applicable; instead, the food bases are grouped into trophic "components" equivalent to compartments in the model. Under this concept a compartment may be a donor or recipient of an unlimited number of other compartments. Models constructed in this manner are useful to describe complexity, but can be used to calculate energy flow only at fixed points in time — since no accommodations are made for changing food habits of species during different life history stages

models are developed to account for the flow of energy or nutrients through woodland food chains the obvious trophic levels may be separated, but this requires detailed knowledge of the components of the food chains. The feeding habits of some groups of soil organisms can be determined by studies in the laboratory (e.g., Hartenstein, 1962). Radioactive tracer methods can provide information on feeding habits of invertebrates in the field and reveal food chains (e.g., Odum and Kuenzler,

1962). Such studies yield estimates of feeding habits only under one set of environmental conditions, and animals may change to other foods seasonally.

The rates at which soil invertebrates ingest food usually are calculated from the differences in weight of the food offered and that remaining after a given time inverval, an approach suitable for both saprophagous (BOCOCK, 1963) and predaceous (PHILLIPSON, 1960) species, although coprophagy (WIESER, 1966) must always be considered. The rate of disappearance of litter *in situ* was used by VAN DER DRIFT and WITKAMP (1960) to verify their laboratory measurements of food consumption. Indirect estimates of the food intake can be made when experimental measurement is impracticable, but these depend on knowledge of the energy balance of the organisms. If growth, reproduction, respiration and assimilation are known, the amount of food necessary to satisfy these energy requirements can be predicted.

The rates at which radiosotopes are consumed in food also can be used to determine the food consumed by organisms. Radiosotopes can be used in two ways to measure the food consumption: 1) when there is no loss of isotope from the consumer, the rate of whole-body accumulation is related directly to the rate of assimilation into tissues, e.g., radioiodine (VAN HOOK and CROSSLEY, 1969), and 2) for most radioisotopes which are excreted, the organisms can be allowed to attain isotopic equilibrium with their food base (REICHLE, 1967). At equilibrium, isotope input = isotope loss

$$\text{and } I = \frac{kQ_e}{a}$$

where I = feeding rate (units of radioactivity), a = proportion of ingested isotope assimilated, k = the elimination coefficient of the radioisotope in time^{-1} (0.693 / the biological half-life), and Q_e = the equilibrium whole-body burden of radioisotope (units of radioactivity). The rate at which isotope is ingested can be converted to food consumption, if the concentration of isotope in the food is known.

Several attempts have been made to relate the amounts of food eaten by soil invertebrates to some dimension of the consumer, such as weight, volume or surface area. VAN DER DRIFT (1951), GERE (1956) and DUNGER (1958) showed that the amount of litter consumed by millipedes and isopods was proportional to their surface area or live body weight to the 0.67 power. REICHLE (1968) found that food intake correlated better with dry body weight for eleven species of forest floor arthropods, and calculated a regression equation ($Y = 0.063 X^{0.68}$) relating food consumption (Y) to the 0.68 power of body weight (X); this relationship, however, probably differs between taxa with different food habits, for it is not the total food intake but its assimilation which should correlate with either surface area or some other power of body weight. Data relating ingestion to assimilation by forest floor invertebrates are summarized by REICHLE (1967).

Assimilation

The amount of material or energy actually assimilated, as opposed to that ingested, is difficult to measure. Net assimilation can be estimated from

$$A = C - F$$

where A is assimilation, C is consumption and F is defecation (PETRUSEWICZ, 1967), or from some other equivalent expression. A budget of both consumption and defecation must be obtained for a suitable period of time. Animals such as soil insects may contaminate food with feces so this also must be considered.

Nonassimilated indicator dyes have been used to measure both the amount of food which passes through the digestive tract and the amount assimilated (Milner, 1967). Although such techniques have been used primarily for large mammals, they can be used with arthropods. Suitable dyes should not be assimilated in the gut and must be readily detectable in the feces; bromophenol blue (Krishna and Singh, 1968), Azo Rubrin (Treherne, 1957) and chromic oxide (McGinnis and Kasting, 1964) have been successful. In such experiments, percentage dry weight assimilated was calculated (Southwood, 1966) from:

$$\text{Assimilation} \ (\% =) \ \frac{1 - \text{conc. indicator in food})/(\text{unit dry wt})}{(\text{conc. indicator in feces})/(\text{unit dry wt})} \times 100 \ .$$

There are difficulties in determining the concentration of dye in the feces and accounting for the slight assimilation of the marker. Radioactive tracers appear to estimate better the assimilation of materials from the gut (Odum and Golley, 1963; Reichle, 1967). A single meal of radioisotope may be excreted in two phases with an initial rapid loss by elimination of unabsorbed radioisotope from the gut followed by an excretion of assimilated radioisotope from the tissues (Reichle and Crossley, 1965).

By plotting whole-body radioactivity against time the components can be separated, and the proportion of isotope assimilated from the food calculated. Assimilation of energy also can be estimated using radioactive tracers, because the difference between calories per unit of tracer in food and calories per unit of tracer in feces represents calories assimilated. This method has not been applied to soil invertebrates, but its usefulness has been established for large mammals (Petrides, 1968).

Respiration of Organisms

Laboratory studies of respiration usually are made to calculate metabolic rates, so that these can be extrapolated to populations in their natural habitat. The metabolic rate of poikilotherms (O_2 consumption/unit wt/unit time) is influenced by many factors such as temperature (Wieser, 1965), age (Wiegert, 1964), and season (Phillipson, 1967). Oxygen requirements of poikilotherms at rest may be related to mitochondrial oxygen requirements and are largely independent of rapid fluctuations in temperature (Newall, 1966). Reichle (1967) quotes data which indicate that oxygen consumption doubles for a temperature rise of 10° C (i.e., Q_{10} of two) for forest floor arthropods. Respiration (O_2 uptake/unit time) is proportional to body weight (Zeuthen, 1953; Bertalanffy, 1964) and can be described by the linear power function:

$$Y = a \cdot X^b$$

where Y is respiration, X is body weight and b is a "constant" derived from empirical data (Brody, 1964). Typical values of b for arthropods have been higher than those for mammals, e.g., 0.83 for the isopod *Porcellio scaber* (Will, 1952), *Armadillidium vulgare* and *Cylisticus convexus* (Reichle, 1967); 0.84 for nine grasshopper species (Wiegert, 1965); and 0.84 for 14 species of forest floor arthropods (Reichle, 1968).

To measure respiration in the laboratory, the animal is enclosed in a chamber and evolved carbon dioxide is absorbed in an alkaline solution, so that the pressure in the chamber changes proportionally to the amount of oxygen consumed. In constant-volume respirometers, e.g., the Warburg, the loss of pressure is measured with a manometer. In differential respirometers the pressure is kept constant by changes in volume which can be recorded from a calibrated syringe or by movement of an oil droplet along a capillary tube (ENGELMANN, 1961; WIEGERT, 1964). The same principle is used in ultra-microrespirometers such as the Cartesian diver (KAY, 1964), but for measurements extending over several hours, oxygen must be replaced into the system. Several electrolytic respirometers have been described (WINTERINGHAM, 1959; PHILLIPSON, 1962) in which change in volume when oxygen is consumed activates an electrolytic apparatus that generates oxygen to replace that used. The time the current flows can be recorded, and the apparatus calibrated to give the volume of oxygen used. Compensating respirometers are modified differential respirometers with internal standardization which eliminates most temperature and pressure corrections and are especially useful to measure respiration if temperature should fluctuate.

The amount of CO_2 evolved during respiration can be measured by titration of an alkali absorbant, and this technique has been successful in field studies (ITO, 1964; MENHINICK, 1967). Respiration of intact soil in the field can be measured by an inverted box technique (HOWARD, 1966; KIRITA and HOZUMI, 1966; WITKAMP, 1966) which involves placing an inverted box on the soil surface, sealing the edges and measuring the change in CO_2 content of the air by either titration, gravimetric measurements of alkali-absorbed CO_2 or direct gas analysis (CONWAY, 1962). Infrared spectrophotometric CO_2 gas analyzers which are very sensitive are available commercially.

Radioisotope tracers have been used to estimate the metabolic rates of invertebrates, e.g., MISHIMA and ODUM (1963) related the rate of excretion of [65]Zn to the metabolic rate of snails. REICHLE (1967) correlated oxygen consumption with the elimination rate of [134]Cs by isopods in the laboratory, and attributed accelerated turnover of isotope in the field (25 to 30% increase) to increased activity of the organisms. This may be useful for measuring the metabolic activity of populations in the field, because it is relatively easy to measure isotope lost from individuals by recapturing them, whereas the direct measurement of respiration of free-living organisms in the field can only be approximated.

Table 1 summarizes the respiration of some of the major soil invertebrate groups. Smaller organisms have greater metabolic rates, and the rate of oxygen consumption increases with increasing temperature. Many of these studies, e.g., NIELSEN (1961), SATCHELL (1967), O'CONNOR (1963, 1967), REICHLE (1968), PHILLIPSON (1965), BERTHET (1963) and HEALEY (1967) relate the size of animals and temperature to the rate of oxygen consumed, and some have applied their results to natural populations by compensating for variations in environmental factors.

Calorific Values

All living organisms require a source of energy which is usually the chemical energy of food that can be measured as heat of combustion in bomb calorimeters. The chemical energy can be converted to heat (measured in calories), and calories used to compare the chemical energy of different foods. Studies of energy transfer and trans-

Table 1. *Respiration rates of selected forest litter and soil invertebrates (all values expressed as μl O₂ per gram live body weight)*

Taxa	μl O₂ g⁻¹hr⁻¹	mean live wt (mg)	temp. (C)	authority
Nematoda	582—354	0.000135—0.000196	16	Nielson, 1961
Annelida				
Lumbricidae				
Lumbricus terrestris	36	2700	15	Satchell, 1967
Enchytraeidae				
Cognettia cognettii	625	0.19	20	O'Connor, 1963
Mollusca				
Pulmonata				
Mesodon thyroides	91*	2108*	20	O'Neill (pers. comm.)
Arthropoda				
Crustacea				
Isopoda				
Cylisticus convexus	205	35	20	Reichle, 1967
Diplopoda				
Dixidesmus erasus	140	78	20	Reichle. 1968
Ptyoiulus impressus	56	432	20	Reichle, 1968
Chilopoda				
Otocryptops sexspinosus	73	559	20	O'Neill (pers. comm.)
Phalangida				
Mitopus morio	960—610	3.7—56.1	16	Phillipson, 1964
Arachnida				
Acarina				
Platynorthrus peltifer	925	0.063	15	Berthet, 1963
Araneae				
Schizocosa sp.	251	23	20	Moulder (et al. 1970)
Insecta				
Collembola				
Onychiurus procampatus	1000—320	0.01—0.10	15	Healey, 1967
Orthoptera				
Ceuthophilus gracilipes	263	259	20	Reichle, 1968
Parcoblatta sp.	158	73	20	Reichle, 1968
Coleoptera				
Sphaeroderus stenostomus	194	169	20	Reichle, 1968
Popilius disjunctus	124	1630	20	Reichle, 1968
Isoptera	199	2.1	20	Reichle, 1968
Diptera				
Tipula sp.	146	277—607	13	Bornebusch, 1930

* Live weight includes shell weight.

formation within ecosystems use the calorie or kilocalorie as the basic unit of measurement (Harris, 1966).

The three most common methods of determining calorific values are 1) various types of oxygen bomb calorimeters, 2) modifications of the Phillipson Microbomb

(PHILLIPSON, 1964), or 3) wet dichromate oxidation (MACIOLEK, 1962). These methods usually require at least one gram of dry weight for an estimation, although microbomb calorimeters may require only a few milligrams of sample.

There are usually significant differences in calorific values between different parts of plants or vegetation collected at different seasons and from different sites (GOLLEY, 1961; CUMMINS, 1967). Nevertheless, SLOBODKIN and RICHMAN (1961) have emphasized the constancy of the calorific value of tissues from various animals with normal nutrition, the average being about 5000 cal/g dry weight. The calorific content of plant and animal tissues varies with species, nutrition (LONG, 1934), life cycle (WIEGERT, 1965), season and previous climatic factors (PHILLIPSON, 1967). Calorific measurements should be made in each habitat and only in extensive, general surveys can estimates of energy content from average calorific values be justified.

Soil invertebrates to be used in calorific analyses should not be collected or stored in organic solvents, because they may lose soluble carbohydrates and lipids. Organic detritus can be separated from soil by flotation methods (e.g., MALONE, 1967). Although the dry weights of biological materials are usually measured after drying for 24 hours at 105° C, many workers prefer heating at 65° C or freeze-drying samples to prevent fats and oils from volatilizing. Estimates of the ash content of materials are necessary for calorific values to be expressed on an ash-free dry wt basis.

The amount of energy liberated during the metabolic processes through respiration can be determined indirectly from oxygen consumption or carbon dioxide evolution, once the respiratory quotient (RQ) is known:

$$RQ = \frac{CO_2 \text{ produced}}{O_2 \text{ utilized}} .$$

Assuming that carbohydrate and fat utilization are the only metabolic processes involved (RQ between 0.70 and 1.00), then the heat energy liberated in respiration (from BRODY, 1964) is:

$$\text{cal/ml } O_2 \text{ utilized} = 3.813 + 1.234 \ (RQ)$$

Table 2 summarizes some new data on calorific values of the tissues of many of the more important soil invertebrates and shows how consistent the calorific values are for different organisms, although estimates vary more without correction for ash content. Cummins (1967) reported the mean calorific content of arthropod tissues as 4209 cal/g dry weight and that for insects as 5388 cal/g dry weight. The average of the 14 species in Table 2 is 5788 cal/g ash-free dry weight, with a range from 5344 to 6247 cal/g ash-free dry wt.

Nutrient Contents of the Bodies of Soil and Litter Invertebrates

Saprophagous soil and litter invertebrates decompose and subsequently release nutrients from litter so that these become available for new plant growth. Detritus-feeding invertebrates and their predators redistribute and transfer nutrients in soil, and the chemical composition of the various species must be estimated before the movement of nutrients through food chains can be determined. There are data on the elemental contents of specific tissues of arthropods, but the nutrient contents of whole bodies are poorly known.

Table 2. *Whole-body calorific values for selected forest litter and soil invertebrates (mean + SE (a) or 95% confidence limits (b))*

Taxa	g cal/g ash-free dry wt	ash (% dry wt)	authority
Annelida			
Lumbricidae			
Lumbricus terrestris	5745.7	23.0	French et al., 1957
Mollusca			
Pulmonata			
Mesodon thyroides	5970.9	22.8	O'Neill (pers. comm.)
Arthropoda			
Crustacea			
Isopoda			
Cylisticus convexus	6123 ± 9.7[a]	45.2	Reichle, 1967
Diplopoda			
Brachoria initialis	5453	54.9	Reichle (unpub. data)
Chilopoda			
Otocryptops sexspinosus	5344	7.2	O'Neill (pers. comm.)
Phalangida			
Leiobunum flavum	5732	6.9	Reichle (unpub. data)
Arachnida			
Acarina			
Tyroglypus interna	5808 ± 446[b]		Slobodkin and Richman, 1961
Araneae			
Drassyllus virginiana	5794 ± 89.4[a]	3.8	Moulder (et al. 1970)
Insecta			
Collembola			
Tomocerus sp.	6063	8.7	Moulder (et al. 1970)
Orthoptera			
Parcoblatta sp.	5722 ± 95.2[a]	4.0	Moulder (et al. 1970)
Acheta domesticus	5896	5.7	Reichle (unpub. data)
Coleoptera			
Evarthrus sodalis	5672	3.2	Reichle (unpub. data)
Diptera			
Sarcophaga bullata	5464 ± 243[a]		Slobodkin, 1962
Hymenoptera			
Camponotus americanus	6247	2.8	Reichle (unpub. data)

The methods of chemical analysis available depend on the element which is to be detected and the sensitivity required. Preparation of samples can be critical because many techniques are sensitive to ionic competition by extraneous elements. Methods have been summarized by Wilde and Voigt (1955), Mitchell (1964) and Willard et al. (1965).

Clark (1958) reviewed the amounts of calcium and magnesium in insects and gave some whole-body values. Marek (1961) studied the nutrient content of the cricket, *Acheta domesticus*, in detail and French et al. (1957) determined the nutrients (ash,

Table 3. *Whole body nutrient composition of selected forest litter and soil invertebrates* (after REICHLE et al., 1969)

Taxa	% ash	mg/g ash-free dry wt		
		Ca	K	Na
Mollusca				
Pulmonata				
Mesomphix vulgatus				
shell	83.73	1703.4	10.7	8.8
flesh	22.76	16.7	4.2	2.2
Arthropoda				
Crustacea				
Isopoda				
Ligidium blueridgensis	28.83	108.9	9.4	5.1
Diplopoda				
Dixidesmus erasus	47.87	294.9	12.1	3.7
Scytonotus granulatus	25.25	103.8	3.2	1.4
Chilopoda				
Otocryptops sexspinosus	7.20	0.6	7.1	9.2
Phalangida				
Leiobunum flavum	6.93	7.2	7.6	5.7
Arachnida				
Araneae				
Pachylomerides audouini	8.13	2.1	7.2	7.3
Insecta				
Orthoptera				
Ceuthophilus gracilipes	9.54	2.1	4.6	5.7
Coleoptera				
Geotrupes sp.	4.01	0.8	4.6	3.4
Sphaeroderus stenostomus	3.00	0.5	2.2	3.8
Lepidoptera				
Arctiidae (larvae)	15.80	4.7	50.0	9.2
Diptera				
Sciomyzidae	5.00	2.0	1.2	5.4
Hymenoptera				
Camponotus americanus				
Camponotus pennsylvanicus	2.76	0.7	9.4	2.9

protein, fat and carbohydrate) in earthworms. Amounts of calcium, potassium and sodium in the whole bodies of 37 species of forest floor arthropods are summarized in Table 3 (REICHLE et al., 1969). The amounts of calcium varied (6.4 mg/g ash in staphylinid beetles to 328.4 mg/g ash in xystodesmid millipedes), and large amounts of calcium were correlated with high ash contents which occurred in species characterized by well-developed exoskeletons, e.g., Diplopoda averaged 327.22 mg Ca/g ash-free dry wt; all other arthropods averaged only 1.89 (median = 1.6). The mean sodium concentration for all species was 4.6 mg/g ash-free dry wt (median = 4.3), and mean potassium concentration was 6.2 mg/g ash-free dry wt (median = 4.2).

11*

Net Production by Animals of Different Trophic Levels

We have considered some of the parameters necessary to construct an energy budget. For an individual animal the energy equation may be summarized in Fig. 2. Engelmann (1966) discussed several approaches to the energetics of natural communities. One uses the analysis of maintenance energy (mainly biomass and respiration) to compare populations. Another, the trophic-dynamic approach (Lindemann, 1942), consists of a compartment model of energy flow between organisms with

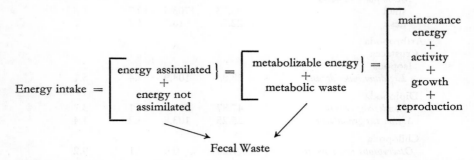

Fig. 2. Energy equation for an individual organism. Population energy budgets are constructed from such individual budgets and the productivity of the population (Note: only the upper term equals the expression to its right)

similar feeding habits (Phillipson, 1966). The energy content of a trophic level is designated by Λ; Λ_1 represents producers, Λ_2 primary consumers, Λ_3 first order predators, etc. The energy exchange per unit time from Λ_{n-1} to Λ_n is designated by λ_n. A portion of the energy received (λ_n) by the organisms constituting Λ_n is dissipated as respiratory heat (R). The heat energy loss per unit time (R) plus the energy (λ_{n+1}) passed per unit time to the next trophic level (Λ_{n+1}) is symbolized as $\lambda_{n'}$. Following Lindemann's nomenclature, the rate of energy change of any trophic level may be written as:

$$\frac{d\Lambda_n}{dt} = \lambda_n + \lambda_{n'}$$

which states that the change in energy content of a trophic level is equal to the difference of the simultaneous rates of energy intake and energy loss.

The efficiency with which a food supply is utilized by a consumer population can be expressed by:

$$\frac{\text{calories of prey consumed by predator}}{\text{calories of food supplied to prey}} \times 100$$

This is termed the *food chain efficiency*. The net production of a trophic level λ_n is a measure of the energy available to the next highest trophic level. Thus, the ratio of λ_n/λ_{n-1} or:

$$\frac{\text{calories of prey consumed by predator}}{\text{calories of food consumed by prey}} \times 100$$

is a measure of how efficiently energy is transferred between trophic levels. This ratio is termed the *gross ecological efficiency*. There are not yet enough trophic studies of soil communities to give a clear picture of community structure or energy dynamics. Data compiled by ENGELMANN (1966) showed a range of ecological efficiencies for invertebrates from 6.8 to 30%. ENGELMANN's (1961) calculations of ecological efficiencies of Oribatei ranged from 8 to 30%. SLOBODKIN (1962, 1964) suggested that the most probable value for gross ecological efficiency in natural ecosystems is about 10%. KOZLOVSKY (1968) reviewed the trophic level concept and evaluated the various expressions used to describe "ecological efficiencies".

Trophic level comparisons of soil invertebrates remain largely dependent upon data expressing the significance of organisms in terms of either density (numbers/unit area), biomass (dry weight/unit area), or occassionally the energy dissipated by the population through respiration. Density estimates in different habitats and ecosystems, e.g., arthropods (MADGE, 1965), earthworms (SATCHELL, 1967), nematodes (O'CONNOR, 1967), show that numbers alone are insufficient for meaningful ecological interpretation. Nematodes at 10^7 individuals/m^2 respire 71 liters of O_2/m^2/yr (NIELSEN, 1961); 1.3×10^5 oribatid mites/m^2 respire 4.5 liters O_2/m^2/yr (BERTHET, 1963); respiration of 7.4×10^4 enchytraeids/m^2 amounts to 32 liters O_2/m^2/yr (NIELSEN, 1961); and 45 lumbricid earthworms/m^2 may respire 22.5 liters O_2/m^2/yr (SATCHELL, 1967). These values are not directly comparable, because they have been calculated from different ecosystems. Comparable data which include all major invertebrate taxa have been summarized for soil habitats by MACFADYEN (1963) for nine grassland, moor, and forest habitats.

The invertebrate populations of soil and litter in a European oak mull (BORNEBUSCH, 1930), a European beech mor (VAN DER DRIFT, 1951) and a cool temperate beech forest of Japan (KITAZAWA et al., 1964) are compared in Table 4. There is a significantly smaller biomass of large decomposers in the mor soil, but the small decomposers compensate for this in terms of annual energy turnover through respiration. Calculations of biomass can be misleading, e.g., in the Japanese beech forest the Collembola biomass is 54% of total biomass but is responsible for 82% of the total energy loss in respiration. The large decomposers dominating mull soils are predominantly earthworms, but in some habitats Isoptera, Isopoda or Diplopoda may be more numerous, and the relative incidence of each may vary greatly. Small decomposers (Nematoda, Enchytraeidae, Acarina and Collembola) range from insignificant to dominant species in different habitats, but the total metabolic activity of the soil animals remains surprisingly constant. Collembola and Oribatei are abundant in fungus-rich raw humus while Nematoda are more common in mineral soils. Parasitic mites are the dominant predaceous species in all types of soils, although Chilopoda are the large active predators in most woodland ecosystems, and spiders are more important in grasslands.

Herbivores constitute from 14 to 30% of total invertebrate biomass in soil and are responsible for 20 to 31% of the respiration. Large decomposers represent between 27 and 82% of biomass and 3 to 33% of total respiration. Small decomposers are most extreme, contributing from 3 to 64% of biomass and 30 to 88% of respiration. Predators constitute from 1.5 to 19% of biomass and 6 to 25% of respiration. We have insufficient data, especially for large and small decomposers, to compare the energy turnover of different soil habitats. MACFADYEN (1963), however, calculated a rather consistent relationship between the ratios of daily metabolic activity to biomass for different ecosystems. Small decomposers range between 10 and 20%, predators between 5 and 20%, and herbivores and large decomposers below 5%. Before the im-

Table 4. *Comparative biomass and metabolism of three forest soil habitats. Oak mull* (Bornebusch, 1930) *and Beech mor* (van der Drift, 1951) *after* Macfadyen, 1963; *Beech forest after* Kitazawa et al., 1964

	Oak mull		Beech mor		Beech (Japan)	
	g/m²	kcal/m²/yr	g/m²	kcal/m²/yr	g/m²	kcal/m²/yr
Herbivores						
Nematoda	2.0	61.1	0.1	3.4		
Mollusca	5.3	32.7	2.6	15.9		
Hemiptera			0.5	3.2		
Lepidoptera	0.6	4.1	0.1	0.7		
Coleoptera	0.2	1.5	0.8	6.7		
Diptera	3.1	19.0	0.6	3.7		
Total	11.2	118.4	4.7	33.6		
Large decomposers						
Oligochaeta	61.0	90.5	2.9	4.3	6.2	7.2
Mollusca					0.3	0.6
Isopoda	0.3	2.1	0.0	0.0	0.2	1.2
Diplopoda	4.7	35.9	1.3	9.6	0.1	0.24
Coleoptera larvae					0.6	2.4
Total	66.0	128.5	4.2	13.9	7.4	11.64
Small decomposers						
Nematoda	1.0	30.5	0.7	21.4	0.2	7.2
Enchytraeidae	0.7	5.8	0.0	0.0	0.8	10.8
Oribatidae	0.4	2.0	1.6	8.2	0.1	0.24
Collembola	0.1	76.3	1.5	46.4	13.0	336.0
Diptera larvae					1.5	9.6
Total	2.2	114.6	3.8	76.0	15.6	363.84
Predators						
Araneae	0.1	0.3	0.5	2.8	0.37	4.8
Phalangida			0.0	0.2		
Chilopoda	0.6	7.8	0.8	9.5	0.7	16.8
Diptera			0.6	3.8		
Coleoptera	0.2	1.4	0.7	5.5	0.6	3.6
Hirudinea					0.3	10.8
Parasitic mites	0.2	12.7	0.3	19.7		
Total	1.1	22.2	2.9	41.5	1.97	36.0
Grand Total	80.5	383.7	15.6	165.0	24.29	411.48

portance of soil animals in the total energy budget can be interpreted, it will be necessary to know the total soil metabolism due to all organisms. Bacteria with $\sim 3 \times 10^{14}$ individuals/m² and respiring 1.6×10^3 liters O_2/m²/yr (calculated from Clark, 1967), soil algae at 1.5×10^9 individuals/m² (Petersen, 1935) and protozoa with 1.7×10^{12} individuals/m² (Stout and Heal, 1967) are undoubtedly important in soil metabolism. While it is recognized that the soil microflora and microfauna may account for as much as 90% of the total energy flow (Phillipson, 1966), invertebrate populations contribute significantly through litter fragmentation, mixing with soil, and creation of suitable fecal substrates for the microbiota as well as adding to energy turnover themselves. Ultimate understanding of the role and importance of soil invertebrates cannot be realized until the interactions and synergistic effects of soil populations are known.

Acknowledgements

Research supported by the U. S. Atomic Energy Commission under contract with the Union Carbide Corporation.

References

ATTIWILL, P. M.: The loss of elements from decomposing litter. Ecology **49**, 142—145 (1968).

BECK, L.: Aus den Regenwäldern am Amazonas. I. Natur. und Mus. **98**, 24—32 (1968).

BERTALANFFY, L. VON: Basic concepts in quantitative biology of metabolism. Helgol. Wiss. Meeresunters. **9**, 5—37 (1964).

BERTHET, P.: Mesure de la consommation d'oxygene des Oribatides (Acariens) de la litiere des forests, pp. 18—31. In: Soil organisms. (DOEKSEN, J., and J. VAN DER DRIFT, Eds.). Amsterdam: North-Holland Publ. Co. 1963.

BOCOCK, K. L.: The digestion and assimilation of food by *Glomeris*, pp. 85—91. In Soil organisms (DOEKSEN, J., and J. VAN DER DRIFT, Eds.). Amsterdam: North-Holland Publ. Co. 1963.

—, and O. J. W. GILBERT: The disappearance of leaf litter under different woodland conditions. Plant and Soil **9**, 179—185 (1957).

BORNEBUSCH, C. H.: The fauna of forest soil. Forstl. Forsögswaesen Copenhagen **11**, 1—224 (1930).

BRAY, J. R., and E. GORHAM: Litter production in forests of the world. Advan. Ecol. Res. **2**, 101—157 (1964).

BRAUN, A.: Die Sukzession der Dipteran-larven bei der Stockhumifizierung. Z. Morphol. & Oekol. Tiere **43**, 313—320 (1954).

BRODY, C.: Biogenesis and growth. New York: Hafner Publ. Comp. 1964.

CHEKALOV, K. I., and V. P. ILLYUVIEVA: Use of the C14 isotope for the study of the decomposition of organic matter in soil. Soviet Soil Sci. **5**, 482—490 (1962). (Engl. transl.).

CLARK, E. W.: A review of literature on calcium and magnesium in insects. Ann. Entomol. Soc. Amer. **51**, 142—154 (1958).

CLARK, F. E.: Bacteria in soil, pp. 15—49. In: Soil biology (BURGES, A., and F. RAW, Eds.). London: Academic Press 1967.

CONWAY, E. J.: Microdiffusion and volumetric error. London: C. Lockwood. 1962.

CROSSLEY, D. A. JR., and M. HOGLUND: A litter-bag method for the study of microarthropods inhabiting leaf litter. Ecology **43**, 571—574 (1962).

—, and M. WITKAMP: Forest soil mites and mineral cycling. Acarologia, 137—145 (1964).

CUMMINS, K. W.: Calorific Equivalents for Studies in Ecological Energetics. Pymatuning Laboratory, Univ. Pittsburg, privately circulated. 52 pp. (1967).

DAVEY, S. P.: Effects of Chemicals on Earthworms: A Review of the Literature. Special Report No. 74., U.S.D.I., Washington. 20 pp. (1963).

DRIFT, VAN DER, J.: Analysis of the animal community in a beech forest floor. Wageningen: Ponsen and Looijen 1—168 (1951).

— The disappearance of litter in mull and mor in connection with weather conditions and the activity of the macrofauna, pp. 125—133. In: Soil organisms (DOEKSEN, J., and J. VAN DER DRIFT, Eds.). Amsterdam: North-Holland Publ. Co. 1963.

—, and M. WITKAMP: The significance of the break-down of oak litter by *Enoicyla pusilla* Burm. Arch. Neerland. Zool. **13**, 486—492 (1960).

DYBAS, H. S., and D. D. DAVIS: A population census of seventeen-year periodical cicadas (Homoptera: Cicadidae: *Magicicada*). Ecology **43**, 432—443 (1962).

DUNGER, W.: Über die Veränderung des Fallaubes in Darm von Bodentieren. Z. Pflanzenernähr. Düng. Bodenk. **82**, 174—193 (1958).

EDWARDS, C. A.: Persistence of insecticides in the soil. New Scientist **19**, 282—284 (1963).
— Effects of soil insecticides on soil invertebrates and plants, pp. 239—261. In: Ecology and the industrial society. Oxford: Blackwell 1965.
— Relationships between weights, volumes and numbers of soil animals, pp. 585—594. In: Progress in Soil Biology (GRAFF, O., and J. E. SATCHELL, Eds.). Braunschweig-Amsterdam: Vieweg 1967.
—, and G. K. A. BUAHIN: The recolonization of sterilized soil by invertebrates, pp. 149—150. In: Rep. Rothamsted Exp. Stn. for 1963, Harpenden, England (1964).
—, and E. B. DENNIS: Some effects of Aldrin and DDT on the soil fauna of arable land. Nature **188**, 767 (1960).
—, and E. GUNN: Control of the glasshouse millipede. Plant Pathol. **10**, 21—24 (1961).
—, and K. E. FLETCHER: Terrestrial arthropod populations. In: Methods for the study of production and energy flow in soil ecosystems (J. PHILLIPSON, Ed.). (in press).
—, and G. W. HEATH: The role of soil animals in breakdown of leaf material, pp. 76—84. In: Soil organisms (DOEKSEN, J., and J. VAN DER DRIFT, Eds.). Amsterdam: North-Holland Publ. Co. 1963.
— E. B. DENNIS, and D. W. EMPSON: Pesticides and the soil fauna. I. Effects of Aldrin and DDT in an arable field. Ann. Appl. Biol. **60**, 11—22 (1967).
—, J. R. LOFTY, and A. R. THOMPSON: Changes in soil invertebrate populations due to some organophosphorus insecticides, Proc. 4th Brit. Insecticide and Fungicide Conf. 48—55 (1968).
—, D. E. REICHLE, and D. A. CROSSLEY JR.: Experimental manipulation of soil invertebrate populations for trophic stidies. Ecology **50**, 495—498 (1969).
ENGELMANN, M. D.: The role of soil arthropods in the energetics of an old field community. Ecol. Monogr. **31**, 221—238 (1961).
— Energetics, terrestrial field studies, and animal productivity. Advan. Ecol. Res. **3**, 73—115 (1966).
FEENY, P. P., and H. BOSTOCK: Seasonal changes in the tannin content of oak leaves. Phytochem. **7**, 87—90 (1968).
FRANZ, H., and L. LEITENBERGER: Biologisch-chemische Untersuchungen über Humusbildung durch Bodentiere. Österr. Zool. **1**, 498—518 (1948).
FRENCH, C. E., S. A. LISCINSKY, and D. R. MILLER: Nutrient composition of earthworms. J. Wildl. Manage. **21**, 348 (1957).
GASDORF, E. C., and C. J. GOODNIGHT: Studies on the ecology of soil arachnids. Ecology **44**, 261—268 (1963).
GERE, G.: The examination of the feeding biology and the humificative function of Diplopoda and Isopoda. Acta Biol. Acad. Sci. Hung. **6**, 257—271 (1956).
— Über einige Faktoren des Streuabbaues, pp. 67—75. In: Soil organisms (DOEKSEN, J., and J. VAN DER DRIFT, Eds.). Amsterdam: North-Holland Publ. Co. 1963.
GHILAROV, M. S.: Distribution of humus, root-systems and soil invertebrates within the soil of the walnut forests of the Ferghana mountain range. Compt. Rend. (Doklady) Acad. Sci. U.S.S.R. **55**, 49—52 (1947).
— Simpozium po pochvennoi fauna v Kieve. Pocvovedenie **1962** (9), 1 (1962).
— Opreditel' obitayushchikh v pochve lichinok nasekomykh, 919 pp. Akademiya Nauk SSR 1964.
GOLLEY, F. B.: Energy values of ecological materials. Ecology **42**, 581—584 (1961).
GRIGOR'EVA, T.: Wirkung des in den Boden eingebrachten Hexachlorans auf die Bodenfauna. Ber. allruss. Akad. Landwirtschwiss. Moskau **12**, 16—20 (1952).
HAMILTON, D. W.: Periodical cicadas, *Magicicada* spp., as pests in apple orchards. Proc. Indiana Acad. Sci. **71**, 116—121 (1961).
HANDLEY, W. R. C.: Mull and mor formation in relation to forest soils. Bull. For. Comm. London **23**, 1—115 (1954).
HARRIS, L. E.: Glossary of energy terms and their biological interrelationships, Washington: National Academy of Sciences 1966.
HARTENSTEIN, R.: Soil Oribatei. I. Feeding specificity among forest soil Oribatei (Acarina). Ann. Entomol. Soc. Amer. **55**, 202—206 (1962).

HEALEY, I. N.: The energy flow through a population of soil Collembola, pp. 695—708. In: Secondary Productivity of Terrestrial Ecosystems (PETRUSEWICZ, K., Ed.). Warsaw: Panstwowe Wydawnictwo Naukowe 1967.

HEATH, G. W., and M. K. ARNOLD: Studies in leaf-litter breakdown. II. Breakdown rate of "sun" and "shade" leaves. Pedobiol. 6, 238—243 (1966).

—, and H. G. C. KING: The palatability of litter to soil fauna, Proc. VIIIth Int. Cong. Soil Sci. Bucharest 979—986 (1964).

—, M. K. ARNOLD, and C. A. EDWARDS: Studies in leaf litter breakdown. I. Breakdown rates among leaves of different species. Pedobiol. 6, 1—12 (1966).

—, C. A. EDWARDS, and M. K. ARNOLD: Some methods for assessing the activity of soil animals in the breakdown of leaves. Pedobiol. 4, 80—87 (1964).

— —, and A. E. WHITING: Studies in leaf litter breakdown. III. The influence of leaf age. Pedobiol. (in press).

HOWARD, P. J. A.: A method for the estimation of carbon dioxide evolved from the surface of soil in the field. Oikos 17, 267—271 (1966).

HUBBELL, S. P., A. SIKORA, and O. H. PARIS: Radiotracer, gravimetric and calorimetric studies of ingestion and assimilation rates of an isopod. Health Phys. 11, 1485–1501(1965).

ITO, Y.: Preliminary studies on the respiratory energy loss of a spider, *Lycosa pseudoannulata*. Res. Population Ecol. 6, 13—21 (1964).

KARG, W.: Die exaphischen Acarina in ihren Beziehungen zur Mikroflora und ihre Eignung als Anzeiger für Prozesse der Bodenbildung, pp. 305—315. In: Soil organisms (DOEKSEN, J., and J. VAN DER DRIFT, Eds.). Amsterdam: North-Holland Publ. Co. 1963.

KAY, R. H.: Experimental biology: Measurement and analysis. New York: Reinhold Publ. Corp. 1964.

KENDRICK, W. B.: The time factor in the decomposition of coniferous leaf litter. Can. J. Botany 37, 907—912 (1959).

—, and A. BURGES: Biological aspects of the decay of *Pinus sylvestris* leaf litter. Nova Hedwigia 4, 313—342 (1962).

KEVAN, D. K., McE.: Soil animals. London: Witherby 1962.

KING, H. G. C., and G. W. HEATH: The chemical analysis of small samples of leaf material and the relationship between the disappearance and composition of leaves. Pedobiol. 7, 192—197 (1967).

KIRITA, H., and K. HOZUMI: Re-examination of the absorption method of measuring soil respiration under field conditions. I. Effects of the amount of KOH on observed values. Phys. and Ecol. 14, 23—31 (1966).

KITAZAWA, Y., S. SAITO, and M. NAKAMURA: Animal communities of primeval ecosystems of Mt. Tanzawa (In Japanese), pp. 255—291. In: Scientific Report Tanzawa Mts. Yokohama: Kanagawaken. 1964.

KOLLMANSPERGER, F.: Lumbriciden in humiden und ariden Gebieten und ihre Bedeutung für die Fruchtbarkeit des Bodens. Rapp. VI Congr. Int. Sci. Sol. (Paris) C, 293—297 (1956).

KOZLOVSKY, D. G.: A critical evaluation of the trophic level concept. I. Ecological efficiencies. Ecology 49, 48—59 (1968).

KRISHNA, S. S., and N. B. SINGH: Sugar-dye movement through the alimentary canal of *Odontotermes obesus* (Isoptera: Termitidae). Ann. Entomol. Soc. Amer. 61, 230 (1968).

KÜHNELT, W.: Soil-inhabiting arthropoda. Ann. Rev. Entomol. 8, 115—136 (1963).

KURCHEVA, G. F.: The role of invertebrates in the decomposition of the oak leaf litter. Pocvovedenie 1960 (4), 16—23 (1960).

— Wirbellose Tiere als der Faktor der Zersetzung von Waldstreu. Pedobiol. 4, 7—30 (1964).

LINDEMAN, R. L.: The trophic-dynamic aspect of ecology. Ecology 23, 399—418 (1942).

LLOYD, M., and H. S. DYBAS: The periodical cicada problem. I. Population ecology. Evolution 20, 133—149 (1966).

LONG, F. L.: Application of calorimetric methods to ecological research. Plant Physiol. 9, 323—337 (1934).

MACFADYEN, A.: Control of humidity in three funnel-type extractors for soil arthropods, pp. 158—168. In: Progress in soil zoology (P. W. MURPHY, Ed.). London: Butterworth 1962.

Macfadyen, A.: The contribution of the microfauna to total soil metabolism, pp. 3—17. In: Soil organisms (J. Doeksen, and J. van der Drift, Eds.). Amsterdam: North-Holland Publ. Comp. 1963.

Maciolek, J. A.: Limnological organic analyses by quantitative dichromate oxidation. Rept. Bur. Sport Fish. Wildlife 20, 1—61 (1962).

Madge, D. S.: Leaf fall and litter disappearance in a tropical forest. Pedobiol. 5, 273—288 (1965).

— How leaf litter disappears. New Sci. 113—115 (1966).

Malone, C. R.: A rapid method for enumeration of viable seeds in soil. Weeds 15, 381—382 (1967).

Marek, M.: Gesamtstoffwechsel der Insekten. 10. Änderungen der chemischen Zusammensetzung der Hausgrille (Gryllus domesticus L.) während der Postembryonalentwicklung. Acta Soc. Zool. Bohemosl. 15, 15—69 (1961).

McGinnis, A. J., and R. Kasting: Chronic oxide indicator method for measuring food utilization in a plant-feeding insect. Science 144, 1464—1465 (1964).

Menhinick, E. F.: Structure, stability, and energy flows in plants and arthropods in a Sericea lespedeza stand. Ecol. Monogr. 37, 255—272 (1967).

Milner, C.: The estimation of energy flow through populations of large herbivorous mammals, pp. 147—162. In: Secondary productivity of terrestrial ecosystems (Petrusewicz, K., Ed.). Warsaw: Panstwowe Wydawnictwo Naukowe 1967.

Mishima, J., and E. P. Odum: Excretion rate of Zn^{65} by Littorina irrorata in relation to temperature and body size. Limnol. and Oceanogr. 8, 39—44 (1963).

Mitchell, R. L.: The spectrochemical analysis of soils, plants, and related materials. Techn. Comm. No. 44, Commonwealth Agricultural Bureaux, England, 1964.

Moulder, B. C., D. E. Reichle, and S. I. Auerbach: The significance of spider predation in the energy dynamics of forest floor arthropod communities. Oak Ridge National Laboratory, USA, ORNL 4452. (1970).

Murphy, P. W.: Extraction methods for soil animals. I. Dynamic methods with particular reference to funnel processes, pp. 75—114. In: Progress in Soil Zoology (Murphy, P. W., Ed.). London: Butterworth 1962.

— The biology of forest soils with special reference to the mesofauna or meiofauna. J. Soil. Sci. 4, 155—193 (1953).

Nakamura, M.: Bio-economics of some larval populations of pleurostict Scarabaeidae on the flood plain of the River Tamagawa. Japanese J. Ecol. 15, 1—18 (1965).

Newell, R. C.: Effect of temperature on the metabolism of poikilotherms. Nature 212, 426—428 (1966).

Nicholson, P. B., K. L. Bocock, and O. W. Heal: Studies on the decomposition of the faecal pellets of a millipede [Glomeris marginata (Villers)]. J. Ecol. 54, 755—766 (1966).

Nielsen, C. O.: Respiratory metabolism of some populations of enchytraeid worms and freeliving nematodes. Oikos 12, 17—35 (1961).

O'Connor, F. B.: Oxygen consumption and population metabolism of Enchytraeidae, pp. 32—48. In: Soil organisms (Doeksen, J., and J. van der Drift, Eds.). Amsterdam: North-Holland Publ. Co. 1963.

— The Enchytraeidae, pp. 213—257. In: Soil biology (Burges, A., and F. Raw, Eds.). London: Academic Press 1967.

Odum, E. P., and E. J. Kuenzler: Experimental isolation of food chains in an old-field ecosystem with the use of phosphorus-32, pp. 113—120. In: Radioecology (Schultz, V., and A. W. Klement Jr., Eds.). New York: Reinhold Publ. Co,, and Washington: AIBS 1962.

—, and F. Golley: Radioactive tracers as an aid to the measurement of energy flow at the population level in nature, pp. 403—410. In: Radioecology (Schultz, V., and A. W. Klement Jr., Eds.). New York: Reinhold Publ. Co., and Washington: AIBS 1963.

Olson, J. S.: Energy storage and the balance of producers and decomposers in ecological systems. Ecology 44, 322—331 (1963).

—, and D. A. Crossley Jr.: Tracer studies of the breakdown of forest litter, pp. 411—416. In: Radioecology (Schultz, V., and A. W. Klement Jr., Eds.). New York: Reinhold Publ. Co., and Washington: AIBS 1963.

OVINGTON, J. D.: Quantitative ecology and the woodland ecosystem concept. Advan. Ecol. Res. 1, 103—192 (1962).

PARIS, O. H., and A. SIKORA: Radiotracer demonstration of isopod herbivory. Ecology 46, 729—734 (1965).

— — Radiotracer analysis of the trophic dynamics of natural isopod populations, pp. 741—771. In: Secondary productivity of terrestrial ecosystems (PETRUSEWICZ, K., Ed.). Warsaw: Panstwowe Wydawnictwo Naukowe 1967.

PARK, O., and S. I. AUERBACH: Further study of the tree-hole complex with emphasis on quantitative aspects of the fauna. Ecology 35, 208—222 (1954).

PATTEN, B. C., and M. WITKAMP: Systems analysis of ^{134}Cesium kinetics in terrestrial microcosms. Ecology 48, 813—824 (1967).

PETERSON, J. B.: Dansk. botan. Ark. 8, 183 (1935).

PETRIDES, G. A.: The use of 51-chromium in the determination of energy flow in mammals. Proc. Symp. Recent Adv. Trop. Ecol. pp. 25—31 (1968).

PETRUSEWICZ, K.: Concepts in studies on the secondary productivity of terrestrial ecosystems pp. 17—49. In: Secondary Productivity of Terrestrial Ecosystems (PETRUSEWICZ, K., Ed.): Warsaw: Panstwowe Wydawnictwo Naukowe 1967.

PHILLIPSON, J.: The food consumption of different instars of Mitopis morio (F.) (Phalangida) under natural conditions. J. Anim. Ecol. 29, 299—307 (1960).

— Respirometry and the study of energy turnover in natural systems with particular reference to harvest spiders (Phalangida). Oikos 13, 311—322 (1962).

— A miniature bomb calorimeter for small biological samples. Oikos 15, 130—139 (1964).

— Respiratory metabolism of the terrestrial isopod Oniscus asellus L. Oikos 16, 78—87 (1965).

— Ecological energetics. New York: St. Martin's Press 1966.

— Studies on the bioenergetics of woodland Diplopoda, pp. 679—685. In: Secondary productivity of terrestrial ecosystems (PETRUSEWICZ, K., Ed.). Warsaw: Panstwowe Wydawnictwo Naukowe 1967.

RAW, F.: Estimating earthworm populations by using formalin. Nature 184, 1661 (1959).

— Studies of earthworm populations in orchards. I. Leaf burial in apple orchards. Ann. Appl. Biol. 50, 389—404 (1962).

— Current work on side-effects of soil-applied organophosphorus insecticides. Ann. Appl. Biol. 55, 342—343 (1964).

REICHLE, D. E.: Radioisotope turnover and energy flow in terrestrial isopod populations. Ecology 48, 351—366 (1967).

— Relation of body size to food intake, oxygen consumption, and trace element metabolism in forest floor arthropods. Ecology 49, 538—542 (1968).

—, and D. A. CROSSLEY JR.: Radiocesium dispersion in a cryptozoan food web. Health Phys. 11, 1375—1384 (1965).

— — Investigations on heterotrophic productivity in forest insect communities, pp. 563—587. In: Secondary productivity of terrestrial ecosystems. (PETRUSEWICZ, K. Ed.) Warsaw: Panstwowe Wydawnictwo Naukowe 1967.

— — Trophic level concentrations of cesium-137, sodium, and potassium in forest arthopods. pp. 678—686. In: Symposium on radioecology. (NELSON, D. J., and F. C. EVANS Eds.) AEC-CONF 670503 (1969).

—, M. H. SHANKS, and D. A. CROSSLEY JR.: Calcium, potassium, and sodium content of forest floor arthropods. Ann. Entomol. Soc. Amer. 62, 57—62 (1969).

RITCHER, P. O.: Biology of Scarabaeidae. Ann. Rev. Entomol. 3, 311—334 (1958).

SAITO, S.: Productivity of high and low density populations of Japonaria laminata armigera (Diplopoda) in a warm-temperate forest ecosystem. Res. Pop. Ecol. 9, 153—166 (1967).

SANDNER, H., and D. COCHY: Research on the effectiveness of fungal and bacterial insecticides. Ekologia Polska-Seria A, 15, 325—333 (1967).

SATCHELL, J. E.: Lumbricidae, pp. 259—322. In: Soil biology. (BURGES, A., and F. RAW Eds.) London: Academic Press 1967.

—, and D. G. LOWE: Selection of leaf litter by Lumbricus terrestris, pp. 102—120. In: Progress in soil biology. (GRAFF, O., and J. E. SATCHELL Eds.) Amsterdam: Braunschweig 1967.

Sheals, J. G.: Soil population studies. I. The effects of cultivation and treatment with insecticides. Bull. Ent. Res. 47, 803—833 (1956).

Slobodkin, L. B.: Energy in animal ecology. Advan. Ecol. Res. 1, 69—101 (1962).

— Experimental populations of Hydrida. J. Anim. Ecol. (Suppl.) 33, 131—148 (1964).

—, and S. Richman: Calories/gm in species of animals. Nature 191, 299 (1961).

Southwood, T. R. E.: Ecological methods with particular reference to the study of insect populations. London: Methuen and Co. 1966.

Stout, J. D., and O. W. Heal: Protozoa, pp. 149—195. In: Soil biology. (Burges, A., and F. Raw, Eds.) London: Academic Press 1967.

Thomas, W. A.: Accumulation and cycling of calcium by dogwood trees. Ecol. Monogr. 39, 101—120 (1969).

Tracey, M. V.: Cellulose and chitinose of earthworms. Nature 167, 776 (1951).

Treherne, J. E.: Glucose absorption in the cockroach. J. Expt. Biol. 34, 478—485 (1957).

Van Hook, R. I. Jr., and D. A. Crossley Jr.: Assimilation and biological turnover of cesium-134, iodine-131, and chromium-51 in brown crickets, *Acheta domesticus* (L.). Health Phys. 16, 463—467 (1969).

Waksman, S. A.: The origin and nature of the soil organic matter of soil humus. Soil Sci. 22, 123 (1936).

Went, J. C.: Influence of earthworms on the number of bacteria in the soil, pp. 260—265. In: Soil organisms. (Doeksen, J., and J. van der Drift Eds.) Amsterdam: North-Holland Publ. Comp. 1963.

Wiegert, R. G.: Population energetics of meadow spittelbugs (*Philaenus spurmarius* L.) as affected by migration and habitat. Ecol. Monogr. 34, 217—241 (1964).

— Intraspecific variation in calories/g of meadow spittelbugs (*Philaenus spumarius* L.). BioScience 15, 543—545 (1965).

Wieser, W. von: Untersuchungen über die Ernährung und den Gesamtstoffwechsel von *Porcellio scaber* (Crustacea: Isopoda). Pediobiol. 5, 304—331 (1965).

— Copper and the role of isopods in degradation of organic matter. Science 153, 67—69 (1966).

Wilde, S. A., and G. K. Voigt: Analysis of Soils and Plants for Foresters and Horticulturists. Ann. Arbor, Michigan: J. W. Edwards, Publ., Inc. 1955.

Will, A.: Körpergröße und O_2-Konsum bei Schaben und Asseln (Isopoden). Z. Vergl. Physiol. 34, 20—25 (1952).

Willard, H. H., L. L. Merritt Jr., and J. A. Dean: Instrumental Methods of Analysis 4th Ed.). Princeton: D. van Nostrand Co. Inc. 1965.

Winterinhgam, F. P. W.: An electrolytic respirometer for insects. Laboratory Practice 7, 372—376 (1959).

Witkamp, M.: Decomposition of leaf litter in relation to environment, microflora and microbial respiration. Ecology 47, 194—201 (1966).

—, and D. A. Crossley Jr.: The role of arthropods and microflora in breakdown of white oak litter. Pedobiol. 6, 293—303 (1966).

—, and M. L. Frank: Retention and loss of cesium-137 by components of the groundcover in a pine (*Pinus virginiana* L.) stand. Health Phys. 13, 985—990 (1967).

—, and J. S. Olson: Breakdown of confined and nonconfined oak litter. Oikos 14, 138—147 (1963).

Wittich, W.: Untersuchungen über den Verlauf der Streuzersetzung auf einem Boden mit Mullzustand. II. Forstarchiv 19, 1—18 (1943).

Zachariae, G.: Zur Methodik bei Geländeuntersuchungen in der Bodenzoologie. Z. Pflanzenernähr. Düng. Bodenk. 97, 224—233 (1962).

Zeuthen, E.: Oxygen uptake as related to body size in organisms. Quart. Rev. Biol. 28, 1—12 (1953).

Micronutrients: Forest Ecology and Systems Analysis

J. A. C. FORTESCUE and G. G. MARTEN

Forest plants are composed of a large number of chemical elements which are usually divided into four groups:

1. the gaseous elements (H, O, N, Cl);
2. macronutrients (C, K, Ca, Mg, P, S);
3. micronutrients (Mo, Cu, Zn, Mn, Fe, B); and
4. nonessential elements (Na, Al, Ba, Sr, Rb, Pb and others).

In this paper we are concerned with the six micronutrient elements in forest plants and forest ecosystems, although occasional reference is made to others. The role of micronutrients in the growth of plants is only a part of the whole subject of plant nutrition and should not, strictly speaking, be separated from that of others. In practice this separation is often made because of the special experimental difficulties involved in the study of micronutrients.

The information on micronutrients in forest plants is small compared with that available either for cultivated plants or for macronutrients in forest plants. As OVINGTON (1962) pointed out in respect to forest ecosystem studies as a whole, the problem for ecologists reviewing the data on micronutrients in forests is largely that of synthesizing fragmentary and incomplete knowledge gained in different disciplines. These include botany, ecology, agriculture, horticulture, biogeochemistry, geochemical prospecting and forestry.

The mathematical description of plant nutrition has been developed largely in agriculture, where problems are somewhat different from those in forestry. Even in agriculture, little attempt has been made to express mathematically the function of the micronutrients in relation to crop growth; consequently any theoretical attempt to apply the "systems analysis approach" to the role of micronutrients in forest nutrition is somewhat speculative at the present time. There is, however, much potential value in such speculation owing to its relationship to the concepts of the International Biological Program. It is important that data from expanding IBP programs in forest nutrition are not fragmentary and incomplete but so designed to be fed into a system analysis approach aimed at the development or testing of mathematical models for forest nutrition.

Our initial objective is to summarize information available on the geochemistry and physiology of the micronutrients and to introduce the mathematical description of nutrition. Following this, we describe important aspects of what is known about the role of micronutrients in forests and to propose two systems models which can be used as starting points for the better interpretation of nutrition data and the fuller understanding of forest nutrition. Data from the growth chamber, or the forest ecosystem, level of detail can be utilized. The chapter ends with some conclusions regarding the evolution of forest nutrition studies.

General Information

Geochemistry of the Micronutrients

It is well established that small amounts of the six micronutrients are needed for healthy growth of all plant species. The essential nature of manganese was determined as long ago as 1905 (Stiles, 1958a) and, by 1939, it had been shown that iron, boron, copper, zinc, and molybdenum were also essential. The relative proportions of these elements required for the healthy growth of plants has been calculated by Epstein (1965) and are shown on the right-hand side of Table 1. These data focus attention on the small amounts of each element needed as well as the large variation in relative proportions of the elements within this group of nutrient elements.

Geochemistry may be viewed as the study of the synthesis and decomposition of natural materials. Before one can study the dynamic aspects of geochemistry, however, it is important to know the relative abundance of chemical elements in natural materials. Consequently, geochemists have gathered together information on the total content of elements in rocks, soils, waters or plants (Rankama and Sahama, 1950; Goldschmidt, 1954; and Bowen, 1966). Information of this kind for the micronutrients, the macronutrients, and the gaseous nutrients is given in Table 1. Other elements (including sodium, silicon, fluorine, iodine, selenium, bromine, rubidium, strontium, cobalt, aluminium, beryllium and barium) are also present in plant material and some evidence has been collected suggesting that one or more of these are essential for particular species (Bollard and Butler, 1966). Small amounts of lead, nickel, chromium and silver can also be found in plants, without any apparent adverse effect on growth (Stiles, 1958b).

The geochemical data on the occurrence of elements in soils and rocks given in Table 1 are of limited value for the study of forest nutrition because they do not indicate what percentage of each element is *available* for uptake by a plant at a given time. This is important in forest nutrition especially at planting time and at other times during the tree's life cycle. The determination of short and long-term availability of micronutrient elements is particularly important in forest areas where nutrition is likely to be a limiting factor in growth.

Some idea of the relative concentrations of micronutrients which are required in available form for optimum growth of plants can be obtained from the study of the constitution of nutrient solutions used for growing plants under strictly controlled greenhouse conditions (Hewitt, 1966). An example of such a solution is included in Table 1 where the content of each micronutrient can be compared with the content of the same element in the natural material. General geochemical information of the type included in Table 1 is of marginal interest in relation to the nutrition of a particular forest because each plant species has a variable requirement for micronutrients during a particular season, and through its life span. If too little of the element is available, growth will slow down and the plant will develop visual *deficiency* symtoms. Similarly if too much of an element is available to a plant, growth also may be inhibited and eventually *toxicity* symptoms develop. Hence, the aim of forest nutrition is to provide trees with amounts of each nutrient element permitting optimum growth during their whole life span.

From the viewpoint of dynamic geochemistry plants participate in the cycling of chemical elements within the ecosystem. Cycling involves the uptake of elements

Table 1. *The concentration (ppm) of essential plant nutrient elements in natural materials*

Element	Rocks[a]				Soil[a]	Fresh water[a]	Nutrient solution[d]	Plant matter (oven dry basis)		Relative number of atoms with respect to molybdenum (plant)	
	Igneous[e]	Shale[e]	Sandstone[e]	Limestone[e]				all plants[b]	Conifers[c]		
Molybdenum	1.5	2.6	0.2	0.4	2[e]	0.00035[e]	0.05	0.1	1	1	micronutrients
Copper	55	45	5	4	20	0.01	0.064	6	4	100	
Zinc	70	95	16	20	50	0.01	0.065	20	30	300	
Manganese	950	850	50	1100	850	0.012	0.55	50	200	1000	
Iron	56300	47200	9800	3800	38000	0.67	5.6	100	200	2000	
Boron	10	100	35	20	10	0.13	0.50	20	20	2000	
Chlorine	130	180	10	150	100	7.8	3.5	100	1000	3000	macronutrients
Sulphur	260	2400	240	1200	700	3.7	48	1000	800	30000	
Phosphorus	1050	700	170	400	650	0.00541	41	2000	1300	60000	
Magnesium	23300	15000	10700	2700	5000	4.1	36	2000	1000	80000	
Calcium	41500	22100	39100	302000	13700	15.0	134—300	5000	2500	125000	
Potassium	21000	26700	11000	2700	14000	2.3	130—295	10000	8000	250000	
Nitrogen	20	—	—	—	1000	0.23	140—284	15000	15000	1000000	
Oxygen	464000	483000	492000	497000	490000	889000		450000		30000000	
Carbon	200	15000	14000	11400	20000	11000		450000		35000000	
Hydrogen	1400	5600	1800	860	15000	111000		60000		60000000	

a Data from Bowen (1966).
b Data from Epstein (1965).
c Data from Keay (1964) for conifers.
d Data for "Long Ashton Solution" (Hewitt, 1966).
e All values for concentration of elements are total content in samples.

by plant roots, their translocation and incorporation within the plant for a period of time, and their return to the soil again. Cycles of this type are called "biogeochemical cycles" and vary from element to element and from plant to plant within the same forest ecosystem. Biogeochemical cycles are of two types, "normal cycles" and those that occur in accumulator plants. In normal cycles different plants may take up different quantities of elements from the same substrate. Uptake of elements may vary with 1) *genotype* (e.g., OUELETTE and DESSUREAUX (1958) showed that two genotypes of alfalfa have different rates of uptake of the same elements under controlled conditions); or with 2) *species* (e.g., COLLANDER (1941) showed that different herb species grown in solutions containing known amounts of essential and non-essential elements took up elements in different proportions), but in either case the magnitude of the variation is relatively small.

GERLOFF (1963) collected species of common plants from different localities in Wisconsin and analysed them for zinc. He found that most species contained less than 50 ppm zinc, but one species *Nemopanthus mucronata* contained consistently higher amounts of zinc, one individual containing as much as 711 ppm. This is an example of an accumulator plant. BOLLARD and BUTLER (1966) reviewed information available on accumulator plants and divided them into two types:

1. those — like *Nemopanthus mucronata* — which accumulate elements from soils containing normal amounts of an element, and

2. those which become adapted to living in soils with particularly high amounts of certain elements — e.g., soils near mineral deposits.

An example of the second type is *Astragulus bisulcatus* which accumulates selenium. According to ROSENFIELD and BEATH (1964) the normal content of selenium in soil is <1 ppm but on a soil containing 5 ppm of selenium they found a specimen of *A. bisulcatus* containing 2640 ppm. Other information on accumulator plants is given by CANNON (1960) and MALYUGA (1964). Accumulator plants may be important in forest nutrition for two reasons:

1. where a micronutrient is scarce they can remove it from the soil in amounts large enough to upset the nutritional balance of a forest, or

2. accumulator plants may be planted as a "negative" treatment in a forest in an attempt to lower toxic levels in other plants.

Plant Physiology of the Micronutrients

All elements essential for plant nutrition are thought to play a number of biochemical roles within a plant. The study of the biochemistry of particular elements in plants is complex and only recently has received systematic study by plant physiologists. Some information can be obtained from texts by STEWARD (1963), BOWEN (1966), and BONNER and VARNER (1965) and from papers by NICHOLAS (1961), McELROY and NASON (1954) and EPSTEIN (1965). Briefly, all six micronutrients (with the possible exception of boron) have been shown to play an important part in particular enzyme systems. McILRATH and PALSER (1956) and SCOTT (1960) consider boron to perform a protective function in plants preventing the excessive polymerization of sugars at sites of sugar synthesis. Useful reviews of the function of micronutrients as revealed by visual examination of cultivated plants are included in books by CHAPMAN (1966), CHILDERS (1966), and WALLACE (1961).

BARROWS (1959) and others have stressed the need for the study of the "critical range of concentration" of elements available to a plant. The relation between growth and increasing foliar content of an element is shown in Fig. 1. A response curve of this type may be related to the study of deficiency and toxicity symptoms of plants as described above. The study of micronutrients in plants would be relatively simple if elements could be studied one at a time by such curves. Unfortunately this is not possible because both the level of one element in a plant and its effect on growth are linked to the levels of other elements.

SCHÜTTE (1964) described a good example of linking between a macronutrient (potassium) and a micronutrient (boron) in soya bean plants growing under controlled

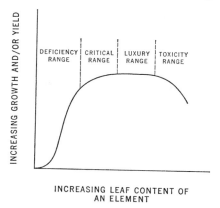

Fig. 1. Diagram designating the different ranges in the mineral element content of leaves in relation to the corresponding positions of the response curve (from BARROWS, 1959)

conditions in a greenhouse. Without boron, over 2% of potassium in the substrate restricted the growth of the plants, but if a small amount of boron was added, healthy growth was obtained even if 20% potassium was present in the substrate. This is a clear example of the *interaction* of one element with another and demonstrates that the boron requirement of the plants is not absolute, but is linked to that of potassium. In another example, SCHÜTTE (1964) described how the addition of phosphates to sour-orange seedlings depressed the uptake of copper, boron and zinc, but increased the uptake of manganese, although it did not seriously influence the uptake of iron. Interactions of this type are sometimes called *antagonisms* and should always be looked for when fertilizer applications are applied to plants.

Empirical evidence, which has been reviewed by MAULDER (1950), LAL and RAO (1954), GAUGH (1957) and, more recently, by BROWN (1963), shows that there are many kinds of interactions which can occur between nutrient elements in plants. A further complication is that interactions may vary with time and space within the same plant. For example CAIN (1959) working with apple trees, described the interaction of nitrogen, potassium and magnesium in different organs of the same plant. He showed that these elements are unevenly distributed through the plant and that changes in the chemical composition of one organ did not always reflect changes in the whole plant. LAL and RAO (1954) noted that manganese deficiency in a plant is inseparable from iron toxicity, and the symptoms of one are identical with those of the other. For example, they noted that for soya bean the optimum active iron:manganese ratio was 1.5:2 *regardless of concentration*. This indicates that both the level of concentration *and* the relative proportions of available elements are important for the healthy growth of plants.

Much empirical research has been directed at the nutrition of fruit trees, details of which can be obtained from a number of texts (e.g., REUTHER, 1961; CHAMPMAN, 1966; and CHILDERS, 1966). BOULD (1966) reviewed information available on the use of leaf analysis for the determination of the nutrient status of fruit trees. He pointed out that experiments can be set up under controlled conditions (in which the nutrition of a single element is the only limiting factor) in order to determine:

1. the best leaf position and stage of plant development to reflect nutrient status,
2. the optimum concentration of nutrient concentrations associated with optimum growth and crop yield, and
3. leaf nutrient levels associated with deficiency and toxicity.

He suggested that leaf analysis offers the only means of applying directly to field work the results of sand- and solution-culture experiments. This is because the interpretation of nutritional status on the basis of foliar analysis is not concerned with the way in which plants are grown but only with the effect on growth of changes in leaf composition. HEWITT (1966b) stressed that the nutrition of forest trees can be investigated in the same way as for herbaceous plants by using culture methods, provided that they can be grown to adult stages and that proper climatic conditions are provided.

Mathematical Description of Plant Nutrition

Early attempts to express mathematically the nutrition process in agricultural plants have been reviewed by RUSSELL (1961). For example, he discussed LIEBIG's "Law of the Minimum", which states that the amount of plant growth is regulated by the factor (i.e., nutrient) present in minimum amount, and that growth rises or falls as this factor is increased or decreased in amount. (Fig. 1 illustrates this type of relationship graphically.) However, because of the diverse interactions described above, this law is not generally applicable to nutrition; one element will interact with another, and it has been found that if several elements are low, but not too low, in concentration, an increase in any one will increase yield. Further discussion of this principle is provided by PRÉVOT and OLLAGNIER (1961).

More recently, largely as a result of theoretical work by FISHER and others after 1910, the nutrition of plants has been studied by the biometric-statistical approach (FISHER, 1958). This utilizes factorial experiments designed to test certain treatments (e.g., fertilizer elements singly and in various combinations). Experimental data are analyzed by statistical methods and a regression formula derived. This is essentially an empirical approach and no detailed hypothesis need be set up taking account of underlying causes before commencement of the experiment. Limitations of this biometric approach are the need for a large number of treatment combinations and a difficulty in generalizing beyond the particular experimental materials and conditions used.

This approach was used by FUEHRING (1960), who carried out a factorial experiment with corn plants involving growth and mineral content under 27 treatments. Each treatment was a combination of five micronutrient elements (Zn, Mn, B, Cu, and Fe) applied to the plants for a period of 100 days. He found that a second-order multiple regression equation fitted the data fairly well and this enabled him to show that, as the amount of some elements was increased above the deficiency range, other elements held constant in the environment showed a decreasing concentration in the plant tissues, apparently being diluted by growth. A similar experiment using the same elements in 81 combinations added to lucerne, oats, and tomato plants was carried out by EL KHOLI (1961). The data from this large number of species and treatment combinations revealed few consistent patterns of statistical significance.

So far this approach has rarely been tried to investigate the interrelations of the six micronutrient elements in relation to forest plants, although unpublished data of this type exist (CARLISLE, 1968, personal communication). This is probably due to the practical difficulties involved in working with trees as opposed to herbs. The biometric approach has, however, been used in experiments or in surveys of macronutrients and forest plants (CARLISLE and WHITE, 1964; and BAKUZIS, 1968).

Recently a third approach towards quantifying the study of plants and ecosystems has been evolving — the "systems approach". During the past few decades systems analysis theory has been developed for a variety of applications in science and engineering (ASHBY, 1956; HILLIER and LIEBERMAN, 1967; MILSUM, 1966; BELLMAN et al., 1969; BIRKHOFF et al., 1962). It should be noted, however, that nearly all these applications have involved systems far simpler than ecological ones. [For introductions to the application of systems analysis to ecology see WATT (1966), WATT (1968) and VAN DYNE (1966).]

This approach has also been applied to some aspects of forest nutrition but not those involving micronutrient elements. Such investigations have included radiotracers of macronutrients (e.g., THOMAS, 1969, who used ^{45}Ca) or nonessential minor elements (e.g., OLSON, 1965, who used ^{137}Cs).

The application of systems analysis to ecology may involve either or both of two trends:

1. a systems theory is tailored to suit the complexities of ecology, or
2. the ecology is simplified to match existing techniques of systems analysis.

Systems analysis in ecology to date has followed the second trend. In the second part of this paper we suggest an approach towards simplifying the description of an ecological system — by means of a flow diagram — before the application of existing mathematical techniques to the system.

In theory the application of systems analysis to ecological systems is justified because of inherent simplifications in their organization due to the finite information-processing capabilities of their biological components. Cybernetic considerations (QUASTLER, 1964) indicate that mathematical principles and patterns may be expected which will explain the workings of such complex biological systems. The problem of mathematical modeling is the discovery of these natural principles and patterns.

Unlike the biometric approach discussed above, the mathematics of the systems approach as suggested here is always based on a hypothesis derived from real-world processes. The procedure of the systems approach involves the description of a system (in this case a single plant, or a woodland ecosystem) by means of a flow diagram that embraces the different components of the system together with the possible pathways which connect the various components. Although this has not yet been done for the micronutrition process in forests, other similar systems in ecology and natural resource management have been thoroughly described in this manner (Hufschmidt and Fiering, 1966; Royce et al., 1963; and Morris, 1963). A mathematical model is then set up to mimic the system, based upon transfer equations devised to describe the flow of energy or materials (e.g., nutrients) between components. The effectiveness of the model cannot surpass either the validity of its transfer equations or the realism of the interactions between components, as shown by the flow diagram on which the model is based.

A variety of imaginary experiments, similar to actual experiments that might be carried out in the real world, may then be conducted on the model by means of simulation. [See Naylor (1966) and Mize and Cox (1968) for descriptions of simulation techniques.] The findings of model experiments are then used to indicate an experimental approach to the study of the real system (Bellman, 1962) and eventually to process data from experiments on the system. During this phase a dialogue is developed between the model-building process on the one hand and the real world on the other. The dialogue involves a series of successive approximations each of which should result in the model approaching more nearly the processes in the real world.

One advantage of the systems approach is that as the performance of the model improves, experiments may be conducted on the model to discover transfer equations which are most important to the system as a whole. These transfer equations may then be the object of more intensive research. Another advantage is that the systems approach can cope with the joint action of many variables, each being incorporated alone as it affects the process.

When a systems model has been constructed successfully it can then be applied quite generally to solve ecological problems. For example, Holling's (1965) discussion of the predation process suggests that a realistic model of a process can be applied to a variety of situations involving that process. This is due to a general uniformity of biological pattern despite diversity of particular biological mechanisms.

Forest Information
Micronutrients in Forest Plants

Greenhouse or nursery investigations have focused attention on visible symptoms of micronutrient deficiency or toxicity in forest plants. For example, Smith (1943) showed that the growing of root tips and extended cork layers of *Pinus radiata* were due to boron deficiency; Kessell and Stoate (1936) that "rosetting" in *P. radiata* was due to zinc deficiency; Benzian and Warren (1956) that seasonal "needle tip burn" of *Picea sitchensis* was due to a deficiency in copper; and Smith (1943) that a chlorosis of needles at the ends of shoots of *P. radiata* was due to manganese deficiency. More information on symptoms of micronutrient deficiencies in forest plants may be obtained from Hewitt (1966), Tamm (1964) and a recent article by Stone (1968).

Little systematic information is available on the growth of forest plants under controlled conditions in a greenhouse beyond the seedling stage. General information on the growing of plants under the strictly controlled conditions needed for the study of micronutrients may be obtained from Hewitt (1966b) who lists the composition of nutrient solutions in which seedlings and small trees have been grown successfully.

Other information on the growing of forest plants in a greenhouse may be obtained from ADDOMS (1937), MITCHEL (1939), KRAJINA (1958, 1959), HACSKYLO (1961), INGESTAD (1957, 1959, 1960, 1962), and INGESTAD and JACOBSON (1962).

A pioneer investigation by PETERSON (1962) on Kauri seedlings (*Agathis Australis* Salisb.) is important because he worked with all six micronutrient elements, one at a time, on the same species under controlled greenhouse conditions. The seedlings were grown for a period of 100 days in solutions containing all required nutrient elements in optimum amounts for growth except one micronutrient. Known amounts of the missing element were added at the start of experiments designed to determine the strength of nutrient solution required to induce deficiency (or toxicity) symptoms in the seedlings. The relationship between the concentration of the treatment element in the nutrient solution and the dry matter production of the seedlings also was studied. The experimentally produced deficiency symptoms were described in relation to the whole plant, or to particular organs. Only zinc deficiency modified the growth of the whole plant; in other cases one or more organs were affected. Toxicity symptoms were induced for all elements except molybdenum. Details for the concentration of

Table 2. *Deficiency, normal and toxic concentrations for micronutrient elements in culture solutions containing all elements (except the test element) at optimum concentration in which Kauri trees were grown for 100 days* (Data from PETERSON, 1962)

Yield of dry matter	Element ppm in solution					
	Boron	Copper	Iron	Manganese	Molybdenum	Zinc
Deficiency	0.001	0.001	0.001	0.001	0.0001	0.001
Normal	0.500	0.010	1.200	0.500	0.5000	0.050
Toxic	5.000	0.200	12.000	2.500	—	0.500

each element required to produce symptoms are given in Table 2. In order to place foliar analysis of forest plants on a rational basis (as described by BOULD (1966) above), many more controlled experiments of the type pioneered by PETERSON (1962) are needed.

Micronutrients in Forest Ecosystems

The uptake of macronutrient elements by forests of different kinds has been calculated by RENNIE (1955). More recently, OVINGTON (1962) has provided extensive summaries of information available on the quantity of macronutrient elements in the components of forest ecosystems, and TAMM (1964) has summarized information available on the nutrient requirements of forest stands. Unfortunately, these articles provide little information regarding micronutrients.

A comprehensive study of the minor-element content of forest soils and plants growing in Finland was carried out by LOUNAMAA (1956). He determined the amounts of five micronutrient elements (B, Mn, Zn, Cu, and Mo) and nine non-essential elements (Cr, Co, Ni, Ga, Y, Zr, Ag, Cd, and Pb) in soils and plants of different kinds growing on silicic, ultrabasic, or calcarous rocks in Finland. Although this paper may well be used as a starting point for the study of minor elements in forests, the data presented are of somewhat limited value from a nutritional point of view because

Table 3. *The total element content of some Finnish rocks and associated soils and plant ashes* (Lounamaa, 1967)

N = Number of samples, ppm = Mean content of element, ± one standard deviation, range = Range in concentration of individual samples

Rock Type	Material	Manganese			Copper			Zinc			Molybdenum		
		N	ppm	range	N	ppm	range	N	ppm	range	N	ppm	range
Silicic	**Deciduous trees and shrubs**												
	Leaves	80	12000±1200	1000—30000	27	150±21	30— 300	36	2100±330	<100—6000	75	5.4±1	< 3—30
	Twigs	80	9000± 950	1000—30000	27	270±47	30—1000	36	4100±340	600±6000	75	6.8±1.5	< 3—60
	Coniferous trees												
	Needles	64	14000±1300	1000—30000	58	240±26	<30—1000	34	1500±250	100—6000	64	2.3±0.6	< 3—30
	Twigs	64	10000±1100	300—30000	58	510±71	30—3000	34	1600+220	100—6000	64	1.6±0.3	< 3—30
	Soil	94	1500± 150	30— 6000	91	110±10	<30— 300	84	320± 50	<100—1000	90	3.4±0.2	<10—20
	Rock	177	610± 59	—30— 6000	173	50± 4	<30— 300	159	180± 17	<100—1000	184	3.3	<10—20
Ultra-basic	**Deciduous trees and shrubs**												
	Leaves	14	15000±3200	3000—30000	14	94±10	30— 200	12	1200±330	100—3000	14	28±9	<3—100
	Twigs	14	11000±2800	3000—30000	14	130±20	60— 300	12	2700±540	300—6000	14	23±6	< 3—60
	Coniferous trees												
	Needles	10	16000±3800	1000—30000	10	59±12	30— 100	10	950±250	300—3000	10	15±8	< 3—60
	Twigs	10	5200±1100	1000—10000	10	82±10	30— 100	10	1200±310	300—3000	10	12±6	< 3—60
	Soil	23	1200± 150	300— 3000	23	22± 5	30— 100	23	270± 49	<100—1000	23	3.0	<10± 0
	Rock	50	1400± 110	100— 3000	50	27± 4	<30— 100	50	190± 29	<100— 600	50	3.0	<10± 0
Calcar-eous	**Deciduous trees and shrubs**												
	Leaves	23	5000± 130	300—30000	16	99±26	30— 300	14	1000±470	<100—6000	23	7.7±2.0	3—30
	Twigs	23	7400± 160	300—30000	16	96±15	30— 300	14	1700±540	<100—6000	23	14.0±4	< 3—60
	Coniferous trees												
	Needles	9	7800±1200	1000—10000	9	130±34	30— 300	8	860±330	100—3000	9	2.2±1	3—10
	Twigs	9	6100±1400	1000—10000	9	290±48	100— 600	8	1200±400	300—3000	9	7.4±4	3—30
	Soil	17	1800± 370	300— 6000	17	42±10	< 30— 100	17	220± 64	<100—1000	17	3.0	
	Rock	34	2300± 350	100—10000	28	34± 7	< 30— 100	28	230± 44	<100— 600	34	3.2	<10—10

1) the total rather than the available content of each element in each rock or soil type was determined, and 2) the data for plants are expressed on an ash rather than an oven-dry weight basis. In a later paper LOUNAMAA (1967) summarized geochemical data of this type available for four micronutrient elements in plants, rocks, and soils from Finland (Table 3). Some observations of general interest can be made from this table; for example, there is always more manganese in the needles than twigs of coniferous plants.

Table 4. *The distribution of six micronutrient elements in the organs of individuals of four species of coniferous trees growing in the University Forest, Stillwater, Maine* (data from YOUNG and GUINN, 1966)

Species	Organ	Distribution of micronutrients (ppm oven dry wt.)					
		Boron	Iron	Manganese	Zinc	Copper	Molybdenum
Red spruce	Needles	31	100	1400	45	7.0	4.8
	Small branches	14	284	580	46	11.5	2.6
	Branches (Wood)	2	6.9	221	14	6.0	0.69
	Branches (Bark)	15	159	855	74	17.8	8.0
	Trunk (Wood)	0.1	14	144	8	4.2	0.9
	Trunk (Bark)	10.5	82	612	50	8.0	4.1
	Roots (Wood)	1.0	8	130	14	4.6	0.6
	Roots (Bark)	11	112	762	72	8.0	10.0
	Small roots	23	300+	460	67	7.5	4.8
Balsam fir	Needles	15	111	1155	50	7.5	5.1
	Small branches	11	166	930	67	8.8	3.7
	Branches (Wood)	3	12	156	15	8.8	0.68
	Branches (Bark)	14	187	815	81	10.0	6.1
	Trunk (Wood)	2	13	127	11	17.1	0.4
	Trunk (Bark)	11.7	135	600	45	8.0	3.6
	Roots (Wood)	1	11	60	26	4.4	0.4
	Roots (Bark)	12	193	340	36	7.4	4.3
	Small roots	16	300+	310	40	5.3	3.4
Hemlock	Needles	44	127	1500	10	4.5	3.0
	Small branches	14	130	720	29	9.0	3.3
	Branches (Wood)	1	8	155	4	3.8	0.18
	Branches (Bark)	19	176	1305	52	8.7	9.9
	Trunk (Wood)	1	6	145	2	4.5	0.3
	Trunk (Bark)	12.1	65	794	15	9.0	5.4
	Roots (Wood)	1.3	19	86	2	5.1	0.31
	Roots (Bark)	16	300	453	13	6.4	3.1
	Small roots	18	300+	300	13	6.3	1.7
White pine	Needles	13	72	375	52	5.3	1.4
	Small branches	11	88	260	68	6.9	3.5
	Branches (Wood)	3	7.0	67	12	5.7	0.48
	Branches (Bark)	15	120	332	98	7.4	5.9
	Trunk (Wood)	1	10	28	11	5.5	0.3
	Trunk (Bark)	11.5	68	185	65	5.6	2.4
	Roots (Wood)	0.6	10	17	6	4.4	0.3
	Roots (Bark)	14	229	46	19	9.0	0.7
	Small roots	22	300+	64	22	5.5	1.3

In the remainder of this section information on the distribution and amount of micronutrients obtained at an instant in time will be described first. Some of this information is expressed on a weight basis, and some on an area basis. The last part of the section includes information obtained on the concentration of micronutrient elements in plants sampled at intervals over a season.

Young and Guinn (1966) provided detailed information on the distribution and amount of all six micronutrients in particular organs of individual trees of three deciduous and four coniferous species growing in the University of Maine forest at Stillwater, Maine. Information from this interesting paper has been summarized in Tables 4, 5 and 6 and may be used as a guide. Several general points of interest are apparent from these tables; e.g., the greatest concentration of micronutrients is found in the extremities of the trees and the lowest in the stemwood. Estimates for the total content of each element in each tree (Table 6) indicate that uptake for boron, iron, copper and molybdenum are similar for each species but that uptakes for manganese and zinc vary from species to species. It would be interesting to know the cause of this

Table 5. *The distribution of six micronutrient elements in the organs of individuals of four species of deciduous trees growing in the University Forest, Stillwater, Maine* (data from Young and Guinn, 1966)

Species	Organ	Distribution of micronutrients (ppm oven dry wt.)					
		Boron	Iron	Manganese	Zinc	Copper	Molybdenum
White birch	Leaves	31	72	315	77	6.8	4.0
	Small branches	13	27	129	91	7.0	3.1
	Branches (Wood)	—	7.7	40	17	10.0	0.4
	Branches (Bark)	12	42	233	100+	7.0	7.6
	Trunk (Wood)	1	10	34	28	4.1	0.63
	Trunk (Bark)	12.9	55	275	99	6.0	6.2
	Roots (Wood)	—	8	5	10	3.5	0.4
	Roots (Bark)	17	300	117	83	7.1	5.6
	Small roots	12	300+	55	100+	125.0	3.0
Red maple	Leaves	30	133	765	41	9.0	3.6
	Small branches	16	58	475	49	6.8	3.8
	Branches (Wood)	2.4	19	102	30	9.0	0.5
	Branches (Bark)	17	59	728	66	8.7	7.4
	Trunk (Wood)	1.3	11	72	29	4.9	0.3
	Trunk (Bark)	14.4	55	668	78	11.2	6.9
	Roots (Wood)	2.0	15	80	31	5.4	0.3
	Roots (Bark)	20	278	577	73	10.9	12.4
	Small (Roots)	13	300+	355	69	10.0	4.8
Aspen	Leaves	30	105	460	100+	8.0	7.0
	Small branches	15	25	118	88	13.0	5.2
	Branches (Wood)	4	24	33	24	11.0	0.5
	Branches (Bark)	16	84	148	100+	14.0	7.1
	Trunk (Wood)	2	12	29	17	6.5	0.4
	Trunk (Bark)	13.4	54	114	97	8.0	6.6
	Roots (Wood)	1.0	32	20	18	5.8	0.2
	Roots (Bark)	27	300	100	94	7.7	6.2
	Small roots	16	300+	45	47	9.0	3.1

(data from YOUNG and GUNN, 1966)

Species	Total height, meters	Complete fresh wt, kgs	Grams of micronutrient elements (complete tree)					
			Boron	Iron	Manganese	Zinc	Copper	Molybdenum
Red spruce	15.8	275.9	0.98	7.66	61.03	3.69	0.99	0.35
Balsam fir	13.7	316.4	1.06	9.64	69.30	4.65	1.71	0.34
Hemlock	13.4	294.6	1.39	7.84	66.13	1.06	0.84	0.23
White pine	15.8	376.9	0.89	7.54	16.42	4.91	1.22	0.18
White birch	18.7	371.9	1.33	8.08	21.96	10.98	1.88	0.47
Red maple	14.7	283.6	1.08	7.28	41.64	7.66	1.32	0.35
Aspen	15.2	232.1	0.91	5.10	9.60	5.62	1.03	0.30

Table 7. *The total weights (grams) of iron and manganese contained in entire tissue categories of ten individuals[a] of 34-year-old red pine collected from Halliday's Field, Petawawa Forest Experiment Station, Ontario, Canada* (unpublished data from RENNIE)

Tree Number	Stem Wood 3—7 (cm diam.)		Stem bark 3—7 (cm diam.)		Dead branches		Twigs		Needles	
	Manganese	Iron	Manganese	Iron	Manganese	Iron	Manganese	Iron	Manganese	Iron
I	0.0878	0.0057	0.1113	0.0210	0.6282	0.7503	0.5759	0.3051	2.7194	0.6148
II	0.1394	0.0030	0.0930	0.0132	0.4539	0.5584	1.1648	0.2314	4.6111	0.5981
III	0.1443	0.0036	0.1168	0.0140	0.3385	0.7722	0.6189	0.2324	3.0300	0.6660
IV	0.1341	0.0058	0.1006	0.0137	0.5209	0.5948	0.6404	0.3784	3.4195	0.6927
V	0.1120	0.0052	0.0810	0.0143	0.1914	0.4747	0.3753	0.2586	2.3393	0.3697
VI	0.1280	0.0047	0.1080	0.0162	0.5240	0.6659	0.4111	0.3674	2.5173	0.6069
VII	0.1249	0.0041	0.0882	0.0100	0.4477	0.4042	0.3836	0.4161	2.5699	0.5253
VIII	0.1543	0.0044	0.1423	0.0115	0.5602	0.3916	0.7398	0.2535	3.8209	0.6523
IX	0.1494	0.0090	0.1598	0.0168	0.2907	0.4743	0.3439	0.2668	2.0327	0.4045
X	0.1575	0.0046	0.1678	0.0118	0.3207	0.4373	0.4770	0.2041	2.6780	0.4439
Mean	0.1332	0.0050	0.1169	0.0142	0.4276	0.5524	0.5731	0.2914	2.9738	0.5574
Coefficient of variation (%)	16.0	33.0	25.7	22.2	32.2	25.3	43.0	24.8	26.1	20.6

[a] Each individual selected because of average basal area for whole stand.

variability, which might be related to the differing needs for these elements by the individual plants, or might be a reflection of differences in the availability of these elements in the soils of the Stillwater forest. [Young and Guinn (1966) did not include detailed information on the magnitude of either sampling or analytical errors in their paper.]

In an earlier paper (Young et al., 1965) tables were provided as guides to the estimation of the amount of twelve elements (N, Ca, K, Mg, P, Mn, Fe, Al, Mo, Zn, Cd, and B) in trees of the same seven species at different diameters and heights. Although this is a promising start to the delineation of the nutrient requirements of forest trees at different stages in their life cycle, for reasons described below these tables may be of rather limited value for micronutrient elements owing to large tree-to-tree variation in the concentration of these elements compared with the macro-nutrients.

Recently Rennie (unpublished data) compared variations in the contents of two micronutrients (Fe and Mn among others) in ten specially selected trees growing in a uniform stand of red pine *(Pinus resinosa)*, growing at the Petawawa Forest Experiment Station, Chalk River, Ontario, Canada. Each tree was selected to have a basal area equal to the average for the 34-year-old stand. The trees were most carefully sampled by methods already described by Rennie (1966). Data for the weights of iron and manganese in comparable samples of stemwood, stembark, dead branches, twigs and needles from each of the ten trees are given in Table 7. It should be stressed that, except in one case, the coefficient of variation for each of the columns of data in Table 7 is greater than 20%. Nitrogen was also determined in all the same samples and gave coefficients of variation of less than 10% for any organ. This indicates the greater difficulty of estimating precisely and accurately the content of micronutrients compared with macronutrients within a given forest stand.

So far we have only been concerned with forest trees and the nutrients they contain. Ovington (1956) has provided information on the total weight of iron and manganese in ground floras under different woodlands at three locations in England. Detailed descriptions of the plant communities from which his data were obtained were given in his 1955 paper. Table 8 includes data on iron and mangenese in these communities — expressed on an area basis. At the time of collection the plantations at West Tofts were 22 years old, those at Bedgebury between 22 and 24 years old and those at Abbotswood from 46 to 47 years old (except those for *Abies grandis, Fagus sylvatica* and the mixed wood which were 24, 39, and 44 years old, respectively). These data focus attention on the relatively large range in content of micronutrient elements in ground floras under plantations of different species but of the same age and in the same general area. It is unfortunate that this interesting study was restricted to iron and manganese and did not include information on the other micronutrient elements.

Rennie (unpublished data) has calculated the total iron and manganese contents of the red pine stand mentioned above on an area basis (Table 9). These are of particular interest because they are expressed on an area basis organ-by-organ. As in Finland, the content of manganese in needles is greater than in the twigs on which they grew and it should be noted that the iron content of the dead branches was relatively high.

So far we have presented information on the composition of plants sampled at an instant in time. Foresters really require information for the nutrient requirements of particular forests to be expressed on an area basis for the duration of the growth cycle. RENNIE (unpublished data) has estimated the budget of 25 elements (Mo, Cu, Zn, Mn, Fe, B, Mg, K, Ca, Si, Na, Al, Ba, Sr, Ti, Pb, Rb, Ag, Zr, Li, Ni, Cr, Co, Sn, and V) which would be removed in logs over 3 inches in diameter from a stand of grand fir 100 years old (Table 10). Although these estimates are first approximations, they do provide information which could be used in relation to other kinds of

Table 8. *The total weights of iron and manganese contained in ground floras under different trees growing in England* (data from OVINGTON, 1956)

Locality	Forest type	Iron content (kg/ha)	Manganese content (kg/ha)
West Tofts,	*Alnus incana*	3.44	0.86
	Betula alba	3.73	1.10
Thetford Chase	*Larix leptolepis*	0.82	0.27
	Unplanted	10.47	2.62
Abbotswood,	*Quercus robur*	2.13	2.13
Forest of Dean	*Quercus* spp.	1.44	1.57
	Castanea sativa	2.02	1.19
	Fagus sylvatica	0.10	0.24
	Pinus sylvestris	1.40	6.32
	Pinus nigra	1.37	6.15
	Larix decidua	5.95	8.69
	Pseudotsuga taxifolia	5.04	6.12
Bedgebury, Kent	*Quercus petraea*	0.19	0.93
	Nothofagus obliqua	0.23	1.15
	Quercus rubra	0.21	0.46
	Larix eurolepis	0.12	1.22
	Chamaecyparis lawsoniana	0.15	0.46
	Pinus nigra	0.04	0.30
	Pseudotsuga taxifolia	0.06	0.47
	Open (1)	25.55	8.29
	Open (2)	31.51	11.16
	Open (3)	10.21	6.56

Table 9. *Estimates for the total weights of iron and manganese contained in a red pine stand at Halliday's Field, Petawawa Forest Experiment Station, Ontario, Canada* (unpublished data from RENNIE)

Organ or category	Iron content (kg/ha)	Manganese content (kg/ha)
Dead branches	1.0621	0.8221
Stem wood (3—7 cm diam.)	0.0096	0.2560
Stem wood (7 cm — stump)	0.2604	9.6777
Stem bark (3—7 cm diam.)	0.0272	0.2246
Stem bark (7 cm — stump)	0.4646	3.4499
Live branches	0.2908	1.3966
Twigs	0.5602	1.1018
Needles	1.0717	5.7170
All aerial organs	3.7466	22.6457

Table 10. *Estimates for the weight of chemical elements taken up by wood of grand fir, Abies grandis (over 3 inches in diameter) in 100 years of growth.* Unpublished data of Rennie based upon analytical values of Ellis (1956)

Elements		kg/acre/100 yrs.
Micronutrients		
	Molybdenum	0.003
	Copper	0.760
	Zinc	0.540
	Manganese	8.0
	Iron	1.8
	Boron	0.180
Macronutrients		
	Magnesium	64.0
	Potassium	148.0
	Calcium	333.0
Non-Essentials		
	Silica	24.0
	Sodium	3.0
	Aluminium	9.0
	Barium	7.4
	Strontium	6.7
	Titanium	1.1
	Lead	1.3
	Rubidium	0.89
	Silver	0.12
	Zirconium	0.10
	Lithium	0.082
	Nickel	0.028
	Chromium	0.015
	Cobalt	0.007
	Tin	0.003
	Vanadium	0.0002

estimates for the element budgets of forests, e.g., those by Young et al. (1965) discussed above.

Relatively little information is available on the role of accumulator plants in forest nutrition. Particular organs of trees are known to accumulate particular elements, e.g., manganese by the needles of conifers, or arsenic in the growing shoots of Douglas fir (Warren, Delavault, and Barasko, 1964). One of the best examples of an accumulator plant in a forest was described by Robinson and Edgeworth (1945) who determined the rare-earth content of hickory leaves. In one experiment samples of different plant species were collected at Falls Church, Virginia and analyzed for total rare-earth content. Although samples of *Lycopodium* collected beneath the hickory trees contained less than 4 ppm of rare earths, the leaves of the hickory trees contained 1,513 ppm. Other experiments in the same area indicated that the accumulation of the rare earths was in the leaves of hickory; e.g., in one tree the leaves contained 981 ppm of rare earths but the fruits contained less than 20 ppm of these elements. As it is likely that other accumulator plants will be discovered in forests, samples of the principal organs of each plant species should be analyzed to determine if accumulator

Table 11. *The effect of sampling height and date of sampling on composition of leaf blades or whole leaves from Craigiebuckler, Scotland (ppm oven dry weight basis)* (from GUHA and MITCHELL, 1966) B = bottom of tree; M = middle of tree; T = top of tree

Plant and plant part		Sycamore No. 3 Leaf blade				Horse chestnut No. 3 Leaf blade					Beech No. 3 Whole leaf				
		May 14	July 3	Aug. 24	Sept. 23	May 19	July 6	Aug. 25	Sept. 29	Oct. 26	May 21	July 14	Aug. 29	Oct. 1	Oct. 20
Molybdenum (ppm)	B	0.32	0.11	0.10	0.14	0.49	0.28	0.26	0.30	0.33	0.16	0.09	0.08	0.09	0.07
	M	0.42	0.12	0.11	0.13	0.83	0.43	0.31	0.33	0.32	0.12	0.06	0.06	0.06	0.05
	T	0.31	0.10	0.07	0.08	0.50	0.44	0.33	0.25	0.28	0.10	0.06	0.06	0.08	0.05
Copper (ppm)	B	15.5	8.2	8.0	7.2	15.6	9.6	7.2	5.9	5.5	13.9	6.9	6.0	4.7	4.2
	M	15.8	8.9	7.4	7.6	16.5	8.3	6.5	5.5	5.0	13.0	4.9	4.4	3.4	3.7
	T	15.1	8.9	6.7	5.7	13.5	8.2	4.6	4.1	3.1	12.3	5.1	4.2	3.5	3.3
Zinc (ppm)	B	31.4	18.7	22.5	22.9	38.0	16.5	15.1	10.4	32.3	33.2	25.5	29.5	30.4	35.1
	M	35.4	23.2	17.7	22.7	42.3	15.9	9.1	12.3	35.4	29.0	21.1	23.1	26.5	32.8
	T	30.5	27.3	20.5	20.8	36.5	13.3	11.9	9.1	20.4	32.5	19.2	23.4	20.5	25.4
Manganese (ppm)	B	55	45	56	63	161	168	179	158	186	240	293	274	337	220
	M	51	48	52	58	146	199	185	226	160	259	194	250	280	242
	T	43	42	36	49	145	210	128	134	115	194	216	236	274	212
Iron (ppm)	B	308	272	328	450	274	276	403	506	462	251	133	161	197	159
	M	253	262	307	438	246	236	351	443	432	174	68	91	130	116
	T	178	152	165	242	206	215	294	286	353	135	72	105	176	120
Boron (ppm)	B	27	41	65	76	16	28	24	26	25	26	38	40	33	26
	M	27	46	65	84	16	22	23	25	26	28	37	37	33	25
	T	30	47	68	86	15	15	14	16	16	25	36	36	35	24

plants are present when undertaking a micronutrient nutrition study. The choice of elements to be included in preliminary investigations of this type should not be restricted to the micronutrients but should also include many nonessential elements as well, as in Table 10.

At the beginning of this section it was noted that seasonal variation was important in any study of the role of micronutrients in forest plants. Seasonal variation in the requirement for copper in nursery plants was studied in Sitka spruce seedlings by Benzian (1965) who discovered that visual symptoms of copper deficiency only occurred late in the season. Vail et al. (1961) described how boron deficiency symtoms in pine disappeared suddenly several years after planting, and suggested that the surface soil in the area was deficient in boron and that deficiency symptoms disappeared when roots penetrated to a part of the soil where sufficient boron was available.

Seasonal variation in the content of micronutrients in leaves of three species of mature deciduous trees collected at different heights above the ground was given by Guha and Mitchell (1966). Table 11 shows that the seasonal variation in elemental content of the leaves of the different species was generally similar for the three species, but differed from element to element. Unlike molybdenum, copper, mangenese, zinc and boron, iron was always lower in concentration in leaves collected at the bottom compared with the top of the tree, and this pattern persisted throughout the summer (Table 11). During the period from June to September the content of an element in a leaf might increase or decrease severalfold compared with the amount at the commencement of the growing seasons. This information focuses attention on the difficulty of choosing a time to sample foliage to determine the amount of particular micronutrients within a given forest stand, because relatively large amounts of micronutrient elements are concentrated in the foliage of trees. For this reason the choice of time of sampling foliage for micronutrient requirement estimation in particular forests should always be considered in relation to the greenhouse approach for the determination of the nutrient requirements of fruit trees described by Bould (1966) and summarized earlier.

As might be expected, the chemical composition of sap as well as that of leaves varies during a growing season. For example, Olsen (1949) studied the seasonal variation of potassium, magnesium, calcium, phosphorus, manganese and iron in the leaves of beech trees and found a steady increase in the content of both micronutrients during the growing season. Parallel investigations were carried out on the macronutrient element content of sap which showed that each of the four elements studied (K, Mg, Ca, and P) behaved differently during the season. The pH of the sap increased from just over 4 in May to over 5 in June and then remained constant till the end of the summer. It seems clear that determinations of the micronutrient, macronutrient, and non-essential element content of sap combined with foliar analysis will provide important information on the rates of movement of individual elements within the plant.

Two Provisional Models for Forest Nutrition

Previously it was pointed out that systems analysis involves the description of a given system in terms of its components, the delineation of the possible transfers of material and energy between components, and the graphing of these features of the system in terms of a flow diagram. The diagram is then translated into a mathematical model in which all the transfers are represented by transfer equations. The geometry of the flow diagram then governs the way in which the equations are arranged in the model.

Systems models are hierarchical in three ways:
1. in level of organization,
2. in time, and
3. in space.

The significance of the hierarchical approach to systems analysis is that instead of having one model of unmanageable complexity, a number of models are allowed, each of which is of manageable complexity. For example, the two models described below are at different levels of organization, the first at the level of the ecosystem and the second at the level of the whole plant. In the case of time, a model might explain the working of a system for one minute, one day, or one year. Spatially, a model might represent one square meter or an entire landscape.

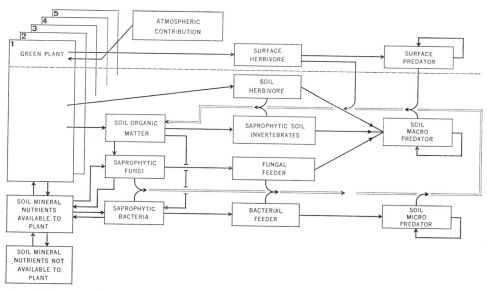

Fig. 2. A generalized flow sheet for the preparation of a systems model of a forest ecosystem, showing the possible paths by which mineral elements may flow between the different components of the system

Flow diagrams have been drawn to represent a forest ecosystem (Fig. 2) and a single plant (Fig. 3). The arrows between components on the diagrams represent possible paths along which atoms of a nutrient element might pass. The overall importance of models of this type is that they can be used to discover general patterns for the transfer of nutrient elements within a system and, once discovered, these patterns can be applied to solve specific problems of plant nutrition.

It is clear from the literature review above that because of 1) the number of plant species present in a forest, 2) the number of micronutrient elements, and 3) the very large number of transfers and interactions which occur between them it is not possible to obtain a complete picture of micronutrient nutrition exclusively from the study of particular instances. What a realistic systems model of a forest ecosystem should do is provide meaningful simplifications of the relationships between the components of the system which can be valuable in the description of the system as a whole.

Unlike an empirical approach to forest ecosystem study the use of a model enables us to predict which field observations are most important for understanding forest nutrition. Once the model is constructed, refining it involves a logical dialogue

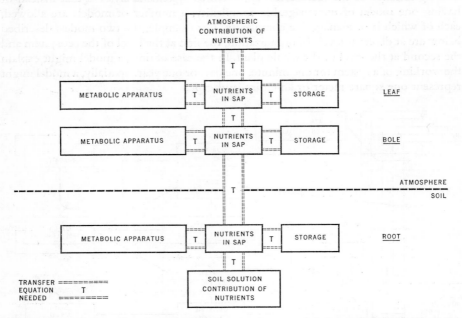

Fig. 3. Flow diagram showing the transfer equations needed for a plant growth model. Note: "storage" includes structural and all other materials which are metabolically inert

between observations on the transfer patterns of the different components in the real world and these same transfers represented by equations in the model. For example, a prediction might be made on the effects of zinc deficiency which could be tested by planting zinc accumulator plants in a forest. Another experiment might be similar to that of Bormann et al. (1968), who reported on a U.S. Forest Service clearcut to observe effects upon nutrient concentrations in streams draining a small basin.

Constructing an ecosystem model requires transfer equations. Fig. 2 demonstrates that there are relatively few basic types of transfers, each with its own typical equation of transfer. Furthermore, some of the transfer equations may be similar to each other, since the nutrition of different ecosystem components — plants, animals, microbes — is a biological process holding some features in common for all organisms. This considerably simplifies the prospect of constructing an ecosystem model.

Selecting one transfer equation as an example, a tree's rate of nutrient uptake depends upon its mineral environment and upon other factors which affect plant growth. This equation could be distilled from a model which describes in some detail the processes of a single plant (Fig. 3). Just as an ecosystem model has transfers of material and energy, a plant growth model has its own internal transfers, and correspondingly its own internal transfer equations. The vertical transfer equations of Fig. 3 are based upon laws of diffusion, active transport, and ionic interference (WILBRANDT and ROSENBERG, 1961; JENNY, 1966). The horizontal equations are based on enzyme kinetics (CLELAND, 1963; WONG and HAINES, 1962), the rate of growth of a particular plant tissue being determined by the concentrations of all nutrients in its solution.

It is appreciated that Fig. 3 alone is not a realistic description of mineral transfers in a plant. The components — leaf, bole, root metabolic storage and sap — are not uniform in function throughout the plant and are therefore subject to subclassification. There are also higher levels of organization in the plant, e.g., hormonal control, acclimation, and endogenous rhythms, which may eventually require attention. The purpose of Fig. 3 is to indicate the transfers and components which require mathematical description.

Just as in the ecosystem model, a logical dialogue is built up between the mathematical model and experimental data from plants growing under controlled conditions in the real world. For example, a sensitive area within the model might be the uptake of nutrient elements by roots, and this transfer might well be studied in detail with excised roots along the lines described by BARNES (1959). Such an experiment might be concerned with the shift in nutrient uptake induced by a sudden change in the root's nutrient environment.

Although model experiments on the computer are in terms of specific numerical values, applications to particular real world situations are often in terms only of principles or qualitative relationships. This is because the functional properties of the real world situation are often too numerous and difficult to measure, to deal with them in terms of precise numerical values.

It is evident from this brief description of two types of systems models that the systems approach to forest nutrition should eventually provide a rigorous and logical foundation for the study of forest nutrition in general and forest micronutrient nutrition in particular. It has been suggested here that the systems approach may be applied at the "single-plant level" of organization — for greenhouse (or growth chamber) studies — or at the "ecosystem level" for the description of mineral cycling in forest. In either case the application of the systems approach will be bound to produce marked changes in the design and planning of micronutrient experimentation upon forest plants or in forest ecosystems.

Summary and Conclusions

This paper has described the role that micronutrient elements play in plants and in forests. The paper has also been concerned with the future application of mathematical methods to nutrition. Fig. 4 summarizes the relationships beween different types of scientific thinking which have contributed to the evolution of the study of forest nutrition. The following conclusions supplement the information summarized on this diagram:

1. The aim of forest nutrition study is to discover general scientific principles (or patterns) which can explain the past, present, or future nutrient requirements of any

forest. Classical investigations concerning the macronutrients in forests (of the type summarised by Rennie, 1955), failed to produce rigorous, logical, explanations for the mechanism of the nutrition of forests which could be applied generally to forests and did not include any reference to the micronutrients.

2. The study of minor elements in forest plants (both micronutrients and non-essential elements) began largely as a result of:

a) parallel studies in plant physiology, agriculture, horticulture and the growth of cultivated fruit trees (which included the development of the statistical-biometric approach for the examination of empirical information obtained during fertilizer trials), and

b) general geochemical and biogeochemical investigations in forests which described the distribution of minor elements in forest ecosystems including the use of radioactive tracers.

3. Recently, the development of the systems approach to ecology has shown promise of uniting these two approaches to the study of forest nutrition. For example,

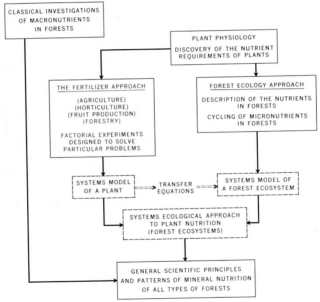

Fig. 4. A flow diagram showing the general evolution of forest nutritional research

in theory, detailed information obtained from growing a plant under controlled conditions can be used for setting up transfer equations which may later be used in a systems model of a forest ecosystem. This beginning should, in time, provide a rigorous logical foundation for the study of *all* chemical elements (including the micronutrients) in forests of all kinds, thus achieving the general objective of the classical investigators (Fig. 4).

The systems approach is now just beginning to be considered in relation to forest nutrition. For this reason there will be a temptation in some quarters to underestimate, or to overestimate, its importance. On the one hand, there are a rapidly growing number of ecologists who are enthusiastic and optimistic over the powerful stimulus which the systems approach is giving to forest ecology in general and forest nutrition in particular; and on the other hand there are a large number of ecologists who consider that novelty is not always synonymous with progress, and require a clear demonstration of the effectiveness of the systems approach in relation to nutrition before accepting the new approach. Surely what is needed by both sides is tolerance and the gradual development of a meaningful dialogue between them.

Acknowledgements

We thank Dr. P. J. RENNIE and Dr. A. CARLISLE who read the preliminary draft of the paper and offered useful suggestions.

References

ADDOMS, R. M.: Nutritional studies on Loblolly Pine. Plant Physiol. **12**, 199 (1937).

ASHBY, W. R.: Introduction to cybernetics. London: Chapman and Hall 1956.

BAKUZIS, E. V.: Forest Synecology. (Lecture notes) School of Forestry, University of Minnesota 1968.

BARNES, R. L.: The study of mineral requirements of excised tree roots, pp. 39—45. In: Mineral nutrition of trees. Duke University, School of Forestry Bull. 15 (1959).

BARROWS, H. L.: Evaluating the micronutrient requirements of trees, pp. 18—31. In: Mineral nutrition of trees. Duke University, School of Forestry Bull. 15 (1959).

BELLMAN, R. E.: Mathematical experimentation and biological research. Federation Proc. **21**, 109 (1962).

—, H. KAGIWADA, and R. KALABA: Inverse problems in biology, engineering and physics. New York: American Elsevier (in preparation).

BENZIAN, B., and R. G. WARREN: Copper deficiency in Sitka Spruce. Nature **178**, 864 (1956).

— Experiments on nutrition problems in forest nurseries. H. M. S. O. London, Forestry Comm. Bull. 1 (1965).

BERMAN, M.: A postulate to aid in model building. J. Theoret. Biol. **4**, 229—236 (1963a).

— Formulation and testing of models. Ann. N. Y. Acad. Sci. **108**, 182—194 (1963b).

BIRKHOFF, G., and G. ROTA: Ordinary differential equations. Boston: Ginn and Co. 1962.

BOLLARD, E. G., and G. W. BUTLER: Mineral nutrition of plants. Ann. Rev. Plant Physiol. **17**, 77—112 (1966).

BONNER, J., and J. E. VARNER: Plant biochemistry. New York: Academic Press 1965.

BORMANN, F. H., G. E. LIKENS, D. W. FISHER, and R. S. PIERCE: Nutrient loss accelerated by clear-cutting of a forest ecosystem, pp. 187—193. In: Symposium on primary productivity and mineral cycling in natural ecosystems (YOUNG, H. E., Ed.). Orono, Maine: Univ. of Maine Press 1968.

BOULD, C.: Leaf analysis of deciduous fruits, pp. 651—688. In: Nutrition of fruit crops, (CHILDERS, N. F., Ed.). New Brunswick, N. J.: Hort. Pub. Rutgers Univ. Press 1966.

BOWEN, H. J. M.: Trace elements in biochemistry. New York: Academic Press 1966.

BROWN, A. H. F., A. CARLISLE, and E. J. WHITE: Nutrient deficiencies of Scots pine (*Pinus sylvestris* L.) on peat at 1800 feet in the Northern Pennines. Commun. For. Rev. **1964**, 43.

BROEN, J. C.: Interactions involving nutrient elements. Ann. Rev. Plant Physiol. **14**, 93—106 (1963).

CAIN, J. C.: Plant tissue analysis. I. Some factors in sampling and analyses for the diagnosis of nutritional status of trees. II. Obervations on antigonistic effects in leaf analysis, pp. 55—70. In: Mineral nutrition of trees. Duke University School of Forestry, Bulletin **15** (1959).

CANNON, H. L.: Botanical prospecting for ore deposits. Science **132**, 591 (1960).

CHAPMAN, H. D.: Diagnostic criteria for plants and soils. Univ. of Calif. Div. of Agric. Sci. 1966.

13*

Childers, N. F.: Nutrition in friut crops. New Brunswick, N. J.: Hort. Pub. Rutgers Univ. 1906.

Cleland, W. W.: The kinetics of enzyme-catalyzed reactions with two or more substrates or products. I. Nomenclature and rate equations. Biochim. Biophys. Acta **67**, 104—137 (1963).

Collander, H.: Selective absorption of cations by higher plants. Plant Physiol. **16**, 691—720 (1941).

El Kholi, A. F.: An experimental study of the influence of microelements on the uptake of macroelements by plants. Centre for agricultural publications and documentation Wageningen 1961.

Ellis, E. I.: The spectrochemical analysis of *Abies Grandis* (Dougl.) Lindl. with particular reference to decay by *Echinodontium Tinctorium* (Ellis) E. and E. Unpublished Ph. D. Thesis, University of Washington, 1956.

Epstein, E.: Mineral metabolism, pp. 438—461. In: Plant biochemistry (Bonner, J., and J. E. Varner, Eds.). New York: Academic Press 1965.

Fisher, R. A.: Statistical methods for research workers, Edinburgh: 13th Ed. Oliver and Boyd 1958.

Fuehring, H. D.: Interrelationships of the trace elements zinc, boron, manganese, and copper on the growth and composition of corn. Ph. D. Thesis. University of Nebraska 1960. Univ. Microfilms Ind. Ann Arbor, Michigan 1962.

Gauch, H. C.: Mineral nutrition of plants. Ann. Rev. Plant Physiol. **8**, 31—64 (1957).

Gerloff, G. C.: Comparative nutrition of plants. Ann. Rev. Plant Physiol. **14**, 107—124 (1963).

Goldschmidt, V. M.: Geochemistry. Oxford: Clarendon Press 1954.

Guha, M. M., and R. L. Mitchell: Trace and major element composition of the leaves of some deciduous trees. II. Seasonal changes. Plant and Soil **24**, 90—113 (1966).

Hacskaylo, J.: Deficiency symptoms in forest trees. 7th Internat. Congr. Soil Sci., Madison, Wisc. Trans. III, 1960.

Hewitt, E. J.: The use of sand and water culture methods in the study of plant nutrition. 2nd. Ed. East Malling, Kent: Commun. Bur. Hort 1966.

— A physiological approach to the study of forest tree nutrition. Suppl. to Forestry, Physiology in Forestry. London: Oxford Univ. Press 1967.

Hillier, F. S., and G. J. Lieberman: Introduction to operations research. San Francisco: Holden Day 1967.

Holling, C. S.: The fundamental response of predators to prey density and its role in mimicry and population regulation. Mem. Entomol. Soc. Can. **45**, 1—60 (1965).

Hufschmidt, M. M., and M. B. Fiering: Simulation techniques for design of water resource systems. Cambridge/Mass.: Harvard Univ. Press. 1966.

Ingestad, T.: Studies on the nutrition of forest tree seedlings. I. Mineral nutrition of birch. Physiol. Plantarum **10**, 418—439 (1957).

— Studies on the nutrition of forest tree seedlings. II. Mineral nutrition of spruce. Physiol. Plantarum **12**, 568—593 (1959).

— Studies on the nutrition of forest tree seedlings. III. Mineral nutrition of pine. Physiol. Plantarum **13**, 513—533 (1960).

— Macro element nutrition of pine, spruce, and birch seedlings in nutrient solutions. Medd. Statens Skogsforskningsinst. **51**, 1—150 (1962).

—, and A. Jacobson: Boron and manganese nutrition of birch seedlings in nutrient solutions. Medd. Statens Skogsforskningsinst. **51**, 1—20 (1962).

Jenny, H.: Pathways of ions from soil into root according to diffusion models. Plant and Soil **25**, 265—289 (1966).

Keay, J.: Nutrient deficiencies in conifers. Scottish Forestry **18**, 22—29 (1964).

Kessel, S. L., and T. N. Stoate: Plant nutrients and pine growth. Australian Forestry **1**, 4 (1936).

Krajina, V. J.: Ecology of forests of the Pacific Northwest for 1957. Appendix B, 1958.

— Natl. Res. Coun. Res. Rep. Univ. B. C. ibid for 1958 Appendix A, 1959.

Lal, K. N., and M. S. S. Rao: Micro-element nutrition of plants. Banaras: Banaras Hindu University Press 1954.

LEYTON, L.: The growth and nutrition of spruce and pine in heathland plantations. Inst. Pap. Imp. For. Inst. Oxf. **31**, 109 (1954).

LOUNAMAA, J.: Trace elements in plants growing wild on different rocks in Finland. Ann. Botan. Soc. Zool. Botan. Fennicæ Vanamo **29**, 1—196 (1956).

— Trace elements in trees and shrubs growing on different rocks in Finland, pp. 287—317. In: Geochemical prospecting in Fennoscandia (KVAUHEIM, A., Ed.). New York: Interscience Publ. 1967.

MALYUGA, D. P.: Biogeochemical methods of prospecting. New York: Consultants Bureau 1964.

MCELROY, W. D., and A. NASON: Mechanism of action of micronutrient elements in enzyme systems. Ann. Rev. Plant Physiol. **5**, 1—30 (1954).

MCILRATH, W. J., and B. P. PALSER: Responses of tomato, turnip, and cotton to variations in boron nutrition. I. Physiological responses. Botan. Gaz. **118**, 43—52 (1956).

MILSUM, J.: Biological control theory. New York: McGraw Hill 1966.

MITCHELL, H. L.: The growth and nutrition of White Pine (*Pinus strobus* L.) Seedlings in cultures with varying nitrogen, phosphorus, potassium and calcium. Black Rock Forest Bull. 9. New York: Cornwall-on-the Hudson 1939.

MIZE, J. H., and J. G. COX: Essentials of simulation. Englewood Cliffs: Prentice Hall 1968.

MORRIS, R. F. (Ed.): The dynamics of epidemic spruce budworm populations. Mem. Entomol. Soc. Can. **31**, 1—325 (1963).

MULDER, E. G.: Importance of copper and molybdenum in the nutrition of higher plants and microorganisms, pp. 41—52. In: Trace elements in plant physiology. Chronica Botanica Co. Waltham Mass. USA.: Intern. Union of Biol. Sci. 1950.

NASON, A., and W. D. MCELROY: Modes of action of the essential elements. Plant Physiol. **3**, 451—536 (1963).

— The metabolic role of vanadium and molybdenum in plants and animals, pp. 269—296. In: Trace elements (LAMB, C. A., O. G. BENTLEY, and J. M. BEATTIE, Eds.). New York: Academic Press 1958.

NAYLOR, T. H. et al.: Computer simulation techniques. New York: Wiley 1966.

NEWBOULD, P. J.: Methods for estimating the primary production of forests. Oxford: Blackwell 1967.

NICHOLAS, D. J. D.: Minor mineral nutrients. Ann. Rev. Plant. Physiol. **12**, 63—90 (1961).

OLSON, J. S.: Analog computer models for movement of nuclides through ecosystems, pp. 121—125. In: Radioecology. (SCHULTZ, V., and A. W. KLEMENT, Eds.). New York: Reinhold Publ. Corp. 1963.

— Equations for cesium transfer in a *Liriodendron* forest. Health Phys. **2**, 1385—1392 (1965).

OLSEN, C.: The mineral, nitrogen, and sugar content of beach leaves and beech leaf sap at various times. Comp. Rend. Trav. Lab. Carlsberg Ser. Chem. **26**, 197—230 (1949).

OUELETTE, G. J., and L. DESSUREAUX: Chemical composition of Alfalfa as related to degree of tolerance to manganese and aluminium. Can. J. Plant Sci. **38**, 206—214 (1958).

OVINGTON, J. D.: Studies of the development of woodland conditions under different trees. III. The ground flora. J. Ecol. **43**, 1—21 (1955).

— Studies of the development of woodland conditions under different trees. V. The mineral composition of the ground flora. J. Ecol. **44**, 597—604 (1956).

— Quantitative ecology and the woodland ecosystem concept. Advan. Ecol. Res. **1**, 103—192 (1962).

PETERSON, P. J.: Mineral nutrition of *Agathis australis* Salisb. The Kauri effects of deficiencies of essential elements on the growth and foliar mineral composition of seedlings. N. Z. J. Sci. Technol. **5**, 141—161 (1962).

PRÉVOT, P., and M. OLLAGNIER: Law of minimum and balanced mineral nutrition, pp. 257—277. In: Plant analysis and fertilizer problems. (RAUTHER, W., Ed.). Washington, D. C.: Am. Inst. Biol. Sci. 1961.

QUASTLER, H.: Emergence of biological organization. New Haven: Yale Univ. Press 1964.

RANKAMA, K., and T. G. SAHAMA: Geochemistry. Chicago: Univ. of Chicago Press 1950.

RENNIE, P. J.: The uptake of nutrients by mature forest growth. Plant and Soil **7**, 49—95 (1955).

Rennie, P. J.: A forest-sampling procedure for nutrient-uptake studies. Commun. For. Rev. **45**, 119—128 (1966).

Reuther, W.: Plant analysis and fertilizer problems. Publ. 3. Washington, D. C.: Amer. Inst. Biol. Sci. 1961.

Robinson, W. O., and G. Edgington: Minor elements in plants, and some accumulator plants. Soil Sci. **60**, 15—28 (1945).

Rosenfield, I., and O. A. Beath: Selenium geobotany, biochemistry, toxicity and nutrition. London: Academic Press 1964.

Royce, W. F. et al.: Salmon gear limitations on Northern Washington waters. Univ. Wash. Publ. Fish. (N. S.) **2**, 1—123 (1963).

Russell, E. W.: Soil conditions and plant growth. London: Longmans 1961.

Scott, E. G.: Effect of supra optimal boron levels on respiration and carbohydrate metabolism of *Helianthus annuus*. Plant Physiol. **35**, 653—661 (1960).

Schütte, K. H.: The biology of trace elements. Their role in nutrition. Philadelphia: J. B. Lippincott Comp. 1964.

Smith, M. E.: Micronutrients essential for the growth of *Pinus radiata*. Australian Forestry **7**, 22 (1943).

Steward, F. C.: Plant physiology. III. Inorganic nutrition of plants. New York: Academic Press 1963.

Stiles, W.: B) Essential micro-(trace) elements. Encyolopedia of plant physiol. IV. Mineral nutrition of plants. Berlin-Göttingen-Heidelberg: Springer 1958a.

— C) Other elements. Encyclopedia of plant physiol. IV. Mineral nutrition of plants. Berlin-Göttingen-Heidelberg: Springer 1958b.

Stone, E. L.: Micronutrient nutrition of forest trees. A review, pp. 132—175. In: Forest fertilization theory and practice. Published by Tennessee Valley Authority National Fertilizer Development Center, Muscle Schoals, Alabama, 1968.

Tamm, C. O.: Nutrient requirements of forest stands, pp. 115—170. In: Int. Rev. For. Res. 1 (Romberger, J. A., and P. Mikola, Eds.). (1964).

Thomas, W. A.: Cycling of calcium by dogwood trees, pp. 661—664. In: Symposium on radioecology (Nelson, D. J., and F. C. Evans, Eds.): AEC CONF 670503, 1969.

Vail, J. W., and W. E. Calton: Boron deficiency die-back in pines. Plant and Soil **14**, 393 (1961).

van Dyne, G.: Ecosystems, systems ecology, and systems ecologists. Oak Ridge National Laboratory TM Report 3957, 1—40 (1966).

Wallace, T.: The diagnosis of mineral efficiencies in plants by visual symptoms. London: H. M. S. O. 1961.

Warren, H. V., R. E. Delavault, and J. Barasko: The role of arsenic as a pathfinder in biogeochemical prospecting. Econ. Geol. **59**, 381 (1964).

Watt, K. E. F.: Systems analysis in ecology. New York: Academic Press 1966.

— Ecology and resource managment. New York: McGraw Hill 1968.

Wilbrandt, W., and T. Rosenberg: The concept of carrier transport and its corollaries in pharmacology. Pharm. Rev. **13**, 109—183 (1961).

Wong, J. J., and C. S. Haines: Kinetic formulations for enzymic reactions involving two substrates. Can. J. Biochem. Physiol. **40**, 763—804 (1962).

Young, H. E., and V. P. Guinn: Chemical elements in complete mature trees of seven species in Maine. Tappi **49**, 190—197 (1966).

—, P. N. Carpenter, and R. A. Altenberger: Preliminary tables of some chemical elements in seven tree species in Maine. Maine Agr. Exp. Sta. Univ. of Maine Tech. Bull. 20 (1965).

14

Biological Cycling of Minerals in Temperate Deciduous Forests

P. Duvigneaud and S. Denaeyer-De Smet

The biological cycle of nutrients is one of the principal processes supporting the production of organic matter. Although energy flow is often useful in describing many ecosystem processes, energy is seldom a limiting factor in the productivity of forest ecosystems. The functioning of most terrestrial ecosystems is typically influenced by the availability of nutrients and water. Mineral cycling in forests is, therefore, an important parameter affecting productivity.

Since the classical data of Albert (in Dengler, 1930, cf. Fig. 1), a recent summary (Ovington, 1962) shows that essentially only even-aged plantations have been studied. A synthetic essay on mineral cycling in temperate deciduous broad-leaved forests was attempted in 1964 by Duvigneaud and Denaeyer-De Smet. Data for beechwoods and coniferous plantations, established in the last century by foresters of central Europe, have been gathered in great detail by Ehwald (1957). Remezov (1963) recently reviewed research accomplished in USSR; other details on mineral cycling in forests of USSR may be found in Sonn (1960) and Sukachev and Dylis (1964). More recently, Bazilevich and Rodin (1966) and Rodin and Bazilevich (1967) have made a panorama of the known biological cycles of minerals in the principal types of ecosystems of the world; they recognized a zonal distribution of biomasses and nutrient content and circulation, parallel to the zonation of plant formations. Research pursued in forests of the USA is summarized in the proceedings of a symposium by Young (1968); these studies include coniferous forests along the West Coast (Cole et al., 1968) and mixed deciduous-coniferous forests along the East Coast (Woodwell and Whittaker, 1968; Whittaker and Woodwell, 1969) and in the Southeast (Olson, Chapter 15). In Japan, similar investigations are being pursued in different types of forest ecosystems by a group from Kyoto University (Tsutsumi et al., 1968). Numerous studies have been made in the UK on nutrient cycling in deciduous forests (principally birch and oakwoods) by Ovington and Madgwick (1959a, 1959b), Ovington (1962), Carlisle et al. (1966, 1967), and others. Many other research programs are underway but cannot be reported here for lack of space.

In this chapter we wish to summarize the results of Belgian research on mineral cycling in deciduous forests (principally oakwoods), and to interpret these data in the context of mineral cycling in European deciduous forests. First, however, it will be necessary to review some basic concepts and terminology of the field.

Nutrient Relationships within Forest Ecosystems

The cycling of minerals in forest ecosystems occurs as a function of time; therefore, annual as well as seasonal and daily cycles can be important and may need to be examined. Nevertheless, the annual cycle has been most intensively studied in temperate deciduous forests; it is a basic measure which may be used for comparison and for study of much longer cycles, e.g., those which develop with aging of the forest (Remezov, 1958). Considering the nutrient relationships within the forest ecosystem, two major pathways of flux must be recognized: the biological "closed" cycle and the geochemical "open" cycle.

The Biological Cycle

The biological cycle involves the more or less cyclic circulation of nutrients between the forest soil and the plant and animal communities. It includes (Figs. 1 and 2) the phenomena of:

uptake or absorption (principally by roots),

retention (in the annual accretion of biomass), and

restitution or losses (leaf litter, organic debris, ground flora, washing by rainwater).

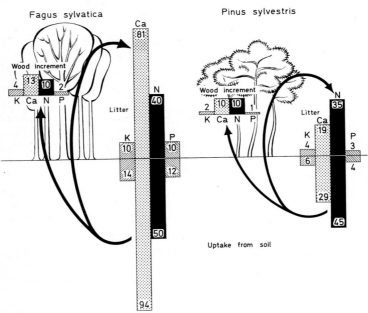

Fig. 1. Annual uptake (values below the soil line), retention (values in the tree crowns) and return (values above the soil line) of macronutrient elements (in kg/ha) by Scotch pine and European beech. Note the large fraction of nutrients returned to the soil in the form of litter — the rotating capital for forest soil fertility (after Albert, in Dengler 1930)

Retention is evaluated from data obtained from measurements of annual biomass change (annual increment) and from the chemical composition of the different organs and tissues comprising that production. Losses are evaluated similarly from the chemical composition of the different components of litter and of rainwater outside and inside the forest. Uptake, which is a means of evaluating the nutrient needs of the forest, is the sum of the two previous values:

$$\text{uptake} = \text{retention} + \text{restitution} .$$

The retained elements, added year after year to the biomass of the growing forest, accumulate in the vegetation and, at a given time, constitute the "mineralomass" (DUVIGNEAUD, 1968 b) of the community. This total content of different mineral elements in the community varies, of course, with age and forest type. Fig. 3 summarizes data gathered from OVINGTON, MADGWICK and others for different broadleaved monospecific forests in the UK (OVINGTON, 1962).

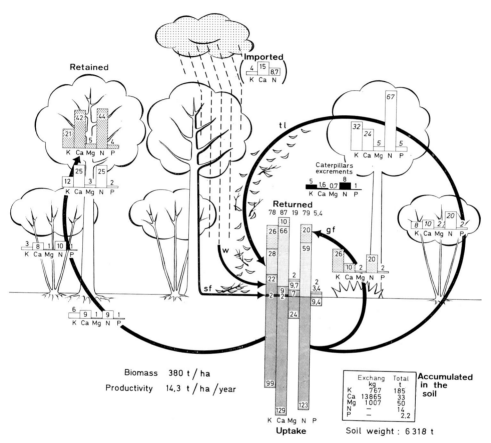

Fig. 2. Annual mineral cycling of K, Ca, Mg and N and P (in kg/ha) in a *Quercus robur — Fraxinus excelsior* forest with coppice of *Corylus avellana* and *Carpinus betulus* at Wavreille-Wève Belgium (after DUVIGNEAUD, 1968a, and DUVIGNEAUD and DENAEYER-DE SMET, 1969). Retained: in the annual wood and bark increment of roots, 1-year-old twigs, and the above-ground wood and bark increment. Returned: by tree litter (*tl*), ground flora (*gf*), washing and leaching of canopy (*w*), and stem flow (*sf*). Imported: by incident rainfall (not included). Absorbed (uptake): the sum of quantities retained and returned. Macronutrients contained in the crown leaves when fully grown (July) are shown on the right-hand side of the figure in italics; these amounts are higher (except for Ca) than those returned by leaf litter, due to reabsorption by trees and leaf-leaching. Exchangeable and total element content in the soil are expressed on air-dry soil weights of particles <2 mm)

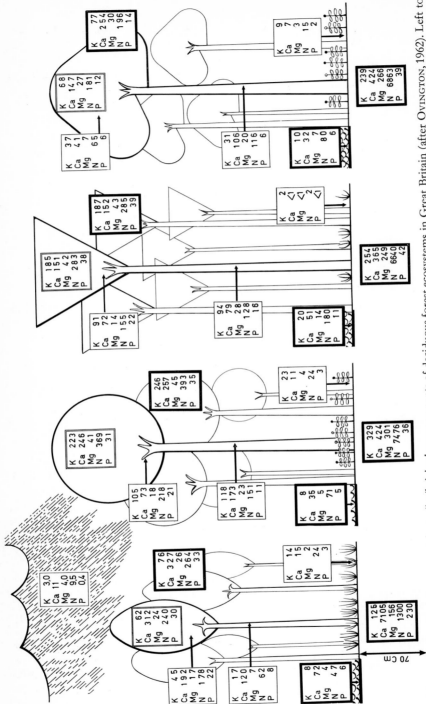

Fig. 3. Distribution of macronutrients (kg/ha) in the components of deciduous forest ecosystems in Great Britain (after Ovington, 1962). Left to right, 22-year-old birch forest, 47-year-old oak forest, 37-year-old beech forest, 47-year-old chestnut forest. Thick-framed compartments, from top to bottom, represent total nutrients in biotic (living plus dead) components of the ecosystem: litter and humus, and soil. Thin-framed compartments, from top to bottom, show respective total nutrient burden in leaves, branches and fruits, boles, ground flora. Double-framed compartments within tree crowns represent tree totals. Annual mineral input in rainfall is taken from Carlisle et al. (1966)

Transfer Pathways

Transfers of mineral elements between the different compartments of the forest ecosystem are often studied separately. Transfers from soil and roots to tree leaves canopy by wood sap are not yet well understood.

Transfers from leaves to soil by rain leaching and shedding of litter have been intensively studied by many authors, e.g., CARLISLE et al. (1966, 1967) see Fig. 4, WITTICH (1939, 1943, 1944), KORNEV (1959), SONN (1960), DENAEYER-DE SMET (1962, 1966, 1969), MINA (1965), ULRICH (1968). For a general summary of litter production in forests of the world, see BRAY and GORHAM (1964).

Transfer of elements from litter to soil is the most intensively studied part of mineral cycling in forests. Losses by litter must be followed by litter humification and mineralization, the rates of which may be very important, before the biological cycle is completed. Many such studies have been performed in European deciduous

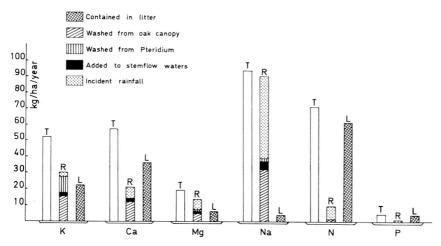

Fig. 4. Comparison of the quantities of nutrients washed from a *Quercus petraea* woodland vegetation by the rainfall with the total nutrient loss (*T*) reaching the soil surface in litter (*L*) and rainfall (*R*), (after CARLISLE et al., 1967)

forests, principally on the types and rates of nitrogen mineralization, e.g., LEMEE et al. (1958), ZÖTTL (1960a, 1960b, 1960c), ELLENBERG (1964), RUNGE (1965), VAN PRAAG and MANIL (1965), LEMEE (1967).

The Geochemical Cycle

The geochemical "open" cycle involves the input and output of mineral elements from the system.

Input of nutrients includes:

Addition by Precipitation. The annual precipitation input of minerals (kg/ha) to ecosystems in temperate regions is (OVINGTON, 1968): K, 1—10 kg; Ca, 3—19 kg; Mg, 4—11 kg; N, 0.8—4.9 kg; and P, 0.2—0.6 kg. The forest canopy is also very efficient in capturing airborne dust particles. Therefore, the action of rain is very complicated. Throughfall (precipitation collected beneath the canopy) contains elements dissolved in the incident rain plus those coming from airborne dust deposited on branches and leaves and then washed by the incident rain. Throughfall precipitation also contains recreted elements accumulated on foliage surfaces during transpiration and washed by incident rain, and elements leached by the rain from leaves and twigs. Mineral input to soil through precipitation also must include measurements of stem flow. Fig. 4 gives some data obtained in an English oakwood by Carlisle et al. (1966, 1967), which quantifies this complex relationship.

Weathering of Parent Rock. This may be a very active process even in mature deciduous forests. Klausing (1956) found erosion of bed rock of 1—2 mm a year for granite and 2.1 mm a year for diorite in German beech forests.

Input of Nitrogen. Gains of nitrogen from the atmosphere may occur through nitrogen-fixation by organisms and electrical discharges in thunderstorms.

Output of nutrients includes:

Losses in Drainage Water. Few data are available for European deciduous forest ecosystems. From lysimetric studies in coniferous forests of Finland (Viro, 1953) and of USA (Cole et al., 1968; Gessel and Cole, 1965) and from studies using small deciduous forest watersheds in the USA (Likens et al., 1967), it appears that mineral losses from forest ecosystems are not large, except for calcium loss which may reach 4—12 kg/ha/year.

Losses through Harvest. These are highly variable and depend, of course, upon the type and degree of exploitation of forest products by man.

Accumulation of Nutrients in Soil

Parallel to and partially dependent on the mineral cycles is the accumulation of nutrients in the soil (Hartmann, 1967) despite the nutrient demand for plant growth. Accumulation takes place in the rooting zone of the soil and may lead to the formation of physiologically enriched horizons. This phenomenon results from the fact that those elements which accumulate in surface litter and humus tend, by their unavailability, to slow down the biological cycle of minerals. The rate of biological cycling, however, must be maintained by uptake from lower soil horizons and the weathering of parent rock. This poorly-understood phenomenon is believed to be the basis of sustained soil fertility (Hartmann, 1967), and may be considered as a basic process in forest perpetuation. The process is not only observed in virgin and well-managed forests, but also in low-productivity forests on very poor nutrient quality parent rock. This process also serves to compensate for the nutrient loss from forest ecosystems through human exploitation (Sonn, 1960). This "active" nutrient balance may give way to a "passive" nutrient balance in case of pathological humus formation.

The accumulation of nutrients at the surface also may be observed during a succession toward the climax. In heliophilic forests such as oakwoods, surface accumulation of nutrients may enhance the development and maintenance of an important ground flora. In evergreen equatorial forests with very high rainfall, mineral leaching from soil is counteracted by storage of nutrients in the vegetation itself and very quick mineral cycling (Nye and Greenland, 1960); here, Hartmann's law is not appropriate.

A particularly important and not yet well understood phenomenon is the accumulation of nitrogen in forest soils—from 4,480 to 13,760 kg/ha in a large range of soils and oak forest types in Belgium (Duvigneaud and Froment, 1969) and from 1,905 to 15,929 kg/ha in a vast range of 44 forest soils on all types of parent rocks in Bavaria (Emberger, 1956). Relation between edaphic conditions, climate and the plant community are considered in both papers.

The degree of accumulation of nutrients of the soil surface can be indicated by the floristic composition of the ground vegetation. This flora may be divided into distinct ecosociological groups (Duvigneaud, 1946; Ellenberg, 1956); more recently, a phytogeochemical classification has been suggested by Schönhamsgruber (1959), Höhne (1962), Duvigneaud and Denaeyer-de Smet (1962). Duvigneaud and Denaeyer-de Smet (in press) have proposed, for Belgian deciduous forest ground flora, a phytochemical system based on nutrient accumulation in photosynthetic organs (e.g., N-P-K-accumulating dicots and monocots, K-accumulating dicots, Ca-nonaccumulating grasses, etc.).

Variations in Chemical Composition of Trees and Ground Flora during the Growing Season

Seasonal parameters affect various compartments of the ecosystem differently; some aspects have been studied in more detail than others: variation of the composition of tree leaves, xylem sap and ground flora following development of successive phenophases.

Seasonal variations in mineral content of tree leaves have been assessed by numerous investigations, e.g., Olsen (1948) and Tamm (1951) for beech and birch leaves, respectively. Young leaves are always richer in N, P and K and poorer in Ca than mature leaves. During active growth, N, P and K content steadily decrease but remain constant when leaves are completely developed. During autumnal yellowing,

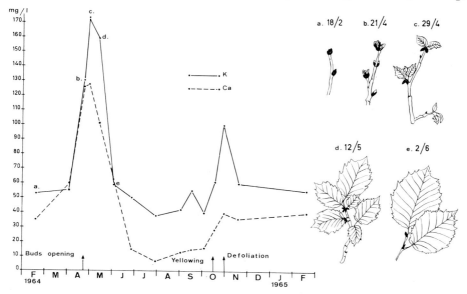

Fig. 5. Seasonal variations in the K and Ca content of the xylem sap of *Corylus avellana* (after Denaeyer-De Smet, 1967a)

N, P and K contents decrease (leaching, return to tree branches, bole and roots), but Ca content increases. Throughout leaf development, Mg level remains constant.

Denaeyer-De Smet (1967a, 1967b, 1968) has shown that in Belgian deciduous forests, K and Ca contents (and also N content, not published) of xylem sap varies during the vegetation cycle. Spring xylem sap has higher concentrations than summer and winter sap (Fig. 5); another peak in sap concentration (primarily for K) appears during autumnal yellowing and presumably downward translocation of part of the foliage nutrient pool.

The shift from vernal to estival phenophases in the ground flora of oak, birch and beech forests, may bring important changes in chemical composition of aerial parts of the herbaceous layer (Duvigneaud and Denaeyer-de Smet, 1964; Duvigneaud, 1968a, 1968b).

Influence of Community Age on the Mineral Cycle

Except for near-climax vegetation, the influence of age of the community on the biogeochemical cycle does not seem to be important. In natural birch stands (6 to 55-year-old, closed canopy and high productivity) on fens in England, the gross annual uptake of nutrients by trees, excluding roots, was relatively constant (Ovington and Madgwick 1959a, 1959b). The same conclusion may be drawn from the studies of Mina (1955) on the mineral cycle in *Quercus-Aegopodium* ("Dubrava") in the Voronezh region of the USSR; he observed nearly the same mineral uptake for oakwoods from 25 to 212 years old. For the same type of forest, however, Remezov (1959) reported a sharp peak in mineral uptake at a forest age of 48 years when productivity, and thus annual increment, was at a maximum.

Mineral Cycling in Belgian Oakwoods

Following initial research on a *Quercus-Betula* forest on very poor soil in the Brussels region (Duvigneaud and Denaeyer-De Smet, 1964), we have now studied mineral cycling in different forest types — principally oakwoods in the Walloon region. Two ecosystems, the Virelles-Blaimont forest (mixed *Quercus-Fagus*) and the Ferage forest (*Quercus-Carpinus*) constitute our maximum program (in the sense of IBP) on productivity and production processes. "Reference" ecosystems, covering a range of very different associations (Schenkler, 1939, 1960) also are being studied; some data on their mineral cycling are available in Duvigneaud and Denaeyer-De Smet (1964) and Duvigneaud and Froment (1969). Another Belgian team is studying a special compartment of mineral cycling turnover of litter and humus) in beech forests of the Ardennes (Manil, 1963; van Praag and Manil, 1965).

In the following section, we shall speak in detail about the mineral cycling (methods and results) in the Virelles-Blaimont forest; the results obtained there and in another Belgian deciduous forest (Wève-Wavreille forest) will be discussed in later sections.

Fig. 6. Organic matter biomass (per hectare) of different living and dead compartments of a mixed oak *(Quercus robur, Carpinus betulus, Fagus sylvatica)* ecosystem in Virelles, Belgium. For the aerial parts, detailed data are given for each species in kg/ha. In the ground flora, H. h. = *Hedera helix* and M. p. = *Mercurialis perennis* populations. Italicized values are informative but not included in totals. Totals are given in t/ha. For further explanation, see text

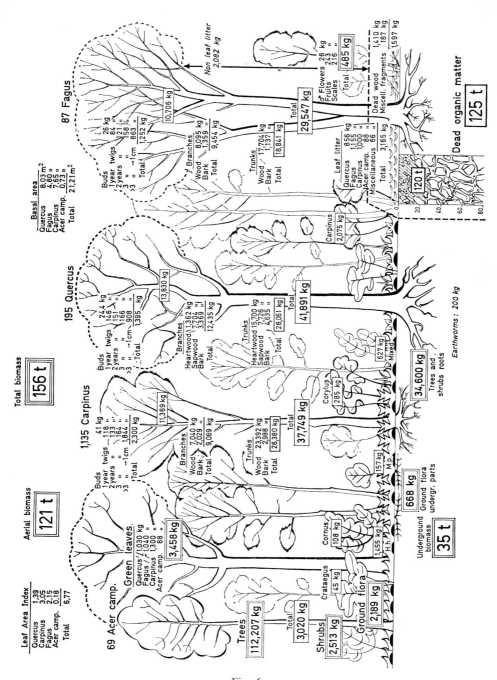

Fig. 6

The Virelles Mixed Oak Forest

Site. This is a young forest on shallow calcareous soil resulting from the reestablishment, since 1936, of a coppice forest of hornbeam (*Carpinus betulus* L.) and beech (*Fagus sylvatica* L.) dominated by older oaks (*Quercus robur* L.) and beech. Dominant trees are 70—75 years old and 15—23 m high; the younger trees, with *Carpinus* dominant, are 30 years old and 11—15 m high. Composition of the forest varies according to soil depth and human activities (Tanghe and Froment, 1967), and some zones are enriched in ash (*Fraxinus excelsior* L.), lime (*Tilia platyphyllos* Scop.) and maple (*Acer pseudoplatanus* L.). Comparative data are being collected in five of these zones; results from one stand representing average forest composition have been selected for intensive effort and will be considered here.

The forest has (1966) a density of 1,500 trees/ha, a basal area of 21.21 m²/ha and a L.A.I. (leaf area index) of 6.8 ha/ha. The image of this young, rapidly growing forest is one of a natural forest, where there are trees of all sizes and age classes (Fig. 6).

The shrub stratum (*Cornus sanguinea* L., *Carpinus betulus* L., *Corylus avellana* L., *Crataegus oxyacanthoides* Thuill., *Rhamnus cathartica* L., and *Euonymus europaeus* L.) is sparse and stunted. In springtime, a vernal stratum of *Anemone nemorosa* L., *Scilla bifolia* L. and *Narcissus pseudonarcissus* L. develops before tree foliation; at the same time, *Primula veris* L., *Cardamine pratensis* L. and *Ranunculus auricomus* L. are flowering. During the growing season, a dense ground flora develops where *Mercurialis perennis* L. dominates, with *Viola reichenbachiana* Jord., *Lamium galeobdolon* (L.) Nath, *Carex digitata* L., *Melica uniflora* Retz, *Potentilla sterilis* (L.) Garcke and a dense moss floor cover. *Hedera helix* L., an evergreen species, is abundant throughout the year.

The shallow soil resulting from decalcification of a calcareous parent rock is of the rendzinoid type, with a dark A_1 horizon extremely rich in roots (often with mycorrhiza). Chemical characteristics of the soil are given in Table 1.

Table 1. *Mineral composition of soil from the Virelles mixed oak forest.*
Mineral content and pH

Horizon	pH	Exchangeable meq/100 g			Total % air dried soil				
		Ca	Mg	K	Ca	Mg	K	N	P
A_1 (0—10 cm)	7.0	50	1.5	0.35	2.8	0.56	0.94	0.53	0.067
A/C (40 cm)	7.8	50	0.40	0.25	16	0.40	2.90	0.15	0.068

Total quantities per hectare

Horizon	Soil (metric tons/ha)	Exchangeable (kg/ha)			Total (metric tons/ha)				
		Ca	Mg	K	Ca	Mg	K	N	P
A_1 (0—10 cm)	640	6400	116	87	18	3.58	6.0	3.40	0.43
A/C (40 cm)	720	7200	35	70	115	2.88	20.8	1.08	0.49
	1360	13600	151	157	133	6.46	2.68	4.48	0.92

The principal meteorological data are:

> total radiation, about 100,000 g cal/cm²/year,
> mean annual temperature, 8.5° C,
> mean length of the vegetation period, 155 days/year,
> precipitation, about 100 cm/year with 45 cm during the vegetation period.

Biomass

It is impossible to quantify the mineral cycle in a forest ecosystem without data on the biomass and the net primary productivity. Although we are not directly concerned with these parameters in this chapter, it is necessary to briefly summarize how biomass and productivity data have been obtained.

Biomass (Fig. 6) has been established using an "allometric method" (KIRA and SHIDEI, 1967; WOODWELL and WHITTAKER, 1968). Forty-eight sampled trees of different DBH (diameter at 1.30 m height) classes species were harvested in winter, adjacent to the experimental plot and weighed separately for trunk and branches. All living branches (diameter > 1 cm) were separated from twigs (diameter < 1 cm) and separated into the following branch diameter classes: 1—3, 3—5, and 5—7 cm classes forming with the twigs the "thin wood"; 7—10, 10—15, 15—20 and 20—25 cm classes forming with the trunk the "strong wood" (German: *Derbholz*). This time-consuming process was necessary to establish the mineral content of various tree parts, since bark and wood have very different mineral contents which vary according to branch diameter. Collected materials were dried in ventilated ovens (85° C) for biomass and productivity estimation and chemical analysis. Dry weight of whole trees or parts thereof, shows good correlation with DBH (SATOO, 1966; WHITTAKER and WOODWELL, 1969); this permits regression analysis predictions of the biomass of trees in the experimental plot.

Total biomass (1966) of woody above-ground parts of the trees was 112.1 metric tons/ha. Fig. 6 gives a breakdown of the total for the different species and for the principal groups of organs ("compartments"). The current year twigs (with buds) were separated from the total twigs. Using a coefficient, different for each species, the biomass of living leaves (3.5 metric tons/ha) was estimated from the biomass of fresh leaf litter.

The biomass of stumps and thick roots was estimated as 13% of the aerial tree biomass (an empirical value used by Belgian foresters and confirmed by root extraction). Smaller roots were collected in five square meters. The total biomass of all roots was about 35 t/ha.

The biomass of the poorly developed shrub stratum was estimated to about 2.5 t/ha, using the same "dimension analysis" method as for trees.

The biomass of aerial parts of ground-stratum vegetation was estimated from plots of 1 to 3 m² surface area sampled during different seasons to be 2.2 t/ha. For each species, a relation was established between aerial and underground parts, which allowed a biomass calculation of 670 kg/ha of underground plant parts.

Summing all the previous data, and adding the biomass of flowers, fruits and bud scales (485 kg/ha), we were able to establish that the total plant biomass of the forest community was 156 t/ha.

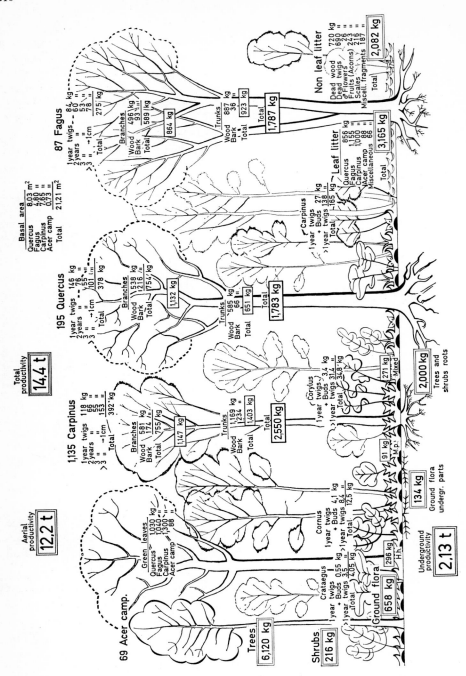

Fig. 7

Net Primary Productivity

Mean productivity of trunks and branches was estimated from the previous years' (1956—1965) increment of wood and bark from the sampled trees, from biomass and density measurements, and using dimension analysis and regression curves for the entire tree layer of the sample plot. This gave an annual tree increment of 6,120 t/ha. For aerial parts of shrubs, a mean annual increment of 216 kg/ha was obtained by dividing biomass by the mean age. The summary of these results is presented in Fig. 7. By using the ratio between aerial productivity and aerial biomass of the trees, the annual productivity of roots and stumps was established at about 2.1 t/ha, which is actually an unsatisfactory approximation. A summary of these data is presented in Fig. 7.

Thus, the annual increment which corresponds to the accretion of OVINGTON (1962) and nears the net ecosystem productivity NEP (Chapter 7) of WOODWELL and WHITTAKER (1968) amounted to 8.4 t/ha. Net primary productivity (Table 2) is the sum of the annual increment, of parts consumed by herbivores (not measured in the present study), of fallen organs and tissues forming the litter (collected in litter traps) and of the ground flora.

Table 2. Net primary production of the Virelles mixed-oak forest

Component	metric tons/ha
Annual increment	8.40
Ground flora (above ground)	0.66
Ground flora (underground)	0.13
Leaf litter	3.16
Other organic debris (non-leaf litter)	2.08
Net primary production	14.43 t/ha

Mineral Cycling

All organs of all plants collected for biomass determinations were analyzed for their contents of K, Ca, Mg, N, P and S. Potassium and Ca were analyzed by flame photometry (Eppendorff photometer) after dry ashing (450° C) and ash solution in HNO_3. Magnesium was analyzed by complexometry (titriplex III) and P by phospho-molybdate colorimetry at 710 $m\mu$. Sulfur was analyzed by gravimetry of $BaSO_4$, after special ashing of the dry sample mixed with a saturated $Mg(NO_3)_2$ solution. Nitrogen was analyzed by NH_3 distillation (PARNAS-WAGNER) after wet mineralization with conc. H_2SO_4, $(CuSO_4+K_2SO_4)$, and catalyzed by Se. Results are expressed as percentage of dry weight. For the leaf litter, it was necessary to use very fresh fallen

Fig. 7. Annual primary productivity (per hectare) of the different compartments of the mixed-oak forest community depicted in Fig. 6. For the aerial parts, data are presented as kg dry organic matter/hectare. In the ground flora, H. h. = *Hedera helix* populations, M. p. = *Mercurialis perennis* populations and Mixed = mixture of several different species. Totals are given in metric t/ha. For further explanation see text

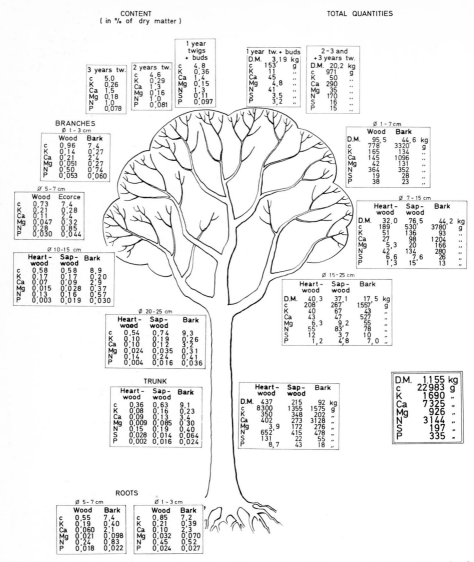

Fig. 8. Distribution of macronutrients in an 89-year-old oak *(Quercus robur)*, from the mixed oak forest of Virelles, Belgium. Left side, macronutrient and ash (c) content in percent dry matter; right side, total quantities (kg) of dry matter (D. M.), ash (c) and macronutrients in aerial parts of the tree

Fig. 9. Phytogeochemical "bisect" of the mixed-oak forest ecosystem at Virelles, Belgium. Foliar ash (c) and macronutrient content in percent of dry matter. Exchangeable K, Ca and Mg in the soil is presented in meq/100 g soil. F. s., *Fagus sylvatica*; Q. r., *Quercus robur*; C. b., *Carpinus betulus*; A. c., *Acer campestre*; T. p., *Tilia platyphyllos*; P. a., *Prunus avium*; H. h., *Hedera helix*; R. sp., *Rubus* sp.; A. n., *Anemone nemorosa*; N. p., *Narcissus pseudonarcissus*; V. r., *Viola reichenbachiana*; S. b., *Scilla bifolia*; M. p., *Mercurialis perennis*; P. c., *Polygonatum officinale*; L. g., *Lamium galeobdolon*; C. d., *Carex digitata*; C. m., *Ctenidium molluscum*

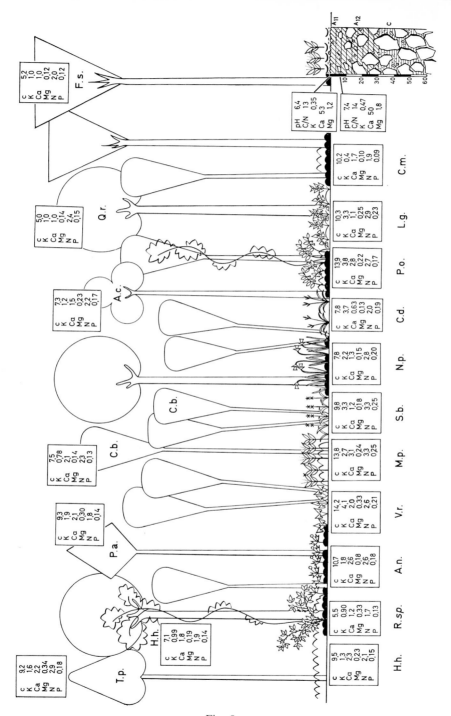

Fig. 9

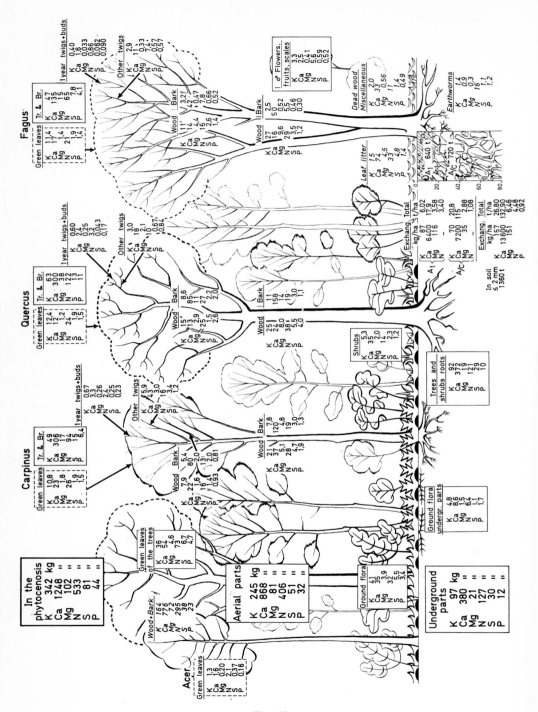

Fig. 10

leaves to avoid losses by leaching. Branches and twigs have been analyzed by diameter classes for elemental contents of bark and wood. Figure 8 is a summary of the chemical content of different parts of an oak tree, including bark, sapwood and eventually hardwood in trunk, branches and roots. Another example is given in Fig. 9 showing data on the mineral content of leaves of all trees and ground flora herbs in a graphic bisect of the ecosystem, which is a more elaborate and site-sensitive means of characterizing a plant association than a floristic listing.

It is possible, but time consuming, to establish the total chemical composition of trees of different size classes and to obtain the total quantities of elements contained in the forest community by multiplying those data times the number of trees of each class and each species. We preferred to establish the chemical composition of every size class category of trunks and branches, to multiply it by the total weight of each category and finally to make a general addition. Therefore, the mineral mass and nitrogen content of the forest community may be calculated from the nutrient content and biomass of each compartment of the system. Some details are given (Fig. 10) where the mineral content of vegetation may be compared to the total and exchangeable nutrient reserve of the soil.

The different pathways of nutrient cycling that we could evaluate are summarized in Fig. 11. The annual retention of elements in vegetation was obtained from the mean annual productivity (during the last 10 years, 1956—1965) of all species and organ categories and their respective mineral contents. The annual restitution of elements from living components of the system were calculated from summing total quantities contained in 1) the different parts of litter, 2) the aerial parts of the ground flora (nearly completely decaying every year), and 3) the leachate of the forest canopy

Table 3. *Total quantities of nutrients in the vegetation of oak forest ecosystems of Virelles and Wavreille* (Belgium)

Forest			Soil t/ha	Aerial parts kg/ha	Under-ground parts kg/ha	Total vegetation kg/ha	Yearly uptake kg/ha
Virelles		K	26.8	245	97	342	69
		Ca	133	868	380	1248	201
Biomass, 156	t/ha	Mg	6.5	81	21	102	19
Productivity, 14.4	t/ha	N	4.5	406	127	533	92
Soil weight, 1360	t/ha	S	—	51	30	81	13
		P	0.9	32	12	44	6.9
Wavreille		K	160	493	131	624	99
Biomass, 380	t/ha	Ca	33.3	1338	310	1648	129
Productivity, 14.3	t/ha	Mg	50.1	126	30	156	24
Soil weight, 6318	t/ha	N	13.8	947	313	1260	123
		P	2.2	63	32	95	9.4

Fig. 10. Distribution (kg/ha) of macronutrients in the different compartments of the mixed-oak forest at Virelles, Belgium. Tree and branch totals (Tr. and Br.) for each species or component are given in boxes. Italicized values and those contained in hatched boxes are informative but not included in totals

(elements containd in throughfall and in stem flow, minus elements contained in incident precipitation). We did not evaluate quantities of minerals released through root secretion and decay; therefore, our estimates of mineral turnover are probably slightly low.

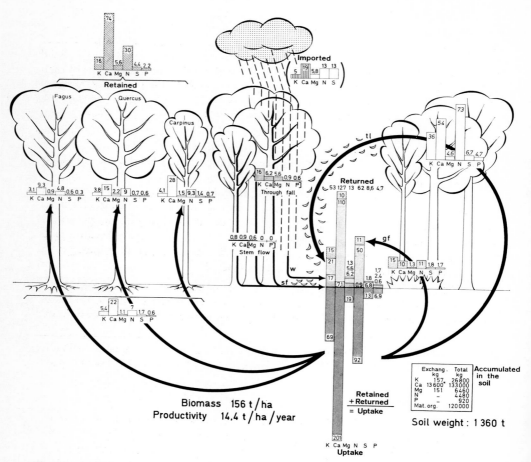

Fig. 11. Annual mineral cycling (in kg/ha) of macronutrients in the "mixed oakwood" ecosystem at Virelles, Belgium. Retained: in the annual wood and bark increment of roots and aerial parts of each species (total is hatched). Returned: by tree litter (*tl*), ground flora (*gf*), washing and leaching of the canopy (*w*), and stem flow (*sf*). Imported: by incident rainfall (not included). Macronutrients contained in the crown leaves when fully grown (July) are shown on the right-hand side of the figure in italics; these amounts are higher (except for Ca) than those returend by leaf litter. Values for Mg, N and P in throughfall and stemflow are taken from Carlisle et al. (1966). Exchangeable and total element content in the soil are expressed on air-dry soil weights of particles <2 mm

Total mineral uptake is considered as the sum of the retained and released elements (Fig. 11, Table 1). The annual elemental balance (kg/ha) shown in Table 4.

Table 4. *Annual element balance (kg/ha) of the Virelles mixed-oak forest*

	K	Ca	Mg	N	P	S
retained (increment)	16	74	5.6	30	2.2	4.4
returned (losses)	53	127	13	62	4.7	8.6
uptake	69	201	18,6	92	6.9	13

These data show that the annual nutrient requirements of an established deciduous forest with a biomass of 156 t/ha amount to only several tens of kilograms; even on a rich soil which could lead to luxury consumption, the nutrient frugality of a forest community can be demonstrated.

The Wève-Wavreille Forest

This forest is a mature oak-ash forest (*Quercus robur* L., *Fraxinus excelsior* L.) with a continuous coppice of hazel (*Corylus avellana* L.) and hornbeam (*Carpinus betulus* L.) Oaks are 115—160 years old and about 24 m in height. The coppice, which is 7—10 m in height, has a rotation period of 22 years. Tree density is about 110 oak and 2 ash per ha and basal area is 23.6 m²/ha. The ground flora, 20—30 cm in height, is characterized in spring by a dense geophytic cover (*Narcissus pseudonarcissus* L., *Ornithogalum pyrenaicum* L., *Anemone nemorosa* L., *Ranunculus ficaria* L.) and in sumer by *Geum rivale* L. *Lamium galeobdolon* (L) Nath. several species of *Rubus*.

The deep and rich soil derived from shale contains calcareous nodules and is of a pseudogley type. Because the forest is established in a large depression, the soil water conditions are favorable, except in very humid (waterlogging) or very dry seasons.

Biomass, productivity and mineral cycling have been studied in the same way as in the Virelles forest. Results are summarized in Fig. 2. More detailed data are available in DUVIGNEAUD (1968a) and DUVIGNEAUD and DENAEYER-DE SMET (1968).

Mineral Cycling in Temperate Deciduous Forests

The cycles of K, Mg, N and P in two oakwood forests (on rich soils) in Belgium (Figs. 2 and 11, Table 3) are of the same order of magnitude, the Ca cycle being on the contrary much different. Nevertheless, clear differences, principally in the uptake of K, Ca and N permit us to distinguish between the two ecosystems, although their net primary productivity is nearly the same (about 14 t/ha/year). The Wavreille *Fraxinus-Quercus* forest (Fig. 2) was mature and, consequently, had a low net relative productivity of 4% ((net productivity/biomass) × 100). Established on a deep, mineral-rich soil, the Wavreille forest had a luxury mineral cycle of K, Ca and N. The Virelle mixed oak forest (Fig. 11) was young and had a high net relative productivity (10%). This forest, however, developed on a shallow, rocky, calcareous soil was relatively frugal in the turnover of K and N, but had luxurious consumption of Ca (201 kg Ca/yr at Virelles compared to 129 kg Ca/yr at Wavreille circulate in the biological cycle of the ecosystem).

These data show that subtypes may exist within a given type of ecosystem, which are conditioned by differences in edaphic chemical characteristics. Comparable to the

Table 5. *Comparison of the macronutrient cycle (kg/ha/year) in several types of temperate forest ecosystems*

Country	Forest type (Association)	Age (years)		K (kg/ha/yr)		
				Restitution[a]	Retention[b]	Uptake
Belgium[c]	*Querceto-fraxinerum*	140	LF	45	21	66
			LT	54	21	75
			LTP	78	21	99
Belgium[d]	*Quercetum mixtum*	30—75	LF	30	16	46
			LT	35	16	51
			LTP	53	16	69
USSR[e]	*Querecto-aegopodietum*	48	LT	46	27	73
USSR[f]	*Querceto-aegopodietum*	55		62	23	85
Germany[g]	*Fagetum* bonity I	—	LF	8.4	8.8	17.2
Germany[h]	*Fagetum* on diorite	115	LF	4.0	4.5	8.5
	Fagetum on granite	125	LF	2.2	2.5	4.7
UK[i]	*Betuletum verrucosae* on fen		LT	25	3	28
USSR[j]	*Picea* forest on gley	120	LF	17	3	20
Germany[g]	*Picea* forest on gley, Bonity I	—	LF	5	6.7	11.7
Germany[g]	*Pinus sylvestris* forest, Bonity I	—	LF	4.4	2.4	6.8

[a] Restitution is equivalent to losses of other authors.
[b] Retention represents that amount in the increment.
[c] Duvigneaud, 1968a.
[d] Duvigneaud et al., 1969.
[e] Remezov et al., 1959.

two Belgian oak forests described here, we may give the example of two German beech forests described by Klausing (1956) on granite and diorite-derived soils (Table 5). The beech forest on diorite takes up, retains and releases about twice the comparable quantities of nutrients as move in the biological mineral cycle of the beech forest on granite. On the other hand, the deciduous oakwood ("Dubrava") *Quercus-Aegopodium* or *Quercus-Carex pilosa* forests) described from the Voronezh region of the USSR (Remezov et al., 1959; Remezov, 1958, 1959, 1963; Mina, 1955) have a mineral biological cycle very similar* to those of Belgium (Table 5).

It is conceivable that the temperate deciduous oak forest may be an ecosystem characterized by its nutrient cycle. Admitting enough light at ground level, this ecosystem favors the development of a dense herbaceous layer, which plays an important role in the biological cycle of nutrients and, consequently, in the processes of soil formation (Neshatayev et al., 1966).

---

* Certain differences in mineral cycles may be due to the fact that there is a strong leaching of fallen leaves and even dead leaves before falling. The exact period of leaf litter collection is therefore very important for analyses of K and N cycles.

(LF, using leaf litter; LT, using total litter; LTP, using total litter plus rainfall restitution)

Ca (kg/ha/yr)			Mg (kg/ha/yr)			N (kg/ha/yr)			P (kg/ha/yr)		
Restitution	Retention	Uptake	Restitution	Retention	Uptake	Restitution	Retention	Uptake	Restitution	Retention	Uptake
52	42	94	9	5	14	55	44	99	4	4	8
76	42	118	12	5	17	79	44	123	5.4	4.0	9.4
87	42	129	19	5	24	79	44	123	5.4	4.0	9.4
84	74	158	6	6	12	44	30	74	3.1	2.2	5.3
120	74	194	7	6	13	61	30	91	4.1	2.2	6.3
127	74	201	13	6	19	62	30	92	4.7	2.2	6.9
81	20	101	12	1	13	48	56	104	12	3	15
86	16	102	13	3	16	59	33	92	3	4	7
69	26	95	0.8	2.5	3.3	29	16	45	3.5	1.2	4.7
23.4	49	72.4	4.5	5.3	9.8	—	—	—	3.3	6.1	9.4
10.7	23.8	24.5	2.1	2.5	4.6	—	—	—	2.8	2.8	5.6
34	10	44	4.7	0.9	5.6	48	8	56	3.6	0.5	4.1
46	6	52	6	1	7	54	8	62	2.4	0.3	2.7
64	22.5	86.5	0.6	2.2	2.8	40	20.6	60.6	3.3	1.8	5.1
35.6	9.0	44.6	0.7	1.2	1.9	22.2	12.1	34.3	2.1	0.9	3.0

[f] MINA, 1955 (cited by OVINGTON, 1968).
[g] EHWALD, 1957.
[h] KLAUSING, 1956.
[i] OVINGTON and MADGWICK, 1959a, 1959b.
[j] PARSHEVNIKOV, 1962.

A comparison of oak forests (Table 2) with the low deciduous oak forest (mixed with pine) on poor soil on Long Island, USA (WOODWELL and WHITTAKER, 1968) shows the importance of edaphic conditions. In this forest, uptake (kg/ha/yr) is 37.5 kg K, 51.5 kg Ca and 10.1 kg Mg which is comparable to European deciduous forests, but with lower values because of the poor soil. The nutrient requirements of deciduous oak ecosystems seem to be much greater, especially for K and N, than those of other forest ecosystems in temperate Europe (Table 5), mainly beech, spruce and pine forests. Nutrient requirements of oak ecosystems approach those of man-made (cultivated) ecosystems of agricultural regions. The mineral cycles of different crops have been evaluated for Germany by EHWALD (1957) (Fig. 12). A comparison between deciduous European beech and oak forests shows that the beech forests (Fig. 1, Table 5) have a K cycle characterized by a reduced uptake of this element, due to an equally low retention and restitution.

Low retention may be explained by a low wood and bark content of K compared to other tree species. Low restitution seems to be due to reduced twig and branch litter (the difference in Table 5 that one might otherwise think was due to the fact that for beech forests only leaf litter (LF) was measured), and principally to the lack of a herbaceous ground layer under the dense beech canopy (herbaceous plant leaves are much higher in K than leaves and other organs of trees, and are important in oakwoods where they grow in great quantities). Nevertheless, leaching of K from leaves by rainwater seems a more important loss in beech forests than in oak forests. Another salient difference between beech and oak forests, and also non-*Pinus* coniferous forests, is the Mg frugality which beech and pine forests share (Table 5).

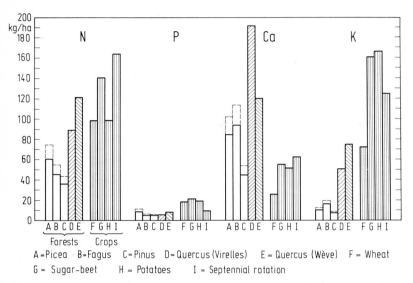

Fig. 12. Comparative annual uptake of N, P, Ca and K (not including restitution by rainfall) in forests and agricultural crops of central and western Europe. Mean uptake values for stands (bonity I) of *Picea*, *Fagus*, *Pinus* and high-yielding crops in Germany are after Ehwald, 1957. Data from Belgian oakwoods (Virelles and Weve) are after Duvigneaud and Denaeyer-De Smet; these data show an intermediate position for N and K and high Ca values due to accumulation in *Quercus* bark. Ehwald's values for German forests take account of leaf litter only; added to these (in broken lines) is the non-leaf litter restitution, evaluated from Belgian oak forests. This permits more exact comparison of data

Birch forests described from English fens by Ovington and Madgwick (1959a, 1959b) occupy an intermediate position between oak and beech forests, and seem to be characterized by very low annual retention and a very high annual restitution of nutrients — especially K and N — partially related to the presence of an important ground flora.

From literature data, Rennie (1955) has extrapolated until the age of 100 years the K, Ca and P content (mineral mass) of three major types of European forest ecosystems:

a) deciduous forests,
b) non-pine coniferous forests, and
c) pine forests.

The data for deciduous forests agree quite well with those for Belgian oak forests and are summarized (bold type) in phytogeochemical bisects in Fig. 13. These data show that regardless of the forest type, Ca is always accumulated in the largest quantities; but one hectare of a 100-year-old deciduous forest generally contains four times more Ca than one hectare of a similar age pine forest and twice as much Ca as a comparable non-pine coniferous forest. The quantities of K and P in a one hectare deciduous forest are nearly the same as in a comparable non-pine coniferous forest; pine forests are much more frugal for K and P. The same situation is found if one considers the total mineral uptake over the 100-year period, which means that losses by human exportation have been added (Fig. 13, numbers in italics).

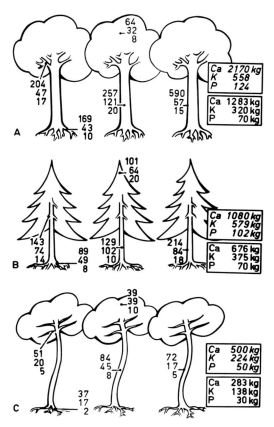

Fig. 13. Distribution (kg/ha) of Ca, K and P in three major types of exploited forests. Total content in standing crop of 100-year-old forests (normal type) and total uptake (including output by thinnings) of the same elements over the century (italics) are taken from RENNIE (1955). Values in boxes are for the total tree stand; various subtotals for leaves, branches, bark, wood and roots are shown in the figures. A, deciduous hardwoods; B, conifers other than pines; C, pines

The similarity of data obtained by simple but comparable methods by many different investigators (Table 5) permits us to consider a forest ecosystem such as the temperate deciduous oak forest as a structured organism. We can consider its peculiar type of nutrition and the transfers of energy, water, minerals and organic substances among its various components. This complex metabolism at the scale of the forest is no more difficult to study than the metabolism of a cell. The complex forest ecosystem functions adequately with a relatively low demand for nutrients for its biotic components and does not deplete or rarely exhaust the large reserve of mineral elements in the soil.

Acknowledgements

We thank Dr. Reichle warmly for reviewing the Englisch translation of this paper.

References

Bazilevich, N. I., and L. E. Rodin: The biological cycle of nitrogen and ash elements in plant communities of the tropical and sub-tropical zones. Forestry Abstr. **27**, 357—368 (1966).

Bray, J. R., and E. Gorham: Litter production in forests of the world. Advan. Ecol. Res. **2**, 101—157 (1964).

Carlisle, A., A. H. F. Brown, and E. J. White: The organic matter and nutrient elements in the precipitation beneath a sessile oak canopy. J. Ecol. **54**, 87—98 (1966).

— A. Brown, and E. White: The nutrient content of tree stem flow and ground flora litter and leachates in a sessile oak *(Quercus petraea)* woodland. J. Ecol. **55**, 615—627 (1967).

Cole, D. W., S. P. Gessel, and S. F. Dice: Distribution and cycling of nitrogen, phosphorus potassium and calcium in a second-growth douglas-fir ecosystem, pp. 197—232. In: Symposium on primary productivity and mineral cycling in natural ecosystems (H. E. Young, Ed.): Orono: Univ. Maine Press 1968.

Denaeyer-De Smet, S.: Contribution à l'étude du pluviolessivage du couvert forestier. Bull. Soc. Roy. For. Belg. **94**, 285—308 (1962).

— Recherches sur l'écosystemes forêt. La chênaie mélangée calcicole de Virelles-Blaimont. Bilan annuel des apports d'éléments minéraux par les eaux de précipitation sous couvert forestier dans la forêt caducifoliée mélangée de Blaimont. Bull. Soc. Roy. Botan. Belg. **99**, 345—375 (1966).

— Recherches sur l'écosystemes forêt. La chênaie à *Galeobdolon* et *Oxalis* de Mesnil Eglise (Ferage). Contribution à l'étude chimique de la sève du bois de *Corylus avellana* L. Bull. Soc. Roy. Botan. Belg. **100**, 353—372 (1967 a).

— Recherches sur l'écosystème forêt. Forêts de Haute-Belgique. Teneurs en Potassium et Calcium des sèves du bois d'arbres et arbustes appartenant à divers écosystèmes forestiers de Haute-Belgique. Bull. Soc. Franc. Physiol. Végét. **13**, 19—32 (1967 b).

— Recherches sur l'ecosystème forêt. La chênaie mélangée calcicole de Virelles-Blaimont. Quelques données sur les teneurs en K et Ca des sèves xylémiques des racines, troncs et branches des essences forestières en fin de repos hivernal. Bull. Soc. Roy. Botan. Belg. **101**, 155—176 (1968).

— Recherches sur l'écosystème forêt. La chênaie mélangée calcicole de Virelles-Blaimont. Apports d'éléments minéraux par les eaux de précipitation, d'égouttement sous couvert forestier et d'écoulement le long des troncs (1965, 1966, 1967). Bull. Soc. Roy. Botan. Belg. **102**, 355—372 (1969).

Dengler, A.: Waldbau auf ökologischer Grundlage. Berlin: Springer 1930.

Duvigneaud, P.: La variabilité des Associations végétales. Bull. Soc. Roy. Botan. Belg. **78**, 107—134 (1946).

— La productivité primaire des écosystémes terrestres, pp. 37—92. In: Problémes de productivité biologique (Lamotte, M., and F. Bourliere, Eds.). Paris: Masson 1967.

— Recherches sur l'écosystème forêt. La Chênaie-Frênaie à Coudrier du Bois de Wève. Aperçu sur la biomasse, la productivité et le cycle des éléments biogènes. Bull. Soc. Roy. Botan. Belg. **101**, 111—127 (1968a).

— Recherches sur l'écosystème forêt. La chênaie mélangée calcicole de Virelles-Blaimont. Bisect biogéochimique et composition des nappes foliaires en polyéléments biogènes. Bul!. Soc. Roy. Botan. Belg. **101**, 129—139 (1968b).

DUVIGNEAUD, P., and S. DENAEYER-DE SMET: Distribution de certains éléments minéraux (K, Ca et N) dans les tapis végétaux naturels. Bull. Soc. Franc. Physiol. Végét. **8**, 96—103 (1962).

— — Le cycle des éléments biogènes dans l'écosystème forêt. Lejeunia, N. S. **28**, 1—143 (1964).

— — Biomass, productivity and mineral cycling in deciduous mixed forests in Belgium, pp. 167—186. In: Symposium on primary productivity and mineral cycling in natural ecosystems (H. E. YOUNG, Ed.). Orono: Univ. Maine Press 1968.

— — Phytogéochimie des groupes écosociologiques forestiers de Haute-Belgique. Ecol. Plantarum (in press).

—, and A. FROMENT: Recherches sur l'écosystème forêt. Série E: Forêts de Haute-Belgique. Eléments biogènes de l'édaphotope et Phytocénose forestière. Bull. Inst. Roy. Sci. Nat. Belg. 45, 25, 1—48 (1969).

—, S. DENAEYER-DE SMET, P. AMBROES, and J. TIMPERMAN: Recherches sur l'écosystème forêt. La chênaie mélangée calcicole de Virelles-Blaimont. Biomasse, productivité et cycle des éléments biogènes dans un écosystème "chênaie mélangée" (Querceto-carpinetum calcicolum). Essai de phytogéochimie forestière. Mem. Inst. Roy. Sci. Nat. Belg. (in press).

EHWALD, E.: Über den Nährstoffkreislauf des Waldes. Deut. Akad. Landw. Wiss. Sitz. **6**, 1—56 (1957).

ELLENBERG, H.: Aufgaben und Methoden der Vegetationskunde (I. Teil) in Grundlagen der Vegetationsgliederung (IV), pp. 1—136. In: Einführung in die Phytologie. (WALTER, H., Ed.). Stuttgart: Eugen Ulmer 1956.

— Stickstoff als Standortsfaktor. Ber. Deut. Botan. Ges. **77**, 82—92 (1964).

EMBERGER, S.: Die Stickstoffvorräte bayrischer Waldböden. Forstwiss. Cbl., **84**, 156—193 (1965).

GESELL, S. P., and D. W. COLE: Influence of removal of forest cover on movement of water and associated elements through soil. J. Amer. Water Works Assoc. **57**, 1301—1310 (1965).

HARTMANN, F.: Was zeigt uns der Wald über die Naturgesetzlichkeiten im Nährstoffhaushalt? Cbl. Ges. Forstwes. **84**, 2—6, 174—181 (1967).

HÖHNE, H.: Vergleichende Untersuchungen über Mineralstoff- und Stickstoffgehalt sowie Trockensubstanzproduktion von Waldbodenpflanzen. Arch. Forstwes. **11**, 10, 1085—1141 (1962).

KIRA, T., and T. SHIDEI: Primary production and turnover of organic matter in different forest ecosystems of the Western Pacific. Japan. J. Ecol. **17**, 70—87 (1967).

KLAUSING, O.: Untersuchungen über den Mineralumsatz in Buchenwäldern auf Granit und Diorit. Forstwiss. Cbl. **75**, 18—32 (1956).

KORNEV, V. P.: The acculumation in the leaves of trees and bushes of Phosphorus and Potassium and their leaching from the leaf fall during the period autumn-winter. Pochvovedenie **1959** (4), 87—94 (1959).

LEMEE, G.: Investigations sur la minéralisation de l'azote et son évolution annuelle dans des humus forestiers in situ. Oecologia Plantarum **2**, 285—324 (1967).

—, P. LOSSAINT, H. METTAUER, and R. WEISSBECKER: Recherches préliminaires sur les caractères bio-chimiques de l'humus dans quelques groupements forestiers de la plaine d'Alsace. Angew. Pflanzensoziol. **15**, 93—101 (1958).

LIKENS, G. E., F. H. BORMANN, N. M. JOHNSON, and R. S. PIERCE: The calcium. magnesium, potassium and sodium budgets for a small forested ecosystem. Ecology **48**, 772—785 (1967).

MANIL, G., F. DELECOUR, G. FORGET, and A. EL ATTAR: L'humus, facteur de station dans les Hêtraies acidophiles de Belgique. Bull. Inst. Agron. Sta. Rech. Gembloux **31**, 114 (1963).

MINA, V. N.: Cycle of N and ash elements in mixed oakwood of the forest steppe. Pochvovedenie **1955** (6), 32—44 (1955).

— Leaching of certains substances by precipitation from woody plants and its importance in the biological cycle. Soviet Soil Sci. **6**, 609—617 (1965).

Neshatayev, Yu. N., O. G. Rastorova, L. S. Schastnaya, I. A. Tereshenkova, and V. P. Tryplenkov: Entry into the soil of ash elements and nitrogen with leaf fall from trees and grasses in the main types of oak stands in the "Forest on the Vorskla". Soviet Soil Sci. **12**, 1372—1379 (1966).

Nye, P. H., and B. J. Greenland: The soil under shifting cultivation. Tech. Commun. Bur. Soil Sci. **51** (1960).

Olsen, C.: The mineral, nitrogen and sugar content of beach leaves and beach sap at various times. Compt. Rend. Trav. Lab. Carlsberg, Ser. Chim. **26**, 197—230 (1948).

— Quantitative ecology and the woodland ecosystem concept. Advan. Ecol. Res. **1**, 103—192 (1962).

Ovington, J. D.: Some factors affecting nutrient distribution within ecosystems. UNESCO, Natural Resources Research **5**, 95—105 (1968).

—, and H. A. I. Madgwick: The growth and composition of natural stands of birch. I. Dry matter production. Plant and Soil **10**, 271—281 (1959a).

— — The growth and composition of natural stands of birch. II. The uptake of mineral nutrients. Plant and Soil **10**, 389—400 (1959b).

Parshevnikov, A. L.: The nitrogen and ash element cycle in relation to alternation of tree species in the forests of Central Taiga. Trudy Inst. Lesa Drev. Sib. Otd. **5**, 52 (1962).

Remezov, N. P.: The relationship between biological accumulation and the eluvial provess under forest cover (according to investigations in the Voronezh National Forest). Pochvovedenie **1958** (6), 1—12 or Soviet Soil Sci. **1958**, 589—598.

— The method of studying the biological cycle of elements in forest. Pochvovedenie **1959** (1), 71—79 (1959).

— Über den biologischen Stoffkreislauf in den Wäldern des europäischen Teils der Sowjetunion. Arch. Forstwes. **12**, 1—43 (1963).

—, L. N. Bykova, and K. M. Smirnov: Uptake and cycle of nitrogen and ash elements in the Forests of European U.S.S.R., Moscow State University (MGU), 1959.

Rennie, P. J.: The uptake of nutrients by mature forest growth. Plant and Soil **7**, 49—95 (1955).

Rodin, L. E., and N. I. Bazilevich: Production and mineral cycling in terrestrial vegetation. Edinburgh-London: Oliver and Boyd 1967.

Runge, M.: Untersuchungen über die Mineralstickstoff-Nachlieferung an nordwestdeutschen Waldstandorten. Flora (Jena) **155**, 353—386 (1965).

Satoo, T.: Production and distribution of dry matter in forest ecosystems. Misc. Inform., The Tokyo Univ. Forests **16**, 1—15 (1966).

Schlenker, G.: Die natürlichen Waldgesellschaften im Laubwaldgebiet des Württenberger Unterlandes. Veröff. Württ. Landesst. Naturschutz **15**, 103—140 (1939).

— Zum Problem der Einordnung klimatischer Unterschiede in das System der Waldstandorte Baden-Württembergs. Mitt. Ver. Förtsl. Standortsk. v. Forstpfl. **9**, 3—15 (1960).

Schönamsgruber, H.: Mineralstoffuntersuchungen an Waldgesellschaften Baden-Württembergs. Ber. Deut. Botan. Ges. **72**, 220—229 (1959).

Sonn, S. W.: Der Einfluß des Waldes auf die Böden. Jena: Gustav Fischer 1960.

Sukatchev, V., and N. Dylis: Fundamentals of forest biogeocoenology (in Russian), Moscow: Publishing Office "Nauka", 1964. (English translation, Edinburgh-London: Oliver and Boyd 1968).

Tamm, C. O.: Seasonal variation in composition of birch leaves. Physiol. Plant. **4**, 461—469 (1951).

Tanghe, M., and A. Froment: La Chênaie à *Galeobdolon* et *Oxalis* de Mesnil-Eglise (Ferage) — Variabilité du tapis herbacé de la Chênaie-coudraie en fonction de caractéristiques édaphiques superficielles. Bull. Soc. Roy. Botan. Belg. **101**, 245—256 (1967).

Tsutsumi, T., T. Kawahara, and T. Shidei: The circulation of nutrients in forest ecosystems. On the amount of nutrients contained in the above-ground parts of single tree of stand. J. Japan. Forestry Soc. **50**, 66—74 (1968).

Ulrich, B.: Extent and selectivity of nutrient uptake in beech and spruce. Allg. Forstzeitschr. **23**, 815 (1968).

VAN PRAAG, H., and G. MANIL: Observations in situ sur les variations des teneurs en azote minéral dans les sols bruns acides. Ann. Inst. Pasteur **109**, Suppl. 3, 256—271 (1965).

VIRO, P. J.: Loss of nutrients and the nutrient balance of the soil in Finland. Commun. Inst. Forest. Fenn. **42**, 1—50 (1953).

WHITTAKER, R. H., and G. M. WOODWELL: Structure, production and diversity of the oak-pine forest at Brookhaven, New York. J. Appl. Ecol. **57**, 155—174 (1969).

WITTICH, W.: Untersuchungen über den Verlauf der Streuzersetzung auf einem Boden mit Mullzustand. I. Forstarchiv **15**, 96—111 (1939).

— Untersuchungen über den Verlauf der Streuzersetzung auf einem Boden mit Mullzustand. II. Forstarchiv **19**, 1—18 (1943).

— Untersuchungen über den Verlauf der Streuzersetzung auf einem Boden mit Mullzustand. IV. Die Streu der Bodenflora. Forstarchiv **20**, 110—114 (1944).

WOODWELL, G. M., and R. H. WHITTAKER: Primary production and the cation budget of the Brookhaven Forest., pp. 151—166. In: Symposium on Primary Productivity and Mineral Cycling in Natural Ecosystems (YOUNG, H. E. Ed.). Orono: Univ. Maine Press, 1968.

YOUNG, H. E. (Ed.): Symposium on primary productivity and mineral cycling in natural ecosystems. Orono: Univ. of Maine Press 1968.

ZÖTTL, H.: Methodische Untersuchungen zur Bestimmung der Mineralstickstoffnachlieferung des Waldbodens. Fortswiss. Cbl. **79**, 72—90 (1960a).

— Dynamik der Stickstoffmineralisation in organischen Waldbodenmaterial. I. Beziehung zwischen Bruttomineralisation und Nettomineralisation. Plant and Soil **13**, 166—182 (1960b).

— Dynamik der Stickstoffmineralisation in organischem Waldbodenmaterial. II. Einfluß des Stickstoffgehaltes auf die Mineralstickstoff-Nachlieferung. Plant and Soil **13**, 182—206 (1960c).

Carbon Cycles and Temperate Woodlands

J. S. Olson

The other chapters of this book have shown many ways of linking different parts of the same system to one another. Among the ways of linking the world's regional and local systems to one another, unifying considerations of the circulation of carbon (and of nitrogen) through a common atmospheric pool have been recognized as important since Dumas and Boussingault (1844; see Riley, 1944). Even now we are not sure how important lands, especially forests, can be in modifying or stabilizing these cycles, but several kinds of information and models suggest that their importance may have been underestimated in the past (Figs. 1 and 2).

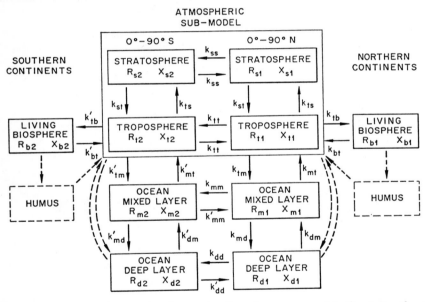

Fig. 1. A model for exchange of carbon and radiocarbon, generated in the stratosphere and redistributed through other pools of the atmosphere, to living and dead organic matter on the continents and to mixed and deep layers of the oceans. R = reservoir magnitude, X = excess over "normal levels" (e.g., due to extra CO_2 from fossil fuels; or extra ^{14}C from thermonuclear explosions beyond that normally produced from cosmic rays acting on ^{14}N. Fractional coefficients of exchange, k_{ij}, have been summarized by Nydal (1968) based on data of Fig. 4 and a review of earlier literature

Burning of fossil carbon left from ancient ecosystems seems to be increasing the CO_2 concentration in the troposphere at a rate of 0.7 ppm/year (Keeling, Harris, and Wilkins, 1968). Yet this is roughly half of the rise expected if all remained in the atmosphere (REVELLE et al., 1965). Can we expect such proportions to be continued? Or would improved mathematical models accounting for the absorption of CO_2 by the oceans and by the continental vegetation help us to predict a change in this proportion — and in the long-range effects on climate and man?

We shall see that some fairly simple mechanistic models have been sufficient to interpolate recent trends of carbon exchange, and to extrapolate a few years into the future. But a better understanding of the processes inside the sea and land ecosystems would seem essential before we can extend analytic or computer predictions to the

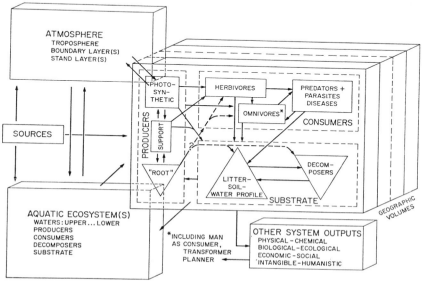

Fig. 2. An expansion of part of Fig. 1 showing interlocking of atmospheric boundary layers into a forest stand (the main pools of organic carbon and other elements and energy) within a "stack" of terrestrial ecosystems. These may exchange substance and influence between one another. Outputs of an idealized landscape contribute to aquatic systems and to variables of interest to human welfare

decades ahead. The possibility of altering the future world balance of CO_2 (and climate, sealevels and coastline habitation) means that far-reaching environmental control may be sought in technology before its consequences are foreseen. Should society hasten the inevitable shift from coal and oil to nuclear fuels for power for these reasons? What new problems are thus invited?

Even if we are presently still unprepared to translate ecosystem predictions into such sweeping recommendations about broad technological-environmental policy, it is at least reasonable to ask whether (or where) the forester already approaches management for the maximum carbon fixation into pulpwood fiber. Can he also afford to manage closely enough to assure that most photosynthate goes to the desired timber species, or at least that not too much goes to fungi or insects? Is a production-oriented policy or model compatible with other values of landscapes and waters for society?

Among those land areas fixing less carbon than the potential suggested by water balances, some forest production rates are known to be improvable by addition of

15*

N and/or P and K. Economic cost frequently makes fertilization for forests more dubious than for agriculture. Yet if forest production can be increased by redistribution or slight adjustment of nutrient pools, then models and field data on nutrient budgets (all major income and loss terms) need to be used as guidance for long-term fertilizer experiments and practices. Can increases in yield be compounded, where native fertility is low? Once established, such experiments involve long-term investment in effort and continued records; but the extra cost of analyses and planning for basic nutrient cycling research can enhance both the scientific and practical value of the work (e.g., TAMM, 1968; COLE, GESSEL, and DICE, 1967; DUVIGNEAUD and TANGHE, 1967; OVINGTON, 1962, 1965; SUKACHEV and DYLIS, 1968; WITTICH, 1953).

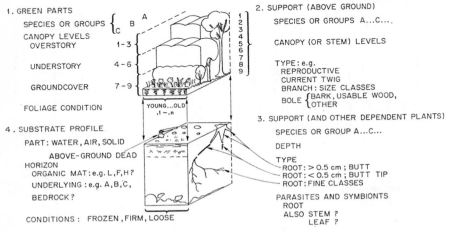

Fig. 3. An expansion of the primary producer and substrate subsystems for a typical forest ecosystem as shown in Fig. 2. Each subsystem is further subdivided into (1) unambiguous components according to species or other trophic groupings (which imply many qualitative aspects of the system), (2) vertical distribution of foliage, support, underground parts and substrate, and (3) variables selected to describe the conditions of each component in ways that are pertinent for purposes of research and model development

Even if we understand the meaning and relations of the main parts in Fig. 2, including the proper place for man, the technical treatment of modeling may demand a further partitioning of the variables in some subsystems. The degree and manner of lumping or subdivision for models follows mainly from the objectives previously discussed.

Because layer structure is so important for a carbon exchange model, canopy structure is chosen as the main basis for subdividing the photosynthetic subsystem (Fig. 3) for compatibility with models which could help distinguish how the canopy strata are adapted to light interception (MONSI, 1960, 1968), leaf area index and photosynthesis-respiration balances (BLACKMAN, 1962, 1968). Leaf age classes can be separated for deciduous dominants; more classes might be separated by year for conifers (MADGWICK, 1968; TAMM, 1968), but many models would find that subdivision less important than vertical position.

Stand Scale

MADGWICK (Chapter 5) and SATOO (Chapter 6) have reviewed the techniques and results which apply to harvest data and the several kinds of corrections which must be added to biomass change in order to estimate the annual production of organic matter. Since tree boles dominate the inventory of older forest stands and ash-poor boles have carbon contents approaching 50%, halving of the tabulated biomass data (cf. OVINGTON, 1962, 1965) approximates the typical magnitudes of carbon for many well-developed temperate forests, both deciduous and coniferous. WOODWELL and BOTKIN (Chapter 7) convert from CO_2 to an organic equivalent representing 45% carbon, a widely used nominal value*. RODIN and BAZILEVITCH (1967) have added many Soviet data hitherto available only in condensed summaries and scattered publications, and place much emphasis as well on problems of cycling of nitrogen and mineral (ash) elements. A specific illustration is given (Table 1, left) for some of the structural components of sedge-goutweed oak forest type planted in the Voronezh area 10 b in the forest steppe zone of European Russia (zone 10 b on the map inside the back cover). Organic content, even excluding nitrogen, ranges from 97.5 to 99.2 percent for woody parts and roots, and 91.3 to 94% for herbs and tree leaves, and would be expected to be high overall for most forests. A special team effort was made to account for the pool magnitude (x_i), annual uptake (judged from peak foliage-period inventory) and approximate losses from (C) herbs** and (B) minor woody species as well as (A) oak dominants (REMEZOV, SAMOYLOVA, SVIRIDOVA, and BOGASHOVA, 1964). Oaks contributed less production (or uptake) and restitution and retention than other species $(B$ and $C)$ in the 12-year-old stand, but overshadowed the others by age 48 years, probably before 20 years (REMEZOV, BYKOVA, and SMIRNOVA, 1959).

Dry mass (and hence carbon) are estimated (Table 1) to have a relative growth rate (= uptake/pool) of 0.05 for oak bole and branches, and almost as high a value for other woody species taken together in a 12-year-old stand. Presumably the equal figure for woody roots is inferred indirectly (cf. REMEZOV, RODIN and BAZILEVICH, 1968). Herbs (root and top) and deciduous foliage are shown as having a turnover of 1.0 in a yearly step, whereas conifer forests in general have ratios from 0.5 to 0.2 or lower in the northern Boreal Forest (RODIN and BAZILIVICH, 1967, 1968).

Losses from branch pruning and bole mortality of the crowded 12-year stand were slightly over half the uptake, leaving less than half the uptake as estimated retention in dry mass increment of woody parts. The annual physiological growth here, as in other forests, was maximum at a pole stage between 20 and 40 years and averaged 1600 g/m² (or metric tons/km²) dry matter — about 760 g/m² of carbon, *omitting* the ground cover. Neglecting possible losses due to animal consumption, REMEZOV et al. (1964) have carefully provided comparable woody growth estimates of 676 g/m² or (738 g/m² *including* herb tops and roots: (Table 1) for the 12-year stand, and 791 g/m² (879 g/m²

---

* WITTICH (1953) tabulates C, N and P contents for numerous woody leaf species. Grasses rich in silica and other herbs of high ash or sugar content are the types whose carbon contents fall below the 45—50% range (LIETH, 1962, 1963, 1965). WESTLAKE's 1963 summary of 46—48% carbon for "most entire plants" is exceeded mainly by fat or protein-rich seeds or animal components of low total mass per unit area.

** Incorrectly translated as "grasses" in English versions of the paper cited, and of RODIN and BAZILEVICH (1968). Centers or "hundred-weight" is misconstrued as 50 kg instead of metric 100 kg in the latter.

Table 1. *Partial organic matter budget of green plants in Voronezh oak forests*[a]
(from Remezov, Samoylova, Sviridova, and Bogashova, 1964)

Age / Live plant parts		Pool size in g/m² (= tons/km²)				% C+H +O+S	Net primary production- or uptake (NPP)	Restitution (annual)	Retention (annual)	$\frac{NPP}{pool}$
		Dry mass	N	Ash	C+H +O+S					
12-year-old stand										
Oak leaves	1A	150	3.76	5.06	141.2	94.0	150[b]	150[b]	0	1+[b]
Other lvs.-woody	1B	126	1.87	6.84	117.3	93.0	126	126	0	1+
Herbs	1C	54	1.08	3.54	49.4	91.5	54	54	0	1
Oak branches	2A1	270	1.43	4.67	263.9	97.8	13	7	6	0.05
Oak bole	2A2	1460	2.19	8.53	1449.3	99.0	72	43	29	0.05
Other branches	2B1	2441	5.74	38.25	2397.0	98.2	92	51	41	0.04
Other bole	2B2	348	0.75	3.37	345.4	99.3	17	10	7	0.05
Roots-woody	3A,B	2160	9.67	40.07	2110.3	97.6	106	60	46	0.05
Roots-herbs	3C	108	1.26	2.38	104.4	96.8	108	36	72	1+
Total		6917	27.75	112.71	6783.4	98.6	738	537	201	0.11
48-year-old stand										
Oak leaves	1A	230	6.00	7.46	216.5	94.0	230	230	0	1+
Other lvs.-woody	1B	79	1.12	5.08	72.8	92.2	79	79	0	1+
Herbs	1C	53	1.12	3.45	48.4	91.3	53	53	0	1+
Oak branches	2A1	3220	12.56	37.67	3169.8	98.5	61	20	41	0.02
Oak bole	2A2	12650	39.22	73.30	12537.5	99.2	247	78	169	0.02
Other branches	2B1	670	2.36	13.88	653.8	97.5	10	3	7	0.01
Other bole	2B2	2160	4.91	22.91	2132.8	98.8	42	13	29	0.02
Roots-woody	3A,B	6910	26.67	80.11	6803.2	98.6	122	43	79	0.02
Roots-herbs		106	1.23	2.32	102.5	96.7	35	35	0	0.33
Total		26078	95.2	246.18	25736.6	98.7	879	554	325	0.03

[a] Based on methods described by Remezov. 1958, 1959, and (in Russian) by Rodin, Remezov, and Bazilevich, 1968.
[b] Excluding not only consumption by secondary producers, but also rain leaching.

including herbs) for a 48-year stand. These values (without herbs) are equivalent to about 317 and 370 g/m² of carbon, respectively. Estimated losses of dry mass were almost the same (537 and 554 g/m² for 12 and 48-year old stands), and remained so for dry mass and element content even to 93 years (Remezov, et al., 1964, Fig. 2).

It is interesting to compare such a forest, with short growing season in a continental climate (having frequent June droughts), with the Brookhaven forest where respiration estimates are available. In the latter, production is limited by poor sandy soil, subject to days of low photosynthesis due to drought or undetermined causes. Woodwell and Botkin (Chapter 7) rely primarily on allometric relations to estimate net primary production near 1195 g/m² dry matter per year, which (for their assumed 45% carbon) is near 540 g/m² per year of carbon. Respiration of the photosynthetic system (leaves and small twigs) was not measured separately from photosynthesis, so the small chamber method approximated "surplus production" (Monsi, 1960) more closely than photosynthesis proper and gave estimates of 1550 g/m²/yr of carbon, which the authors considered very uncertain (even for the three dominant tree species that were measured).

Two alternative methods of estimating respiration of bole and branch (surface area 1.5 m² stem/m² ground) gave annual released *carbon* equivalents of 427 g/m² derived from annual curves, and 560 g/m² based on temperature (excluding shrubs < 1 m in height). Lower shrubs, total groundcover and a combination of soil and root respiration were estimated from annual curves as 470 g/m²/yr carbon.

Woodwell and Marples' (1968) estimate of 621 g of dry organic matter decayed per m² per year (about 280 g C/m²/yr) is not far different from 350 g C/m²/yr income estimated from litterfall (Woodwell and Whittaker, 1968, Table 1). If the slight difference between these loss and income estimates is the realized litter increment, then the difference 470 − 280 = 190 g C/m²/yr is a fair estimate of respiration by roots (mostly) plus total vegetation below 1 m height. Because the respiration of the photosynthetic system itself was not separately measurable, the respiration estimates of 617 or 730 g C/m²/yr are lower bounds for total green plant respiration. Adding these estimates to 560 g C/m²/yr net primary production gives 1157 or 1270 g C/m²/yr for a *part* of gross primary production. Then division by 540 g C/m²/yr indicates a rate of presumably 2.1 to 2.3 times the net primary production for *part* of true gross primary production. This is not far below the near-maximum ratio of 2.9 which was given cautiously by Woodwell and Botkin in Chapter 7 for the two dominant oaks and pitch pine (considering favorable days only).

Such provisional estimates, and related ones on the allocation of fixed carbon by Woodwell and Whittaker (1968) are subject to adjustment as technique and data improve; yet already they confirm a strikingly high throughput of carbon. Annual transfers (the integrals for one year of y_{ij} of Smith, Chapter 2, Table 1) are estimated in Table 2 (see footnotes!). The top *row* summarizes the prompt allocation of photosynthate; the second row some delayed translocation from leaf storage, and remaining transfers also are indexed by row (source) and column (destination).

The first *column* of data are the returns of carbon to the air in respiration; the second column includes presumed initiation of leaf growth from stem storage, in addition to photosynthate by each season's (mostly new) new leaves, etc. Translocation estimates probably could be increased, in view of the strong capability of these oaks and pitch pine to sprout and grow rapidly from underground reserves. Such transfers

Table 2. *Estimated cumulative annual transfers of carbon (g/m² per year)[a] in an oak-pine forest at Brookhaven National Laboratory, Long Island* (Forest Zone 10c on map inside back cover, after WOODWELL and WHITTAKER, 1968; WOODWELL and BOTKIN, Chapter 7)

| | | | | Destination | | | |
| | Pool index | 1 Atmosphere | 2 Leaves | 3 Flowers and Fruit | 4 Bole and Branch | 5 Underground Parts | 6 Substrate | 7 Consumers |
Source								
Atmosphere	1	—	166[b]	~ 0	627—720[c]	371[c]	~ 0	0
Leaves	2	?[c]	—	0	50[d,e]	0[e,f]	122	14
Flower + fruit	3	?	0	—	0	0	11	?
Bole + branch	4	27—510	20[e]	11[d]	—	0[e]	32	?
Underground parts	5	190—	?[x]	?[x]	?[x]	—	140	0
Substrate + decomposers	6	280	0	0	0	0	—	?
Consumers (aboveground)	7	12	0	0	0	0	0	—

[a] Assuming 45% C for tree and herb leaves, roots and organic substrate; 47% for stems; and 50% for flowers and fruits. Slight discrepancies may results from wrong selection of these factors, and from differences in dry matter estimates between WOODWELL and WHITTAKER, 1968, and WOODWELL and BOTKIN, Chapter 7.

[b] Leaf litter 122 + consumptions 14 + loss before abscission 14 + reproductive 11 = 186, but 20 g/m²/year of this is assumed to consist of translocation from stems: see c below.

[c] Carbon for stem and root growth, while photosynthesized in leaves is assumed to be promptly translocated to stem and root cambium and to a labile storage pool. The assumed 20 g in stems seems minimal for the spring flush of growth.

[d] Carbon for reproduction is shown as temporarily translocated into stems and then into flowers and fruit, with a probable value of 11 g shown here. (Actual amounts should vary yearly as a function of prior fruiting and weather.)

[e] Loss of 39 g carbon by leaves before abscission is here shown going to storage in stems, but might be partly due to leaf respiration (not estimated here) and to leaching loss: 50 = 39 + 11 (from d).

[f] Additional stem and root storage might be expected in order to balance upward spring translocation during normal flushing of new growth, most conspicuously during sprouting after removal of tops. Estimates are not available for these cumulative transfers.

among all possible pools (Table 2) are not displayed when sources of uptake and loss are summed, as they were in Table 1.

If such exchanges, and the great turnover of respired carbon, could be added to the dry matter changes in the Russian, Belgian and other forests summarized by DUGINGEAUD and DENAEYER-DE SMET (Chapter 14), the role of terrestrial ecosystems in the world carbon cycle would be even more impressive than most previous authors have thought (OLSON, 1964). Both inside cover mops may give low estimates.

World Scale

During the same centennial anniversary seminar at which HUTCHINSON reviewed the conceptual model of DUMAS and BOUSSINGAULT (in RILEY, 1944), these authors summarized SCHRÖDER's (1919) and NODDACK's (1937) estimates of carbon fixation per unit of land area. RILEY used the 19th century data of LIEBIG and EBERMEYER primarily, multiplying by 4/3 to make a supposed correction for respiration. Our foregoing discussion suggests first that the approximations to net primary production were almost certainly underestimates, even for the data source areas (in Germany). While gas exchange or radiocarbon studies of gross production are very limited, examples like that just mentioned suggest that factors of at least 1.5—3 or more could ultimately become validated in adjusting net primary production data upwards to approximate gross primary production.

DUVIGNEAUD and TANGHE (1967, Tables 10—11) raised the estimates of production per km², and adjusted representative area factors of the earth's surface by which they might be multiplied (even allowing for less favorable environmental factors in many zones) to calculate world-wide productivity. This increased the estimates of annual carbon fixation on the continents. A slightly different areal breakdown (subject to improvement with better maps) and still higher estimates of net primary production in metric tons/km²/year (the same number as g/m²/yr) are used in Table 3. This current estimate of annual primary production (*not* biomass increment) of about 36 billion tons of carbon is almost twice as great for woodlands or forests as for all other landscapes considered together, three times the forest estimate of SCHRÖDER (1919), and four times that of NODDACK (1937). Furthermore, these estimates may be nearer net primary productivity than gross productivity which these authors purported to estimate.

We are still far from having a satisfactory weighted average of total live and dead biomass (and their subdivisions) for the broad types of woodlands or other ecosystems. Totals of well-developed stands from RODIN and BAZILEVICH (1967, 1968) appear satisfactory for a world perspective. Lower values are used in Table 3 because many younger stands affect averages over wide areas. The turnover fraction, even in the sense of *net* primary production of carbon is fairly high (last column). We have seen that gross photosynthesis could be 2—3 times as great.

Conceptual models and mathematical models of the carbon cycle will be important in predicting the fate of ^{14}C and of excess carbon dioxide generated by fossil fuels and other sources. Changes of climate and sea level that are expected by some geophysicists from CO_2 pollution could be so drastic that man cannot afford to wait for hard experience in order to judge their effects (REVELLE, et al., 1965). The mathematical models of ERIKSSON and WELANDER (1956) first related the world subsystems of green land plants, dead organic material and atmosphere in terms of nonlinear

Table 3. *Provisional geographic estimates for ecosystem areas, annual net primary carbon production and carbon budget of the world's landmasses*

	Area 10^6 km^2	Net primary Production of Carbon NPP ton/km^2/yr	10^9 tons	Live Carbon Pool $\dfrac{\text{tons}}{\text{km}^2}$	10^9 tons	$\dfrac{\text{NPP}}{\text{pool}}$
Woodland or forest						
temperate "cold-deciduous"	8	1000	8	10,000	80	0.10
conifer: boreal and mixed	15	600	9	8,000	120	0.075
rainforest: temperate	1	1200	1.2	12,000	12	0.10
rainforest: tropical, subtropical	10	1500	15	20,000	200	0.075
dry woodlands (various)	14	200	2.8	5,000	70	0.04
Subtotal	48		36		482	
Nonforest						
agricultural	15	400	6	1,000	15	0.4
grassland	26	300	7.8	700	18.2	0.43
tundra-like	12	100	1.2	600	7.2	0.17
other "desert"	32	100	3.2	600	19.2	0.17
glaciers	15		0			
Subtotal	100		18.2		79.6	
Continents	148		54.2		561.6	
Average/continents		366		3,800		
Average/earth surface (continental share)		106		1,100		

differential equations and concluded that these subsystems dominated changes in atmospheric CO_2 over time scales of years and decades. They considered the sea primarily as a "damping factor" of possible greatest importance for oscillations over periods of centuries or millenia. Geochemical investigations by Craig (1957) and many others have used stable and radioactive isotope ratios to evaluate the compartment model exchange between atmosphere and shallow and deep layers of the ocean.

Various refinements of an atmospheric compartment submodel were recently reviewed by Nydal (1968). Instead of assuming a single well-mixed atmospheric compartment, he and several authors have distinguished the reservoirs R_s and R_t (stratosphere and troposphere, respectively, separated by the tropopause) and the northern and southern hemispheres (Fig. 1). Arrows show transfers with the living matter of the biosphere on northern and southern continents, and the exchanges with longer-lived humus which had been treated by Craig (1957, 1958). Approximate quantitative rate coefficients k_{ij} for each arrow can be inferred indirectly from Nydal's exchange times since $k = 1/T$. His estimates of uncertainty are partly due to heterogeneity of each pool and partly to a variety of methods for estimation by investigators he cited. The total turnover fraction (1/exchange time) is the sum of the transfer coefficients out of each pool (Fig. 1). After allowing for transfers from the stratosphere of radiocarbon from thermonuclear tests within the northern stratosphere, Nydal's best estimate from recent changes in redistribution of the resulting excess ^{14}C over natural levels suggests an exchange time, from the troposphere into the *combined* ocean and biosphere, of 4.0 ± 1.0 years. The reciprocal of this time is what he calls

$k_{tu} \cong 0.25$ — the fractional rate of the troposphere transfer to these two destinations each year. Symbolically,

$$k_{tu} \equiv k_{tm} + k_{tb} = 0.25 \tag{1}$$

where k_{tm} and k_{tb} are troposphere transfers to ocean (mixed layer) and living biosphere, respectively.

Data from different ocean regions gave estimates of k_{tm} from 0.2 (subtropical) to 0.1, straddling the early estimate of 0.14 from the natural ^{14}C balance by CRAIG (1957). Subtracting the second of NYDAL's estimates from Eq. (1) leaves a preliminary estimate of $k_{tb} = 0.15$ as the fraction per year of the tropospheric carbon pool (adjusted for isotope fractionation between ^{14}C and ^{12}C) moving to the continental biosphere—a relatively greater contribution than for the ocean! The first estimate of k_{tm} (based only on data from latitudes 5° N to 35° N leaves a relatively smaller estimate of $k_{tb} = 0.05$ for subtropical regions and this is not surprising, since the oceans are most extensive here.

A source of evidence independent of either natural or artificial ^{14}C comes from sensitive measurements of seasonal oscillations in CO_2 concentrations, which are not great near the equator or the southern poles but which are marked above latitude 45° N. For this region BOLIN and KEELING (1963) inferred a net carbon assimilation by land plants of at least 4.1 billion metric tons from June through August alone. Accepting JUNGE's (1963) estimate of 0.32 as the stratosphere/troposphere ratio of CO_2 content and 640 billion tons for the atmospheric total, implies a northern troposphere reservoir (all latitudes included) of carbon of about 243 billion tons. Dividing 4.1 billion tons assimilated by 243 billion tons = 0.017 as a fractional loss in slightly less than 3 months. On an annual basis, this implies a temporary *net* fixation *coefficient* for northern ecosystems alone of 0.012 for the summer (followed by a net CO_2 *release* rate in early autumn, when litter freshly killed by frost begins to decompose rapidly). This exceeds NYDAL's lower bound, even neglecting carbon assimilation below latitude 45° N. If the troposphere above latitude 45° N were taken as a denominator, this fractional coefficient would approximate NYDAL's upper estimate for exchange coefficients of carbon. Below latitude 45° N higher, not lower, rates seem likely!

Because some exchange is going on between the atmosphere and biosphere in both directions at all times, it seems probable that all these are conservative estimates of the biosphere coefficients for models like that given in Fig. 1. They agree, independently, in suggesting that improved models for carbon exchange will require explicit attention to the continental living and dead organic materials, preferably subdivided according to major biological-climatic zones, instead of being pooled into a single continental biosphere pool, or two pools for the two hemispheres as Fig. 1 implies. Furthermore, the slowly exchanging humus pool needs to be treated explicitly, since it could influence equations like those used here (cf. ERIKSSON and WELANDER, 1956). Meeting this need requires modeling of the dynamic exchange between substrate, roots and decomposers (Figs. 2 and 3).

The authors quoted above generally recognize the limitations inherent in use of box models or compartment models where heterogeneity and lack of equilibration can be important. The word *pool* is used here to avoid implication of perfect mixing. The main point, however, is that use of models makes clear that mixing within some pools

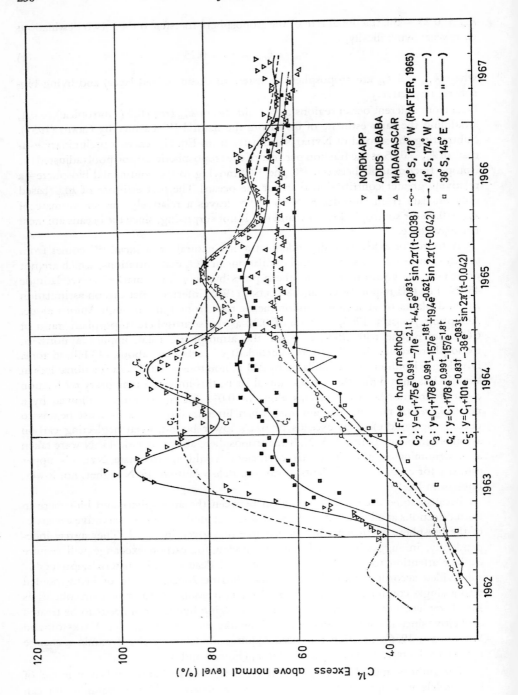

C¹⁴ Excess above normal level (%)

⊳ NORDKAPP
■ ADDIS ABABA
△ MADAGASCAR
-o- 18° S, 178° W (RAFTER, 1965)
-•- 41° S, 174° W (——)
-□- 38° S, 145° E (——)

c_1: Free hand method
c_2: $y = c_1 + 75 e^{-0.99t} - 71 e^{-2.1t} + 4.5 e^{-0.83t} \sin 2\pi(t - 0.038)$
c_3: $y = c_1 + 178 e^{-0.99t} - 157 e^{-18t} + 194 e^{-0.62t} \sin 2\pi(t - 0.042)$
c_4: $y = c_1 + 178 e^{-0.99t} - 157 e^{-1.8t}$
c_5: $y = c_1 + 101 e^{-0.83t} - 38 e^{-0.83t} \sin 2\pi(t - 0.042)$

(within the atmosphere or subdivisions like those of Fig. 3) is very much faster than that in some others like the ocean (especially the deep layer) and the continental biosphere (including humus, peat and long-lived forests). Geochemists suggest refinements for their own purposes (e.g., dividing each half of the troposphere near 30° lat.; dividing the thermocline into four intermediate layers between well-mixed upper ocean and nearly stagnant deep oceans), and ecologists presumably will do likewise for the biosphere on a regional and stand-component basis.

NYDAL's lower and upper estimates, 0.05 and 0.15 of the northern tropospheric reservoir of 227 *billion* tons, if valid, would allocate 11 to 34 billion tons of carbon to a continental area of 110.9 million km² of land in the northern hemisphere. A corresponding mean production rate of 118 metric tons $C/km^2/yr$ to 352 metric tons $C/km^2/yr$, is considered a crude average between unproductive icecaps and bare desert to the best forest or agriculture. The latter figure is remarkably close to the estimate of 366 metric tons $C/km^2/yr$ from Table 3, which was compiled entirely independently, but both seem conservative because biases mostly tend to be on the low side for all methods.

Each of these independent methods (extrapolation from local stand and ecosystem area estimates; radiocarbon behavior; CO_2 behavior) has enough uncertainties that we could still remain cautious about raising terrestrial production estimates so much from the previous estimates. Yet even the highest values (for forests, or for the continents or hemispheres in general) seem more likely to underestimate rather than overestimate the true values of net primary production.

While the models to date, on both a stand scale and world scale, have made no attempt to incorporate all the mechanistic detail which would interest the physiologist or the physicist, a further test arises from the capability of fitting available data on pollution problems, like those concerning ^{14}C and CO_2 changes in the atmosphere. Since excess radiocarbon was injected in the northern stratosphere, it is not surprising that NYDAL (1968, Fig. 8) could approximate a negative exponential decline of the difference between southern and northern stratospheres as the two became mixed, and the latter lost ^{14}C downward to the troposphere.

The sharp rise to 1963 maxima in the tropospheric burden of ^{14}C, followed by a decline, fits the general system of linear differential equations which NYDAL used as a first approximation. A second approximation neatly takes account of the sinusoidal terms (Fig. 4) to represent the well-known seasonality of the downward radiocarbon transport from the stratosphere. Whether the mathematical functions which seem satisfactory for *interpolation* can be *extrapolated* well into the future, or for new cases of contamination, remains to be seen. Predictions of ^{12}C and ^{14}C transfer into woody communities (of dimensions and turnover rates something like the chapparal) has been attempted only in a preliminary manner (NEEL and OLSON, 1962; OLSON, 1963). A first-order approximation at this stage would suggest that smoothing of oscillations

---

Fig. 4. Interpolation curves and seasonally varying equations of NYDAL (1968) for describing radiocarbon fluctuations, which are due to input from the stratosphere and depletion primarily to the continental ecosystems and oceans. The equations have some capability for extrapolation for future atmospheric conditions, and may serve as input for worldwide carbon monels to evaluate contamination of organic matter by photosynthetic intake of radiocarbon by plants (very much like that of Fig. 4) is to be expected as radiocarbon, or other aerial pollutants, are exchanged with and accumulated in woody communities having fairly rapid turnover rates (Tables 1 and 3). Cases like the Brookhaven forest of Table 2 would suggest high and prompt return of most photosynthesized carbon, but the fraction continuing to be stored in wood rings should respond in specific activity with the oscillations in the atmosphere like those shown in Fig. 4.

Conclusions

The analysis of primary producers (above and below ground), consumers and decomposers (Chapters 1—12 of this book) can be translated into budgets and cycles of carbon (Vernadsky, 1926, 1930, 1944). Knowledge of these cycles for all types of ecosystems must involve more insight into the quantitative aspects of cycles of nitrogen, mineral nutrients and water (Chapters 13, 14).

Preceding examples of cycling on the scale of a forest stand, subsystems (like the soil layers) and the world system illustrate how so many ecosystems are linked together. It will be an endproduct of the International Biological Program, and of continuing research beyond, to understand these linkages and, hopefully, to improve man's ability to measure and perhaps alter them. The studies summarized, and others still underway, suggest a relatively greater role for forests in the world's cycles of carbon and perhaps other elements than has been assumed previously by most geochemists or biologists, but the assumptions upon which such a broad generalization are based urgently require study. The importance of relating the large-scale dependence between all living matter, the whole atmosphere, the oceans and solid earth add to the many economic reasons which motivate us to understand and regulate the productivity of particular ecosystems.

At one stage of ecosystem analysis it may be necessary to narrow the aims of a particular project, i.e., its data collection or modeling. For example, (a) an improved *descriptive accounting* and "explanation" for the conservation of mass (or energy) during the various transfers in an ecosystem may be quite sufficient as a near-term objective of some environmental investigation. However, (b) *predictions* may well be demanded of us, whether or not we feel ready to make, justify or explain them. (c) Even if *management goals* for control seem beyond the scope of one person's initial obligation, he may find (and preferably design) his work to have some bearing on man's own role within ecosystems.

A research program (even more than a single project) will usually involve *all three aims* just listed: description, prediction and control. Their importance may be very different from one stage of study to another. Hence, I view systems analysis as requiring not only modeling related to the most important aim of the moment. Ideally, it would look ahead and help us to anticipate the shifting aims of investigation and action — not merely the changing conditions and values of the system chosen for study.

The chosen system may vary in scale: from the world system of Fig. 1, through regional or local land-water complexes as shown in Fig. 2, to a local ecosystem or subsystem. At least the three dimensions of subdivision are suggested in Fig. 3. These are: vertical position (or structure, even in three dimensions), composition by species or natural groupings (with particular priority to "target" organisms or age classes), and condition or type of component (within foliage, above-ground support, roots or below-ground support, substrate; also the decomposers and above-ground consumers shown in Fig. 2). The individual components and variables may have to be defined closely for unambiguous analysis, even though the purposes of modeling at one stage of investigation can very well tolerate broad grouping into heterogeneous pools like those of Fig. 2 or even Fig. 1.

Acknoledgements

Research supported partly by the U.S. Atomic Energy Commission under contract with the Union Carbide Corporation, and Ford Foundation grant no. 67-448 to Oak Ridge Associated Universities. Contribution 330 from the University of Tennessee Department of Botany also supported by National Science Foundation grants GB 6103 (in cooperation with Dr. EDWARD CLEBSCH) and G2-395 (Systems Ecology Seminar Project).

References

BLACKMAN, G. E.: The limit of plant productivity. Annual report of the East Malling Research Station for 1961. (1962).

BLACKMAN, G. E.: The application of the concepts of growth analysis to the assessment of productivity, pp. 243—259. In: Functioning of Terrestrial Ecosystems at the Primary Production Level (F. E. ECKARDT, Ed.), Paris: UNESCO 1968.

BOLIN, B., and C. D. KEELING: Large-scale atmospheric mixing as deduced from the seasonal and meridional variations of carbon dioxide. J. Geophys. Res. 68, 3899—3920 (1963).

COLE, D. W., S. P. GESSELL and S. F. DICE: Distribution and cycling of nitrogen, phosphorus, potassium and calcium in a second-growth Douglas-Fir Ecosystem, p. 197—232. In: Symposium on primary productivity and mineral cycling in natural ecosystems (H. E. YOUNG, Ed.), Orono: University of Maine Press 1967.

CRAIG, H.: The natural distribution of radiocarbon and the exchange time of carbon dioxide between atmosphere and the sea. Tellus 9, 1—17 (1957).

—, A critical evaluation of radiocarbon techniques for determining mixing rates in the oceans and air using carbon-14. In: Second U.N. Geneva Conference on Peaceful Uses of Atomic Energy, 18, 358—363 (1958).

DUMAS, M. J., and M. BOUSSINGAULT: The chemical and physiological balance of organic nature. An essay. Philos. Mag. 1844 Translation from Third French Edition of 1841 lecture, Paris 1844.

DUVIGNEAUD, P., and MARTIN TANGHE: Ecosystèmes et Biosphère — L'Écologie, Science Moderne de Synthese (Vol. 2). (2nd Ed.) Documentation 23, Ministere de L'Education Nationale et de la Culture, Rue de la Loi, 155 Brussels 1967.

ERIKSSON, E., and P. WELANDER: On a mathematical model of the carbon cycle in nature. Tellus 8, 155—175 (1956).

JUNGE, C. E.: Studies of global exchange processes in the atmosphere by natural and artificial tracers. J. Geophys. Res. 68, 3849—3856 (1963).

KEELING, CHARLES D., THOMAS HARRIS, and E. M. WILKINS: Concentration of atmospheric carbon dioxide at 500 and 700 millibars. J. Geophys. Res. 73, 4511—4528 (1968).

LIETH, H.: Die Stoffproduktion der Pflanzendecke. Stuttgart: G. Fischer 1962.

— The role of vegetation in the carbon dioxide content of the atmosphere. J. Geophys. Res. 68, 3887—3898 (1963).

— Versuch einer kartographischen Darstellung der Produktivität der Pflanzendecke auf der Erde. Geographisches Taschenbuch, 1964—1965. Wiesbaden: Steiner 1965.

MADGWICK, H. A. I.: Some factors affecting the vertical distribution of foliage in pine canopies, pp. 233—245. In: Symposium on Primary Productivity and Mineral Cycling in Natural Ecosystems (H. E. YOUNG, Ed.). Orono: University of Maine Press 1968.

Monsi, M.: Dry-matter reproduction in plants. Schemata of dry-matter or reproduction. Bot. Mag. Tokyo, **73**, 81—90 (1960).
— Mathematical models of plant communities, pp. 131—148. In: Functioning of Terrestrial Ecosystems at the Primary Production Level (F. E. Eckardt, Ed.) Paris: UNESCO 1968.
Neel, R. B., and J. S. Olson: Use of analog computer models for simulating the movement of isotopes in ecosystems. Oak Ridge National Laboratory Report, ORNL-3172 (1962).
Noddack, W.: Der Kohlenstoff im Haushalt der Natur. Z. Schl. Angew. Chem. **50**, 505—510 (1937)
Nydal, R : Further insertigation on the transfer of radiocarbon in nature. J. Geophys. Res. **75**, 3617—1635 (1968).
Olson, J. S.: Analog computer models for movement of nuclides through ecosystems, pp. 121—125. In: Radioecology, Proc. First National Symposium on Radioecology (V. Schultz, A. Klement jr., Eds.), 1961, New York: Reinhold Publishing Co. 1963.
— Gross and net production of terrestrial vegetation. J. Ecol. **62**, 99—118 (1964).
Ovington, J. D.: Quantitative ecology and the woodland ecosystem concept. Adv. Ecol. Res. I: 103—192 (1962).
— Organic production, turnover and mineral cycling in woodlands. Biol. Rev. **40**, 295—336 (1965).
Rafter, T. A.: Increase in the C^{14} activity in the atmosphere of the southern hemisphere from the testing of nuclear weapons. N. Zealand J. Sci **8**, 472 (1965).
Remezov, N. P.: Relation between biological accumulation and eluvial process under forest cover. Soviet Soil Sci. **1958**, 589—598 (1958).
— Method of studying the biological cycle of elements in forest. Soviet Soil Sci. **1959**, 59—67 (1959).
—, L. N. Bykova, and K. M. Smirnova: Consumption and circulation of N and ash elements in forests of European Russia. Published by Moscow State University, Moscow (In Russian) 1959.
Remezov, N. P., Y. M. Samoylova, I. K. Sviridova, and L. G. Bogashova: Dynamics of interaction of oak forest and soil. Soviet Soil. Sci. **1963**, 222—232 (1964).
Revelle, R., W. Broecker, H. Craig, C. D. Keeling, and J. Smagorinsky: Atmospheric carbon dioxide, pp. 111—133. In: Restoring the Quality of Our Environment. Environmental Pollution Panel, President's Science Advisory Committee, The White House, Washington, D. C. 1965.
Riley, G. A.: The carbon metabolism and photosynthetic efficiency of the earth as a whole. Amer. Sci. **32**, 129—134 (1944).
Rodin, L. E., and N. I. Bazilevich: Production and mineral cycling in terrestrial vegetation. London: Oliver and Boyd (Translation from Russian): Dynamics of the organic matter and biological turnover of ash elements and nitrogen in the main types of the world vegetation. Leningrad: Publishing house 'Nauka', 1965 (1967).
— — World distribution of plant biomass, pp. 45—52. In: Functioning of Terrestrial Ecosystems at the Primary Production Level (F. E. Eckardt, Ed.), Paris: UNESCO 1968.
—, L. E. Rodin, and N. I. Bazilevich: Methodical Information on the study of dynamics and biological circulation in plant communities. Leningrad: Publishing house 'Nauka', Leningrad Branch 1968.
Schroeder, H.: Die jährliche Gesamtproduktion der grünen Pflanzendecke der Erde. Naturwissenschaften **7**, 8—12 (1919).
Sukachev, V., and N. Dylis: Fundamentals of forest biogeocoenology. London: Oliver and Boyd (Translation from Russian), Publishing Office 'Nauka' Moscow: 1966 (1968).
Tamm, Carl Olaf: An attempt to assess the optimum nitrogen level in Norway spruce under field conditions. Studia Forestalia Suecica, No. 61 (1968).
Vernadsky, V. I.: Outlines of Geochemistry. Akad. Sci. USSR, Moscow (Fourth Edition, 1934, In Russian) 1926.
— Geochemie in ausgewählten Kapiteln. Leipzig: Akad. Verlagsges. 1930.
— Problems of biogeochemistry. The fundamental matter energy differences between the living and the inert bodies of the biosphere. Trans. Conn. Acad. Arts and Sci. **35**, 483—517 (1944).

Westlake, D. F.: Comparisons of plant productivity. Biol. Rev. **38**, 385—425 (1963).

Wittich, W.: Untersuchungen über den Verlauf der Streuzersetzung auf einem Boden mit starker Regenwurmtätigkeit. Reprinted from Schriftenreihe der forstlichen Fakultät der Universität Göttingen und Mitteilungen der niedersächsischen forstlichen Versuchsanstalt, 9, 33p (1953).

— Die Melioration streugenutzter Böden. Forstw. **73**, 193—256 (1954).

Woodwell, G. M., and T. G. Marples. The influence of chronic gamma irradiation on production and decay of litter and humus in an oak-pine forest. Ecology **49**, 456—465 (1968).

—, and R. H. Whittaker: Primary production and the cation budget of the Brookhaven Forest, pp. 151—166. In: Symposium on Primary Productivity and Mineral Cycling in Natural Ecosystems (H. E. Young, Ed.). Orono: University of Maine Press 1968.

The Water Flux in Temperate Forests: Precipitation and Evapotranspiration

Gerald Stanhill

Until recent times man's exploitation and manipulation of the temperate forest ecosystem was limited to harvesting its biological productivity by hunting and forestry. As the density and requirements of the human populations living in these areas increased, the nature of man's exploitation changed to include the physical attributes, and the temperate forest is increasingly looked to as an important source of water and recreational space.

This changing emphasis also is seen in forest water relations studies which are now more concerned with the influence of biological factors on hydrological productivity than the influence of hydrological factors on biological productivity.

This contribution will consider the effect of biological factors on precipitation and evapotranspiration: the main sources and sinks for water in the temperate forest. Attention will be focused on those measurements and functional relationships needed in the systems analysis approach.

The salient features of the precipitation and evapotranspiration processes have been tabulated in Table 1.

Precipitation

Table 1 shows that at present it is both easier and more accurate to measure precipitation than to estimate it from past climatological records or to forecast it from current weather data. Even so, the absolute accuracy of rainfall measurements at a given point is unknown, due to the lack of any absolute method of measurement. It will be seen later that this fact has important implications in hydrological research.

Some estimate of the probable error in rainfall measurements can be obtained from the very few studies in which alternative methods of precipitation measurement were available. A recent example is McGuiness's study at the lysimeter installation of the North Appalachian Experimental Watershed (McGuiness, 1966). The amount of rain reaching soil surfaces of large weighing lysimeters was 6% more than that measured in an adjacent, standard raingage.

Another example is the Lake Hefner study, where Harbeck and Coffay (1959) showed that the mean rain catch in 22 standard raingages surrounding the 10 km² lake was 10% less than the amount falling on the lake itself as calculated from detailed water budget measurements. The discrepancies between the two estimates for individual rains increased with increasing wind speed, confirming conclusions derived from comparison of precipitation gages exposed at standard heights with those at ground level surrounded by anti-splash devices (Koschmeider, 1934; Stanhill, 1959; and Struzer, Nechayev, and Boganova, 1965). The Lake Hefner

Table 1. *Some features of the precipitation and evapotranspiration processes*

	Precipitation	Evapotranspiration
Methods of measurement and estimation	Absolute accuracy of standard measurements unknown but thought to be about 10%, relative accuracy about 1%	Absolute physical methods of estimation exist, but no standard methods of measurement are available as controls
Data available	Many comparable measurements over large areas and long periods	Very few accurate measurements available for either long or short time periods or for different land surfaces
Distribution	Discrete and very variable in space and time	Continuous in time and space: not enough data to characterize its variability but thought to be less variable than rainfall
Theoretical understanding	Physical theory inadequate to explain its occurrence quantitatively	Physical theory adequate for quantitative explanation in simplest cases; biological control factors cannot be assessed quantitatively
Surface influences	Of minor significance except in special cases	Of major importance especially in periods of water deficit
Predictability	Probability of occurrence can be predicted with fair (80%) accuracy for a few days in advance, longer range or more quantitative forecasts unsatisfactory	Quantitative estimates of high (90%) accuracy as long as the necessary meteorological parameters can be predicted. Biological control factors cannot be predicted at present
Control possibilities	Varies greatly but increases of about 10% are attainable by cloud seeding over a wide range of conditions	Can be controlled to some extent by forest management practice; in recreation areas, control by biochemical, genetic and weather control practices may prove feasible

results are particularly relevant to forest watershed studies as in both cases precipitation on an area inaccessible to, or unsuitable for, direct measurement has to be estimated from peripheral measurements. Thus, large sampling errors may be introduced in the areal estimate in addition to the instrumental error of point measurements. The extensive literature dealing with sampling errors in precipitation measurements recently has been reviewed by CORBETT (1967) with special reference to forest hydrology. Even under optimum conditions*, short term estimates of precipitation reaching a forest canopy may show considerable random variation about the true values; over longer periods, the measured value may well be between 0.05 and 0.10 below the true amount.

This conclusion is pertinent to the long-debated question of forest influences on precipitation. Obviously, the absolute size of any such influence cannot be clearly established unless it exceeds the error in precipitation measurements. Similarly, the

---

* That is, a dense network of raingauges in a flat area where the precipitation consists of predominantly cyclonic rain falling during calm periods.

relative magnitude of any changes in precipitation brought about by afforestation or forest clearing cannot be demonstrated unless they exceed the annual variation. Even in high rainfall areas, where annual rainfall is normally distributed, year to year variation (coefficient of variation) is seldom less than 10%. The large sampling errors generally present in forest rainfall measurements make it difficult to demonstrate even larger changes in precipitation with any degree of confidence.

Effect of Forest Cover on Precipitation

Three mechanisms have been put forward to explain the greater precipitation which is widely believed to occur over forested areas. One mechanism, thought to be especially important in flat areas, is the greater air turbulence generated by a forest compared with most alternative forms of land cover. This increased turbulence is thought to be especially pronounced where the tree cover is not continuous but, as in wind-break afforestation, planted in wide belts orientated 90° to the prevailing winds.

A second mechanism which can be very important in some mountainous regions is the direct interception by trees of atmospheric moisture in the form of fog, mist or haze. This and the former mechanism are local in effect, increasing rainfall at the leading edge of the forested area at the expense of an equivalent decrease downwind.

A third mechanism often suggested results from an increased atmospheric water content of forested areas, presumed to occur because of their greater evapotranspiration rates. Although such increased water loss is well established (Shachori and Michaeli, 1965), there is still considerable discussion as to its size and its significance in increasing rainfall. Any such increase in rainfall should lead to an increase in the rate of hydrological cycling and, unlike the two previous mechanisms, could lead to a general increase in the areal water balance.

Representative values of the aerodynamic roughness, horizontal interception efficiency and relative evapotranspiration rate of forest areas compared with alternative forms of land use are not available. Even if they were, it still would be impossible to calculate theoretically the effect of forest cover on rainfall in the absence of a quantitative physical model of the precipitation process. Drozdov and Grigor'eva (1965) have reviewed a number of empirical calculations of the effect of afforestation on precipitation in the central steppe region of the USSR. In early calculations, a correlation between the perimeter length of forest strips and summer rainfall was used to estimate that 20% afforestation of these areas had caused an 8% increase in annual precipitation through increased turbulence. The greater evapotranspiration from such areas was calculated, again on an empirical basis, to have led to a 5% increase in summer rainfall. Later research, reviewed by the same authors, suggested a much smaller effect (10% afforestation leading to a 1.5% increase in precipitation); the major portion of which was attributed to the enrichment of atmospheric water content rather than increased turbulence. The more recent Russian work has demonstrated the expected decrease in precipitation on the leeward side of afforested areas and also has shown that some earlier studies did not allow for increased raingage efficiency in calmer forested areas.

In especially favorable mountainous regions, much larger increases in precipitation may occur through the horizontal interception of atmospheric moisture. This, however, is a very local phenomenon (Penman, 1963). In certain northern Pacific regions, forest belts have been planted on the coast to "filter off" this moisture and prevent its deposition from interfering with agricultural crops (HORI, 1953).

PENMAN (1963) also summarizes the strong meteorological arguments against the proposition that the type of vegetation cover can exert any significant, large-scale influence on the amount of precipitation. At the present time, the limited evidence suggests that for large areas, any increases of rainfall brought about by the presence of forests are smaller than the errors in rainfall measurements.

Interception
The Fate of Canopy-intercepted Precipitation

In forest hydrology, interception forms an important link between precipitation and evapotranspiration and illustrates the significance of biological factors in the hydrological cycle. The fate of water intercepted by tree cover has long been the subject of debate. Formerly, the evaporation of intercepted rainfall was regarded as a separate

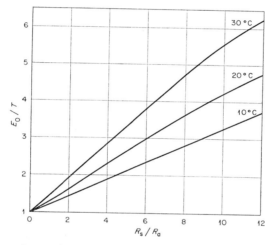

Fig. 1. Relative rate of water loss from wetted and dry canopies (E_0/T) at different ratios of stomatal to aerodynamic resistances (R_s/R_a) (after MONTEITH, 1965)

pathway for water loss additional to, and independent of, losses by leaf transpiration and soil surface evaporation. More recently, evaporation of intercepted water has been viewed as being compensated for by an equivalent decrease in evapotranspiration.

RAESCHKE (1958) and MONTEITH (1965) have developed physically-based equations with which the rate of evaporation from a rain-wetted canopy (E_0) relative to the rate of transpiration from the same canopy in a dry condition (T) may be calcutated. Assuming that the radiation balance of the two surfaces is the same, the ratio E_0/T depends on the air temperature and the relative stomatal and aerodynamic resistances of the forest canopy to the diffusion of water vapor (R_s and R_a, respectively.) Fig. 1, based on MONTEITH's calculations (1965), shows the relationship between E_0/T and R_s/R_a for three different air temperatures.

Unfortunately, few experimental data are available which quantify the diffusion resistances of forest tree canopies. RUTTER (1967) suggests an E_0/T ratio of 4 for a

Scots Pine plantation in southern England, and a similar value has been calculated for a mature orange tree plantation in Israel (Kalma et al., 1968). However, the anatomy and distribution of the stomata in both these evergreen species favor high stomatal resistances, and much lower values may occur for other deciduous forest species.

The hydrological potential for intercepted water loss is dependent upon both the length of time that the canopy remains wetted (i.e., the distribution of rainfall and its correlation with evaporating conditions) and the amount of intercepted precipitation that the canopy can hold (i.e., its storage capacity). Zinke (1967) recently tabulated the results of a large number of forest interception studies. The mean storage capacity expressed in rainfall equivalents, was 2.6 mm for the 15 coniferous forests listed by Zinke with individual values so variable that the coefficient of variation (cv) was 112%. The eight hardwood forests had a mean storage capacity of 0.9 mm, with a cv of 76%; the mean for six understory shrubs listed was 1.3 mm (cv = 28%). The mean storage capacity for two grass species was the same as that for the shrub understory. These figures also suggest that interception is of greater hydrological significance in coniferous than in hardwood forests. Similar conclusions can be drawn from the USSR investigations reviewed by Molchanov (1963).

Interception is also of significance in element cycling, since evaporation of intercepted rain can lead to a considerable concentration of "rainout" elements including radionuclides. These elements, together with leached leaf constituents and leaf deposits from dry "fallout", will subsequently be washed from the forest canopy by later rains. The degree of concentration will, of course, depend on the time sequence of rainfall and evaporation.

The infiltration of rainfall through the forest canopy via stemflow, throughfall and canopy drip also can increase heterogeneity in the distribution of soil water, associated soil nutrients and, consequently, tree growth. Eschner (1967) has reviewed the limited experimental data on this aspect of forest interception.

Despite considerable research, it is not yet possible to predict the amount of precipitation that will be intercepted by a forest canopy nor the pattern of throughfall distribution below the canopy from a knowledge of its physical characteristics. The amount of interception (I) must be calculated for each stand from measurements of above-canopy rainfall (P), stemflow (S) and throughfall (T) using the following equation:

$$I = P - (T+S) \tag{1}$$

The threshold value of P above which T commences is nearly always less than that at which S commences; therefore, the storage capacity of the canopy (C) can be derived from the slope and intercept of a linear regression of T on P using data of individual storms or, more usually, individual days of precipitation. Thus,

$$C = b/a \text{ where } T = aP\text{-}b. \tag{2}$$

Values of I and C are especially necessary when the hydrological influence of forests are the object of study, since reviews of investigations made in America (Zinke, 1967) and elsewhere (Penman, 1963; Molchanov, 1963) have shown that a significant proportion ($\cong 20\%$) of the annual precipitation often is intercepted by the forest canopy.

Evapotranspiration
Empirical and Theoretical Means of Estimation

In contrast to both precipitation and interception, evapotranspiration is much more amenable to calculation on the basis of physical principles than it is to direct measurement. Evapotranspiration is normally measured from changes in soil water content and precipitation. The large horizontal and vertical heterogeneity in soil water content makes it difficult to accurately measure changes in soil water content over a representative area. Even where such measurements are possible, they cannot allow for upward movement by capillarity or downward movement by delayed deep drainage.

The same hydrological balance approach to the measurement of evapotranspiration is often applied to entire watersheds, balancing measured precipitation against surface runoff and attributing the balance to evapotranspiration. There are two important limitations to the accuracy of this method of estimation. Firstly, it is difficult to determine with sufficient precision the true area of the catchment, especially with regard to horizontal outflow beneath the soil surface. Secondly, the previously discussed, limited accuracy in gaging precipitation still remains.

It is possible to measure directly changes in the water content of the soil in which trees are growing with large weighable lysimeters. A number of such instruments large enough to permit forest trees to be grown to maturity under soil conditions closely resembling those of surrounding trees have been described (e.g., HAROLD and DRIEBELBIS, 1958; URYVAEV, 1953). Practical difficulties have limited this method of evapotranspiration measurement: the high cost of the lysimeters and the long period of time elapsing between planting of the lysimeter and obtaining useful results. Nevertheless, such installations are highly desirable, if only to serve as controls by which other methods (including the meteorological) can be evaluated.

The first and most widely used physically-based method of estimating evapotranspiration from meteorological data is the combined energy balance and aerodynamic equation of PENMAN. In its simplest form water loss from the forest under nonlimiting soil moisture conditions (PE_t, potential evapotranspiration) is taken to be a constant fraction (f) of the loss from a hypothetical open water surface, infinitely large in area, small in heat storage capacity and exposed to the same microclimate as the forest (E_0).

E_0 is itself calculated from standard climatological measurements using the following equation:

$$E_0 = (\Delta Q* + \gamma E_a)/(\Delta + \gamma) \qquad (3)$$

where $Q*$ is the energy balance, source strength term. For the hypothetical water surface, this can be taken as its net radiation balance; this value can be obtained in a variety of ways, normally the following expression is used

$$Q* = K\downarrow(1\text{-}\alpha) - L_n . \qquad (4)$$

Here $K\downarrow$ is the total short wave radiation from sun and sky, either measured or calculated from measurements of hours of bright sunshine or degree of cloud cover; α is the short wave reflection coefficient or albedo, which for a water surface can be obtained from tables according to date and latitude (BUDYKO, 1958); and L_n is the net long radiation loss, calculated from one of the many empirical expressions using local measurements of air temperature, humidity and cloud cover.

All energy terms are expressed in evaporation equivalent units. E_a is the aerodynamic sink strength term, and for an open water surface it is written

$$E_a = 0.35 \, (0.5 + u/161) \, (e_a\text{-}e_d) \text{ mm day}^{-1}. \tag{5}$$

Here u is the measured wind speed at a height of 2 m, km day^{-1}; e_a is the saturation vapor pressure of the air at its mean temperature, mm Hg; and e_d is the mean vapor pressure of the air, mm Hg.
The constants of this aerodynamic equation are inversely related to the aerodynamic resistance of the evaporating surface (R_a).

$\varDelta$ and γ are two constants, introduced to combine the energy balance and aerodynamic terms and to keep the units consistent. $\varDelta$ is the slope of the saturation vapor pressure: temperature relationship and is temperature dependent, and γ is the constant of the psychrometer equation. PENMAN's approach has proved very useful under conditions as diverse as N.W. Europe (PENMAN, 1967) and equatorial Africa (PEREIRA, 1967). The mean value of the annual ratios, $f = PE_t/E_0$, obtained from these and other forest studies was 0.85, nearly all of the individual values falling within the range of ± 0.15. The possibility of an independent check on the major evapotranspiration term in the hydrological equation has proven very stimulating in the analysis of classical watershed studies (PENMAN, 1956).

Recently, a more sophisticated form of PENMAN's equation has been used in forest hydrology. RUTTER (1967) replaced the empirical conversion factor (f) relating potential evapotranspiration to evaporation from a hypothetical open water surface by incorporating appropriate values of the radiative (α) and aerodynamic (R_a) characteristics of a Scots Pine forest canopy, together with calculated values of its diffusive resistance (R_s) to water vapor transport.

Actual evapotranspiration (E_t) was calculated for the forest canopy from the following equation, incorporating appropriate values of $Q*$ and E_a

$$E_t = (\varDelta Q* + \gamma E_a)/(\varDelta + \gamma) \, (1/SD) \tag{6}$$

where S is the ratio of aerodynamic to total (aerodynamic plus stomatal) diffusive resistance; i.e., $R_a/(R_a + R_s)$ in the previously defined notation, and D is a daylength factor introduced to allow for stomatal closure during the night. The need for this factor has been questioned on both theoretical and empirical grounds. The estimates of E_t obtained with this form of PENMAN's equation were more highly correlated with values derived from soil water measurements than were the estimates obtained by the simple meteorological approach. RUTTER (1967) has pointed out the practical difficulties in use of the expanded form of PENMAN's equation. The difficulty partly lies in the paucity of data on the radiative and aerodynamic characteristics of various forest canopies and the way in which these are affected by 1) age, 2) environment and management practices, and 3) even more so, the lack of comparable data on the canopy resistances to water vapor diffusion.

In considering the limited data available on these three vegetative characteristics influencing the rate of water loss, it is useful to compare them with those of alternative forms of land use, such as meadows or agricultural crops. Such a comparison enables the hydrological effect of changes in land use to be assessed. Monteith (1965) has estimated in this way that a pine forest would transpire at a rate similar to a short grass cover under the climatic conditions of southern England, but that a tall agricultural crop would use some 20% more water. Assuming similar vegetational parameters and the climatic conditions of California, this same pine forest would transpire at a rate similar to a tall agricultural crop, and a short grass cover would use some 25% less water.

Alternatively, where the water losses of contrasting vegetation have been measured directly, canopy resistances may be calculated. Such data give valuable insight into the probable effect of management policy on water loss.

Although many measurements of the radiative characteristics of forests have been made, they mostly have taken the form of isolated observations. Reifsnyder and Lull (1965) have reviewed much of this data. Budyko (1958) summarized the extensive USSR data and concluded that the short wave (0.3 to 3.0 μ) reflection coefficient (albedo) of coniferous forests lies between 0.10 and 0.15, while for deciduous forests the appropriate range is between 0.15 and 0.20. The albedo of a variety of agricultural surfaces (meadows, cereal crops, potatoes and cotton fields) falls within the range of 0.15 to 0.25.

The few systematic investigations which have been made suggest that these values include a considerable seasonal variation, necessitating periodic observations for each surface type. Thus Grulois (1968) found the albedo of a mixed oak forest in Belgium to range from a maximum decadal value of 0.21 during a period of snow fall to the minimum annual value of 0.11 during late winter and early spring. The seasonal values thus ranged 25% around the mean annual albedo of 0.15. Federer (1968) reports an equally large seasonal variation in the albedo of a mixed birch and maple stand in New Hampshire. The mean annual albedo was 0.17, with a midwinter maximum of 0.23 (with some snow ground cover) preceeded by a late autumn minimum of 0.14. These observations suggest that the seasonal variation in albedo is much larger than the spatial and specific variation. Birkebak and Birkebak (1964) report that the albedo of the upper surfaces of single leaves from 22 North American broad leaf tree species showed a standard deviation of only 11% of the mean value, 0.28.

Federer (1968) also reported the results of an investigation into the spatial variation in net radiation, albedo and surface temperature in a mixed hardwood stand in New Hampshire. Six measurement sites were selected above the uniform but mixed tree canopy, and six additional sites were chosen above adjacent but different cover types, including a pine forest, juniper stand, an old field and a rock ledge. The results of measurements on 5 separate days showed remarkably little spatial variation. Above the forest the range (maximum to minimum) in values was approximately 10% of the mean; in absolute terms, this range was often less than the accuracy of the measurement. Except during the winter, net radiation above all widely differing cover types showed a range in values less than 0.1 cal cm^{-2} min^{-1}. Federer concluded that the short wave (albedo) and long wave (surface temperature) characteristics contributed about equally to the differences in net radiation balance.

In comparison with meadow vegetation or short, agricultural crops, very few data are available on the appropriate aerodynamic characteristics of forest canopies. There are two such parameters, both expressed in units of length. The roughness length, z_o, parameterizes the turbulence generating properties of the canopy while the zero plane displacement height, *d*, indicates the height above the soil surface to which the forest displaces the passing air. Determination of these two characteristics requires accurate measurements of wind and temperature profiles above the canopy at precisely known heights; such measurements are difficult to obtain in a mature forest. In certain situations where the horizontal heterogeneity in stand characteristics is considerable, many sets of measurements are needed. A further complication with many forest types is that the aerodynamic characteristics probably change significantly according to both wind velocity and season. At different stages of leaf development, increases in wind velocity will cause leaf flutter and tree-top swaying which will generate additional turbulence; however, at still greater velocities the roughness elements may adopt streamline positions reducing the turbulence they generate. Yet another complication is caused by the uneveness in topography and canopy height which occurs in many forests. Such sites are unsuitable for aerodynamic measurements or for the use of meteorological methods of calculation which include aerodynamic terms.

In the absence of reliable measurements of the aerodynamic properties of different forest canopies, an approximate estimate of their roughness length can be obtained by extrapolating the linear relationship between log roughness length and log stand height derived by Tanner and Pelton (1960) from data of mainly agricultural crop surfaces.

Values of z_o and d can be used together with measurements of wind velocity above the forest canopy, u to calculate the aerodynamic resistance of the canopy, R_a, using the following equation (Monteith, 1963):

$$R_a = \{\ln[(z\text{-}d)/z_o]\}^2 / uk^2 \tag{7}$$

where, in addition to the previously defined symbols, z is the height of wind measurement above the ground surface and k is Von Karman's constant (0.41). This expression is strictly only valid during isothermal, adiabatic conditions. Under non-isothermal conditions; i.e., lapse and inversions, a correction for forced convection should be introduced based on the Richardson number, a stability parameter calculated from the temperature and wind gradients measured just above the canopy. However, when the wind velocity is measured close to the canopy and when the rate of water loss is not limited by non-meteorological factors, the correction required is very small.

In the absence of the necessary measurements from above the forest canopy, it is possible to calculate the turbulent diffusion coefficient for vegetation of different heights from standard climatological data assuming that both the wind velocity and atmospheric stability in the free air above the forest are the same as for the alternative forms of land use. The relationships involved are given by Businger (1956) and Tanner and Pelton (1960). These transfer coefficients can then be used to calculate evapotranspiration in the absence of any internal canopy resistance; i.e., under potential evapotranspiration conditions.

An example of this approach using calculated values for a mixed oak forest 12 m high and a short grass meadow 0.3 m high will be used to illustrate the significance of the various terms in the evapotranspiration equation. Climatological values of total radiation $(K\downarrow)$, wind velocity (u), air temperature (T_a) and air saturation vapor pressure deficit $(e_a\text{-}e_d)$ were taken from Oak Ridge, Tennessee, a station representative of the oak-chestnut forest region of southeastern USA. The values used and results obtained are listed in Table 2. The calculated annual potential evapotranspiration from the short grass meadow, 886 mm, shows reasonable agreement with some independent estimates of the water loss in a similar region (see discussion in PENMAN, 1956). The calculated annual potential evapotranspiration for the forest was, however, almost four times larger. Such a difference is extremely unlikely, and the experimental evidence from comparable areas (DOUGLASS, 1967) suggests that the annual water loss from such forests should be only 15% greater than that from meadows.

Table 2. *Annual potential evapotranspiration, Oak Ridge, Tennessee, USA (all terms in evaporation equivalents, cm cm^{-2} yr^{-1})*

Evaporation term	Mixed oak forest	Short grass meadow
Energy, $Q*$	136	125
Weighted energy, $Q*(\varDelta/\varDelta + \gamma)$	83.2	76.3
Aerodynamic, E_a	649.7	31.5
Weighted aerodynamic, $E_a(\gamma/\varDelta + \gamma)$	253.4	12.3
Potential evapotranspiration, PE_t(i.e. SD $= 1$)	336.6	88.6

Annual climatological values

$K\downarrow = 133$ Kcal cm^{-2} yr^{-1}, $u = 77$ km day^{-1}, $T_a = 14.5°$ C, $e_a\text{—}e_d = 1.8$ mms.

This gross discrepancy between calculated and experimental values is too large to be attributed to uncertainties in the values of radiative characteristics used. Errors in the aerodynamic characteristics adopted for the forest canopy are of greater significance, and the assumption of similar velocity and stability over the two ecosystems could also introduce significant error. However, to bring the calculated values of water loss into line with the experimental data would require that the roughness lengths of each vegetation type are equal or, alternatively, that very much lower rates of air movement and much more stable atmospheric conditions prevail above the forest canopy than above the meadows. This seems improbable although there is insufficient data to disprove the possibility. A more likely reason for the exagerated calculated values of forest evapotranspiration is thought to be the neglect of canopy resistance in the calculations. RUTTER's (1967) measurements of this resistance in a Scots Pine stand show the term to be of major importance even when soil moisture conditions are nonlimiting and stomata almost fully open. LEE and GATES' (1964) calculations show similar results in that the stomatal resistance of a pine needle (per unit leaf surface) is 2 to 8 times larger than that of an alfalfa leaf, a typical mesophytic species.

Nearly all the evidence suggests that stomata form the major component of internal canopy resistance, although under certain circumstances resistance to water uptake by the roots, or water transport by the soil, stem and leaves, may limit transpiration. The position of the stomata at the interphase between plant and atmosphere makes them, teleologically, the preferred site for a controlling resistance to minimize fluctuations in the plant's internal water status.

Very few measurements of the diffusive resistance of stomata have been reported for forest species even under laboratory conditions. The limitations in current techniques of measurement together with the enormous sampling problem posed by the extreme random, vertical, horizontal, and diurnal variation of stomatal character- istics makes it improbable that representative resistance values could ever be ob- tained by direct measurement for a mature forest canopy. Stomatal resistance can also be calculated from measurements of their dimensions and density, but this method has little to recommend it. The sampling problem is similar to that for direct measure- ments, and the calculations involved are complex and of limited accuracy (Wag- goner, 1965).

Monteith (1963) proposed a meteorological method of calculation whereby the mean stomatal resistance of the entire canopy is derived from temperature and humidity estimates of its idealized, active surface. These surface values are obtained by extrapolation from profiles of wind speed, temperature and humidity measured above the surface of the canopy. This method avoids many of the problems of samp- ling and measurement associated with the methods previously described. However,

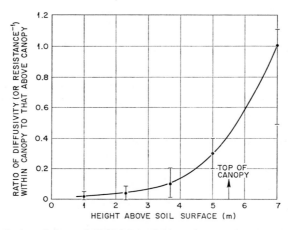

Fig. 2. Mean vertical variation of diffusivity within a young pine forest as a ratio of that within the free air above the canopy (after Denmead, 1964). Brackets include range of ratio values at each height

it rests on the contested assumption that the active surfaces for heat, water vapour and momentum exchange are at the same height within the canopy. Denmead's measurements within a pine forest do not support this assumption (Denmead, 1964).

The alternative meteorological approach used by Denmead (1964) derives values of water vapor diffusivity and evapotranspiration from profiles of net radiation, temperature, humidity and wind speed made within the canopy. Although the measurements involved are more demanding, and in certain types of open, mixed forest stands there may be sampling problems caused by horizontal heterogeneity, this method offers an important advantage. The sources of water vapor and the resistance to its transport within the canopy can be identified. Fig. 2 gives, as an ex- ample of this approach, the results obtained by Denmead in his investigation within a 10-yr-old stand of *Pinus radiata*. The trees were 5.5 m high; the top meter of the canopy consisted of isolated stems and its base extended to within 0.5 m of the soil surface. Diffusivity, or resistance^{-1}, is shown in dimensionless units as a ratio to the values computed for the unobstructed air layer above the canopy during the same period. The lines in Fig. 2 connect the mean value of 8 sets of measurements, each of 40 or 60 min duration and covering a wide range of radiation, wind velocity and atmospheric stability conditions. The range in individual values of the ratio has been shown for each height.

Information of this type is of the greatest practical importance in areas where the reduction of evapotranspiration is an important aim of forest management and land use policy. At present decisions on management practice have to be made on the basis of the often conflicting results of costly and lengthy watershed experimentation which, for reasons previously outlined, is of limited accuracy. DOUGLASS (1967) and GOODELL (1967) in their reviews of such experiments have noted the difficulty in identifying the important operative mechanisms.

The preceeding discussion of the evapotranspiration process suggests that it is unlikely that different management practices could change the radiative characteristics of a forest sufficiently to cause major changes in its rate of evapotranspiration. Such changes are more likely to be caused by alterations in the interception capacity and aerodynamic and canopy resistance characteristics of the forest.

Because of the similarity between the evapotranspiration and photosynthetic process, systematic investigation of the aerodynamic properties and resistances of various forest species under different management practices would be further reward-ing. Both evapotranspiration and photosynthesis are physically diffusion processes — the downward flux of carbon dioxide being controlled by its concentration gradient between free air and the site of photosynthesis within the leaf and the various resistances occurring in series along this diffusion path. In the case of transpiration, the diffusion path is between the free air and the evaporating walls of the substomatal cavity. Many of the resistances are common to both fluxes, and so the effect of the various management practices on the size of the various internal and external resistances is important to an understanding of the rates of dry matter production as well as water loss.

Where forest lands are important as watersheds, as they often are near population centers, timber production is likely to be of secondary importance. Chemical methods currently under development for increasing stomatal resistance and reducing tran-spirational water losses and photosynthesis may prove to be an important means of affecting the hydrologic cycle and maximizing water yield. The first field tests of such methods have not proved successful (WAGGONER and HEWLETT, 1965), because of difficulties in applying the chemicals to those parts of the forest canopy which transpire most.

Current understanding of evapotranspiration from forest canopies is sufficiently advanced to indicate the most promising avenues for future research. Suitable techniques for such investigations are now available. Ultimately, the results of research could provide the necessary data for a systems analysis approach to the problems of forest management policy, where the multiple goals of water yield, recreational access and timber production must be reconciled. This would allow optimization of management practice for a wide variety of ecosystems growing under the full range of their physical environments.

So far attention has been concentrated on the need for the precise data and exact relationships required in detailed hydrological analysis and simulation studies of particular forest ecosystems. The gaps in the data currently available for such purposes have been stressed. However, when attention is directed to the temperate forest on a much larger scale, as for example in the inside back cover map showing the world distribution of this ecosystem, the values of the major terms in the energy balance and hydrological equations found in the main areas of temperate forest show surprisingly good agreement. The approximate, annual values of these terms are presen-

Table 3. *Approximate annual values of energy balance and hydrological terms of the main centers of temperate forests*[a]

Geographical region[a]		Eastern United States (10)	Central Europe (10)	Northeast Asia (10)	Southeast Australia (11)	Southern Andes (10)
Total radiation Kcal cm^{-2} yr^{-1}	$K\downarrow$	130	120	130	130	110
Net radiation Kcal cm^{-2} yr^{-1}	$Q*$	55	40	45	60	50
Convective heat transfer Kcal cm^{-2} yr^{-1}	A	20	15	15	25	20
Latent heat transfer Kcal cm^{-2} yr^{-1}	LE	35	25	30	35	30
Actual evapotranspiration cm yr^{-1}	E_t	60	40	50	60	50
Precipitation cm yr^{-1}		100	60	70	70	200
Runoff — surface and subsurface cm yr^{-1}		40	20	20	10	150

[a] See numbered areas on map inside back cover.

ted in Table 3, interpolated from the Physical Geographical Atlas of the World (SENDEROVA et. al., 1964) for the main center of typical temperate forests in each continent.

The relationships governing the distribution of forests and other ecosystems are notoriously difficult to generalize and quantify. However, the investigations made so far with the previously discussed environmental parameters have shown much promise (BUDYKO, 1958). From the hydrological point of view, such studies along transects of increasing and decreasing aridity between temperate forests and their neighboring associations should be most useful.

Acknowledgement

Research sponsored by the US Atomic Energy Commission under contract with the Union carbide Corporation.

References

BIRKEBAK, R., and R. BIRKEBAK: Solar radiation characteristics of tree leaves. Ecology 45, 646—649 (1964).

BUDYKO, M. I.: The heat balance of the Earth's surface. Translated by NINA A. STEPANNOVA Washington: U. S. Dept. of Commerce 1958.

BUSINGER, J. A.: Some remarks on Penman's equation for evapotranspiration. Neth. J. Agr. Sci. 4, 77—80 (1956).

CORBETT, E. S.: Measurement and estimation of precipitation on experimental watersheds, pp. 107—126. In: International Symposium on Forest Hydrology (SOPPER, W. E. and H. W. LULL, Eds.). London: Pergamon Press 1967.

DENMEAD, O. T.: Evaporation sources and apparent diffusivities in a forest canopy. J. Appl. Meteorol. 3, 383—389 (1964).

DOUGLASS, J. E.: Effects of species and arrangements of forests on evapotranspiration, pp. 451—461. In: Internatioal Somposium on Forest Hydrology (SOPPER, W. E. and H. W. LULL, Eds.). London: Pergamon Press 1967.

DROZDOV, O. A., and A. S. GRIGOR'EVA: The hydrological cycle in the atmosphere. Israel Program for Scientific Translation, Jerusalem: 1965.

ESCHNER, A. R.: Interception and soil moisture distribution, pp. 191—200. In: International Symposium on Forest Hydrology. (SOPPER, W. E. and H. W. LULL, Eds.). London: Pergamon Press 1967.

FEDERER, C. A.: Spatial variation and net radiation, albedo and surface temperatures of forests. J. Appl. Meteorol. 7, 789—795 (1968).

GOODELL, B. C.: Watershed treatment effects on evapotranspiration, pp. 477—482. In: International Symposium on Forest Hydrology. (SOPPER, W. E. and H. W. LULL, Eds.). London: Pergamon Press 1967.

GRULOIS, J.: La variation annuelle du coefficient d'albédo des surfaces supérieures du peuplement. Bull. Soc. Roy. Botan. Belg. 101, 141—153 (1968).

HARBECK, G. E., and ETHEL W. COFFAY: A comparison of rainfall data obtained from raingage measurements and changes in lake levels. Bull. Amer. Meteorol. Soc. 40, 348—351 (1959).

HAROLD, L. L., and F. R. DRIEBELBIS: Evaluation of agricultural hydrology by monolith lysimeters 1944—1955. U. S. Dept. Agr. Tech. Bull. 1179, 1958.

HORI, T. (Ed.): Studies on fogs in relation to fog preventing forest, pp. 399. Hokkaido: Tanne Trading Co., Soppora 1953.

KALMA, J. D., G. STANHILL, and E. URIELI: Rainfall interception and stemflow in an orange plantation. Israel J. Agr. Res. 18, 3—14 (1968).

KOSCHMIEDER, H.: Methods and results of definite rain measurements. III. Danzig Report. Mon. Wea. Rev. 65, 5—7 (1934).

LEE, R. D., and D. M. GATES: Diffusion resistance in leaves as related to their stomatal anatomy and microstructure. Amer. J. Botany 51, 963—975 (1964).

McGUINESS, J. L.: A comparison of lysimeter catch and rain gage catch. October 1966 ARS 41-124 U.S.Dept. Agriculture, 1966.

MOLCHANOV, A. A.: The hydrological role of forests. Translated by A. GOUREVITCH. Jerusalem: Israel Program for Scientific Translations 1963.

MONTEITH, J. L.: Gas exchange in plant communities, pp. 95—112. In Environmental control of plant growth (EVANS, L. T. Ed.): New York: Academic Press 1963.

— Evaporation and environment, pp. 205—236. In: The state and movement of water in living/organism (FOGG, G. E. Ed.). XIX Symp. Soc. Expt. Biol. 1964. Cambridge Univ. Press 1965.

PENMAN, H. L.: Estimating evaporation. Trans. Amer. Geophys. Un. 37, 43—50 (1956).

— Vegetation and hydrology. Tech. Comm. 53. Farnham Royal: Commonwealth Agricultural Bureaux 1963.

— Evaporation from forests: a comparison of theory and observation, pp. 373—380. In: International Symposium on Forest Hydrology (SOPPER, W. E. and H. W. LULL, Eds.) London: Pergamon Press 1967.

Pereira, H. C.: Effects of land-use on the water and energy budgets of tropical watersheds, pp. 435—450. In: International Symposium on Forest Hydrology (Sopper, W. E. and H. W. Lull, Eds.). London: Pergamon Press 1967.

Raschke, K.: Über den Einfluß der Diffusions wide stande auf die Transpiration auf die Temperatur eines Blattes. Flora (Jena) 146, 546—578 (1958).

Reifsnyder, W. E., and H. W. Lull: Radiant energy in relation to forests. U. S. Dept. Agr. Tech. Bull. 1344, 1965.

Rutter, A. J.: An analysis of evaporation from a stand of Scots Pine, pp. 403—417. In: International Symposium on Forest Hydrology (Sopper, W. E. and H. W. Lull, Eds.). London: Pergamon Press 1967.

Senderova, G. M., et al. (Eds.): Physical-geographic atlas of the world. Moscow: Acad. Sci. U.S.S.R. 1964.

Shachori, A. Y., and A. Michaeli: Water yields of forest, maqui and grass covers in semi-arid regions. A literature review. UNESCO and Zone Research 25. Methodology of Plant Ecophysiology. Proc. Montpellier Symposium. Paris: UNESCO 1965.

Stanhill, G.: Rainfall measurements at ground level. Weather 13, 33 (1959).

Struzer, L. R., I. N. Nechayev, and E. G. Bogdanova: Systematic errors of measurements of atmospheric precipitation. Soviet Hydrology 5, 500—504 (1965).

Tanner, C. B., and W. L. Pelton: Potential evapotranspiration estimates by the approximate energy balance method of Penman. J. Geophys. Res. 65, 3391—3413 (1960).

Uryvaev, V. A.: Eksperimental'nye gidrologicheskie issledovaniya na Voldae (Experimental hydrological investigations at Valdai), Leningrad: Gidrometeoizoat 1953.

Waggoner, P. E.: Calibration of a porometer in terms of diffusive resistance. Agr. Meteorol. 2, 317—329 (1965).

—, and J. D. Hewlett: Test of a transpiration inhibitor on a forested watershed. Water Res. 1, 391—396 (1965).

Zinke, P. J.: Forest interception studies in the United States, pp. 137—161. In: International Symposium on Forest Hydrology (Sopper, W. E. and H. W. Lull, Eds.). London: Pergamon Press 1967.

Measurement and Sampling of Outputs from Watersheds

Daniel J. Nelson

The major outputs of materials from drainage basins include water and its associated load of dissolved and particulate material. Since small watershed study areas are manageable research units, the accrual and losses of materials from these prescribed basins may be quantified and related to rate processes associated with hydrologic and geochemical budgets. Accrual and loss processes affect site fertility of the terrestrial environment and losses from the landscape become inputs to associated aquatic environments. Thus, knowledge of the processes affecting the landscape are important for understanding consequent effects on surface waters.

Water Outputs

The output of water from watersheds via stream discharge is measured readily and routinely either in natural control sections or through the use of artificial control structures (Fig. 1). Selection of natural control sections in a stream is dependent upon a stream reach where channel stability is unchanged with time. Each reach must be rated individually, and it often requires years to obtain a satisfactory rating curve, particularly at the higher discharges. The accuracy of discharge measurements in natural control reaches is rarely better than ± 5%. Natural control sections for measuring stream discharge are generally more applicable to large streams where the construction of an artificial control structure would be physically difficult or prohibitive in cost. However, the availability of discharge data from dams constructed for purposes such as for power generation, flood control or navigation should not be overlooked in studies involving large areas of the landscape and major rivers.

Streams draining experimental watersheds are usually small and the construction of artificial control structures such as weirs is feasible. The general types of control structures used in small watershed research are described by REINHART and PIERCE (1964). Water discharge as measured by weirs is generally quite accurate and a carefully installed and maintained sharp-crested weir may yield measurements accurate within 1 or 2 percent (ROTHACHER and MINER, 1967). The accuracy of a weir design may be verified in hydraulic laboratories. Artificial control structures may be built so that previously developed rating formulas can be applied to obtain discharge data. However, the latter approach is not completely infallible, and ROTHACHER and MINER (1967) recommend measurement of discharge to verify ratings in the actual field installations, particularly when the head of water over sharp-crested weirs is less than 0.2 ft (6.1 cm).

Recording of stage height data is accomplished with a variety of water level recording devices. The availability of automated stage height recorders with a paper tape punch or telemetering systems enables investigators to accumulate data at a rate virtually impossible with the older, analog systems. Furthermore, use of automated

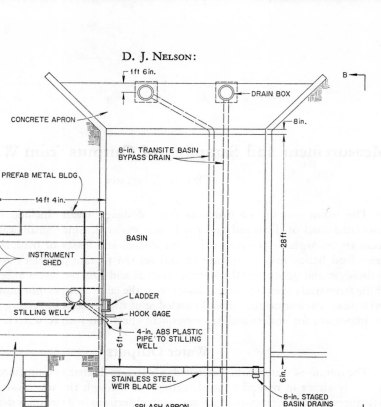

Fig. 1 a

PLAN VIEW

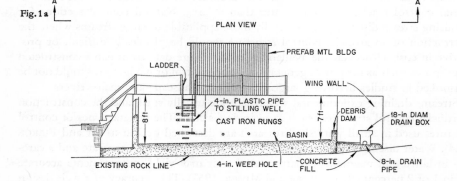

VIEW B-B

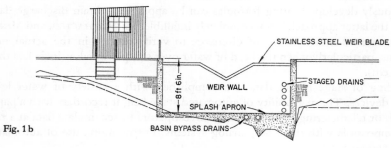

Fig. 1 b

VIEW A-A

systems reduces human errors occurring in the transcription of data. Data may be recorded at 5 min intervals which is desirable when studying hydrologic responses of experimental watersheds.

Use of automatic equipment does not relieve the investigator from the responsibility of visiting the recording installation daily to check on proper operation of equipment. A particular problem in V-notch weirs (Fig. 1) is the entrapment of leaves, twigs or other debris in the notch. At times of low flows, algae will accumulate in the V-notch but they may be removed readily by passing a finger over the edge of the weir blade. A well-maintained control structure is necessary for obtaining accurate data on stream discharge.

Water Sampling

A portion of the discharge stream must be sampled to determine quantitatively the output of materials carried by the water. Sampling methods include continuous proportional samples, continuous nonproportional samples and grab samples. The sampling method chosen will depend on the objectives of the research and the facilities available.

A grab sample is by far the simplest and, in some instances, may be as adequate as samples obtained from more sophisticated sampling equipment. When the analyses of grab samples are related to instantaneous stream discharge, correlations can be developed between discharge and an independent variable such as particulate matter or dissolved materials. Using this approach it is possible to calculate material budgets for dissolved or particulate matter effluent from the watershed. Also, where the chemical quality of the water does not change with stream discharge, grab samples are completely acceptable. Such a circumstance occurs at Hubbard Brook (Thornton, New Hampshire, USA) where BORMAN and LIKENS (1966) observed that total output of dissolved chemical elements was dependent on volume of discharge. Limited data from several Coweeta Hydrologic Station (Franklin, North Carolina, USA) watersheds suggest a similar condition may pertain there.

A continuous nonproportional sampler provides samples having the least value. Chemical quality of the water in nonproportional samples cannot be related to other parameters such as instantaneous stream discharge, and the results are not suited for the development of material budgets for chemical elements leaving the watershed. Nonproportional samplers are available commercially and many of these were developed for special purposes such as sampling water streams in pipes. Attempting to use these samplers in environmental sampling is not recommended unless they are modified to relate the sample to discharge.

A continuous proportional sampler is advisable when water quality changes with discharge. A sampler of this type was developed in connection with the Clinch River Study (STRUXNESS et al., 1967) for obtaining material balances on radionuclides and other chemical elements in the Clinch River. These samplers take water samples in proportion to stream discharge and are suitable for unattended operation. A periodic (weekly) collection of the filled sample container and replacement with an empty con-

Fig. 1. Diagram of Walker Branch Watershed weirs at Oak Ridge, Tennessee, USA. Note the staged drains and bypass drain to be used when the stilling basin is being cleaned

17*

tainer is required. Similar samplers are in use at the Walker Branch Watershed (Oak Ridge, Tennessee, USA) but these were modified with the addition of a fraction collector which permits the collection of up to 250 individual samples during a period manually variable from 1 hour to 16 days (Curlin and Nelson, 1968). A further feature is an electronic sensing device (operational amplifier) which can detect a rising stream hydrograph and automatically initiate operation of the fraction collector. Thus, discrete water samples can be obtained during a storm period when the stream hydrograph rises and recedes. Preliminary grab samples from both forks of Walker Branch show that water quality varies with stream discharge. This new water sampler appears to be a valuable tool for studying changes in water quality as a function of stream discharge or storm flow.

Storm flows account for a large proportion of the sediment carried by streams. The proportion of stream load carried in storm events varies with the type of stream but data from several rivers shows that 50% of the annual load may be carried in from 4 to 95 days (Leopold et al., 1964). Typical of such results are those obtained from the Broad River, Georgia where about 90% of the annual load is discharged in 20% of the year and 50% of the load is carried within 2% of the year (Kennedy, 1963). These relationships are obtained during conditions of normal high discharge and not during catastrophic events. Thus, the importance of sampling streams during high discharge periods is obvious.

Fractionation of Stream Load

Movement of material in the continuing downstream flow of water is in contrast with windblown materials or the activities of animals which may transport matter either into or out of catchment basins. Water-transported materials include the bedload and dissolved and particulate matter which originated either on the watershed itself (allochthonous) or within the streams (autochthonous). Separation of the total stream load into its component parts and derivation of its sources is a task of major importance.

Methods used to sample small streams typical of watershed projects generally are simpler than those required for larger rivers (Interagency Committee on Water Resources 1940, and following at irregular intervals). Bedload of small streams is moved along the bottom, and when weirs with catchment basins are installed it is relatively simple to clean the catchment basin and subsample these materials to determine their component parts. In the Walker Branch weirs we have installed a bypass and staged drains (Fig. 1). The staged drains permit us to lower the water level in the stilling basin gradually to prevent the outwash of trapped particulate material. The basin can be cleaned out while the normal water flow is diverted through the bypass pipes.

The suspended materials in surface streams include dissolved, colloidal and particulate matter. Separation of these into components is usually accomplished either by centrifugation or by filtration (Fig. 2). A standardized separation technique has not been developed for IBP programs (Golterman and Clymo, 1967). Although the importance of the separatory methods for specific elements such as phosphorus has been recognized (Olsen, 1967), it appears that analysts will continue to utilize generally available techniques without regard for classic concepts of dissolved, colloidal, and particulate matter. One of the problems in choosing a separatory method involves application of techniques and equipment which will meet the criteria required to separate the various fractions in water. In actual practice, the colloidal fraction is usually ignored, being considered as part of the particulate fraction or the dissolved fraction or unevenly divided between the two.

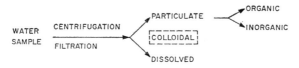

Fig. 2. Schema of fractionation of the particulate and dissolved fractions of the stream load by centrifugation or filtration result in variable separations of the colloidal material. The colloidal and dissolved fractions also include organic and inorganic matter

Filtration is the simplest technique, since it does not require extensive instrumentation. With the widespread use of membrane and glass filters, investigators can readily reproduce experimental methods and filtration has become generally acceptable. The 0.45 μ pore size membrane filter which is used for water filtration is a typical example of a widely used filtration device. The 0.45 μ pore size implies a retention of particles 0.45 μ and larger. Actually these filters retain smaller-size particles and colloids through adsorption and partial occlusion of the pores. While water may be passed through filters having decreasing pore size with the idea of obtaining graded separations, the accuracy of the break points in the separations is more implied than real. Filtration of natural water is convenient and use of this method will continue because of its simplicity.

Centrifugation of water to remove particulate matter has been employed widely by aquatic ecologists since the work on lakes by BIRGE and JUDAY (1934). The earlier centrifuge techniques were relatively uncomplicated and employed either continuous flow or bucket-type centrifuges. Centrifuge methods also suffer from the inability to separate effectively colloidal materials from natural waters, but most low speed centrifuges will allow colloidal matter to remain in the supernatant fraction.

The efficiency of conventional centrifuges in removing the particulate fraction from water varies with the type of particulate material. HARTMAN (1958) recommended at least three centrifugings of water samples containing significant quantities of blue-green algae. JUDAY (1926) observed that one centrifuging removed about 98% of the organisms except for a species of blue-green algae. Subsequent centrifugings, up to 11, removed the remaining algal cells and continued to remove silt and suspended organic matter. A continuous flow centrifuge removed about 70% of the total material caught by 0.45 μ membrane filters (NELSON and SCOTT, 1962). Thus, there may be real differences in the quantities of material removed from water by filtration and centrifugation, but a clearer definition of the magnitude of the problem awaits more intensive research on this subject using ultracentrifuges.

The recently developed centrifugation techniques utilizing continuous flow, density-gradient and zonal ultracentrifugation show a great deal of promise for concentrating, isolating and purifying the colloidal and suspended materials in water (LAMMERS, 1967). Separational procedures include the centrifugation of water in continuous flow, high centrifugal force centrifuges which remove particles as small as colloids. Following collection of the colloidal and particulate material, this fraction is resuspended and subjected to a density-gradient centrifugation through a preformed gradient of CsCl, sucrose or the methylglucamine salt of 3,5-diiodo-4-pyridone-N-acetic acid. Organic and inorganic fractions are separated at their isopynic banding point in the gradient. The method is sufficiently sensitive to separate species of algae having density differences of several hundreths of a gram per cc. These separational methods represent the most sophisticated techniques curently available and their general application should be feasible in the near future.

Stream Outputs

Water is the vehicle by which dissolved and suspended materials are removed from drainage basins. The rate of transport of these materials to the oceans on continental and global scales has received attention from the aspect of chemical denudation rates of terrestrial environments. CLARKE's (1924) estimate of worldwide denudation rates is being refined continuously (DURUM, HEIDEL, and TISON, 1960; LIVINGSTONE, 1963) through more detailed studies from restricted areas. Geographical areas studied recently include Finland (VIRO, 1955), Malay Peninsula (KOBAYASHI, 1959) and the Amazon basin (GIBBS, 1967). Data obtained in these types of investigations are usually obtained from periodic grab samples of water from the respective rivers. In this context the mainstem rivers are considered integrative in that their chemical composition reflects inputs from various physiographic regions as well as varying inputs from cultural sources.

Studies of major rivers have been concerned primarily with the specific ionic content of the water for major cations and anions and in some instances for trace elements. With most small watershed research in the United States the output of streams has been expressed as turbidity units or as parts per million (ppm) of sediment (PACKER, 1967). This unfortunate disparity in reporting research results limits the value of much of the data obtained in watershed research. Consequently, it is seldom possible to compare research results between large geographic areas and experimental watersheds. Perhaps this is not surprising since much research on small watersheds has been directed toward determining water or sediment yield in relation to forestry management practices. The transport of materials by larger rivers has been related to the discharge of specific ions or to the transport of sediments as a function of discharge.

The research on rivers relating stream load to stream discharge is informative with respect to inputs of material to surface waters from the watershed (LEOPOLD and MADDOCK, 1953). The model relating these parameters is usually of the type $Y = aX^b$ where Y is load and X is discharge. Load can refer individually to several components comprising the total load. A diagramatic representation of these relationships is shown in Fig. 3. When the slope, b, equals one, the relationship is actually arithmetic. Slopes of one would be characteristic of streams where each added increment of discharge is accompanied by an equivalent increment of load. This condition would be applicable to constant flowing springs or well-regulated watersheds which might be typical of many forested watersheds. The Hubbard Brook watersheds respond to dissolved cations in this manner (LIKENS et al., 1967). Slopes greater than one are characteristic of stream load-discharge relationships during periods of heavy rainfall when surface run-off occurs. The surface run-off carries suspended sediment and organic debris at a rate greater than the accrual of water. Slopes less than one are typical of situations where there is a dilution of materials in the stream and stream bed by the incoming water.

The organic particulate fraction was constantly 13% of the total particulate fraction in the Middle Oconee River (NELSON and SCOTT, 1962) and a quite similar 15.8% in the Little Miami River (WEBER and MOORE, 1967). Thus, the same factors

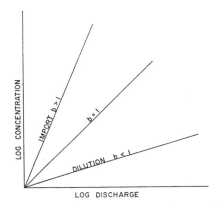

Fig. 3. Stream load-stream discharge relationships. When $b = 1$, the increase in stream load is directly proportional to the increase in discharge. When $b > 1$ the accrual of stream load is greater than the accrual of water and when $b < 1$, the materials within the stream are diluted by incoming water

appear to affect the behavior of organic and inorganic matter in the rivers. This suggested that physical rather than biological factors were initially responsible for the levels of particulate material in the river water. In both rivers there were distinct seasonal differences in the slope of the regression equations relating particulate load to stream discharge. The slopes were greater during winter conditions (November through April) than during summer conditions (May through October). Greater slopes during winter conditions show the effect of greater direct run-off from winter rains.

A comparison of slopes during summer conditions in the two rivers is interesting, because autochthonous primary productivity in the Middle Oconee River was insignificant while productivity in the Little Miami River comprised 38% of the particulate organic matter. The regression slope in the Middle Oconee was slightly less than one showing the dilution of materials present in the stream. The slope in the Little Miami River was negative indicating a significant dilution of the autochthonous algal population by incoming water. A comparison of these results suggests the slope of the regression of particulate organic matter on stream discharge can be used as an indicator of the relationship between autochthonous production and allochthonous organic matter in streams.

The dissolved (probably including colloidal) organic matter in the Middle Oconee River normally was two to ten times as abundant as the particulate fraction. Actual levels of dissolved organic matter ranged from 8 to 47 mg/l with most samples having from 10 to 20 mg/l. Organic matter in the Little Miami River ranged from 5 to 25 mg/l and, on the average, was 2.65 times more abundant than the particulate organic fraction. In both rivers the particulate organic fraction exceeded the dissolved organic fractions during periods of storm discharge. Total dissolved materials carried by base flow of a small stream in Maryland was 1.5 metric tons per year while the suspended material was only 0.28 metric tons per year (BRICKER et al., 1968). Quantitatively, dissolved materials constitute a significant proportion of both the inorganic and organic load carried by streams.

The input of tree leaves to streams constitutes a major source of food for aquatic organisms in the form of organic detritus. The actual mechanism by which this food source is utilized is unknown but it is generally conceded that either bacteria, fungi, or both are important links in the food chain. In addition, leaves may have a significant effect on water quality, especially during low discharges (SLACK 1964; SLACK and FELTZ, 1968). Dissolved oxygen and pH decreased during the peak of leaf fall while conductivity, iron, manganese and bicarbonate concentrations increased. Thus, leaves can have important geochemical implications as well as serving as an energy source for aquatic organisms.

The studies referred to previously have contributed valuable information with respect to the functioning of portions of ecosystems. These researches represent information from streams and rivers in diverse habitats. Techniques and data development in prior work will be useful for preliminary synthesis and testing of models of ecosystems. Ultimately, we can expect to obtain a comprehensive suite of data from a single ecological unit of the type described by BORMANN and LIKENS (1967). The Hubbard Brook watershed is beginning to provide unique information with respect to chemical budgets of a manageable experimental unit (LIKENS et al.,

Table 1. *Chemical outputs from experimental study areas. All data are expressed as kg ha^{-1} year^{-1}*

Area and author	Ca	Mg	Na	K	SO$_4$	Cl	P	N	SiO$_2$
Hubbard Brook									
LIKENS et al., 1967	12.8	2.6	5.9	1.8					
JUANG and JOHNSON, 1967						4.1			
BORMANN et al., 1968								1.8	
Pennine Moorland									
CRISP, 1966	58.6		45.5	10.4			0.86	17.7	
Finnish Rivers (mean)									
VIRO, 1953	12.0	4.35	5.6	4.57	14.2	5.61	0.26	1.87	26.4
New Zealand									
MILLER, 1963	27	13	63	13	40	128	0.03	1.8	

1967). Chemical budgets of small watersheds can be determined with relatively good accuracy and these results will be useful for understanding the maintenance of mineral balance of natural systems. Quantitative estimates of chemical outputs from experimental areas are in Table 1. Because of geologic settings, values for cations may represent lower estimates of loss rates from many watersheds. The extreme range in chloride export reflects the marine influence in New Zealand in contrast to the continental sources at Hubbard Brook. Sufficient data are not available from contrasting areas to make generalizations with respect to biotic effects on chemical cycles.

Small watersheds are ideally suited for studies of the effect of cultural activities on the landscape on mineral movement from land into water. The accelerated loss of calcium, magnesium, sodium, potassium, and nitrate from the catchment basin into the stream draining the area is a significant, immediate effect of clear-cutting of a forest ecosystem (BORMANN et al., 1968). The inputs of nutrient elements to surface water are important for understanding productivity of aquatic habitats. While it may be convenient to consider terrestrial and aquatic habitats separately, they are inextricably related and effective models of ecosystems must consider the relationship.

The importance of land-water relationships was postulated by NELSON (1967) as the result of a comparison of the trace element content of contemporary and pre-Columbian clamshells from the Tennessee River. The pre-Columbian shells came from Indian middens 1000 to 2000 years old. These shells contained concentrations of Sr, Ba and Mn 50 to 100% higher than contemporary shells of the same species. It was suggested that the destruction of forests by an agricultural economy resulted in reduced concentrations of carbon dioxide in the soil and consequently in reduced rates of dissolution of alkaline earth carbonates from the soil matrix. Hence, in prehistoric times river water contained more dissolved minerals than occur presently. Mineral element budgets developed from catchment basin study areas will be useful for testing hypothesis relating the effect of land management practices on water quality in streams.

An input and output budget of several chemical elements and water was developed for a moorland catchment basin in northern England (CRISP, 1966). Since this was a grazed land, estimates were made of the chemical outputs included in sheep removed from the area in normal agricultural practice. These losses were small when compared with losses of sodium, potassium, calcium, phosphorus and nitrogen dissolved in water. Surprisingly, the erosion of peat from the area constituted the most significant loss of nitrogen. Development of data from contrasting areas is an important consideration for the International Biological Program. In this respect Hubbard Brook, on crystalline rocks, will be an interesting comparison with the Walker Branch Watershed (CURLIN and NELSON, 1968) which is on sedimentary rock. As data become available from these and other watersheds, such as those in the Oregon Cascades (ROTHACHER et al., 1967), we will obtain a valuable background of the biogeochemical processes affecting relatively undistrubed catchment basins. This background information is necessary in order to determine the effect of human activities on some of the less obvious but important environmental processes.

Acknowledgements

Research sponsored by the US Atomic Energy Commission under contract with the Union Carbide Corporation.

References

BIRGE, E. A., and C. JUDAY: Particulate and dissolved organic matter in inland lakes. Ecological Monogr. 4, 440—474 (1943).

BORMANN, F. H., and G. E. LIKENS: A watershed approach to problems of nutrient cycling in forest ecosystems. Proc. Sixth World Forestry Congress (Madrid) (1966).

—, and G. E. LIKENS: Nutrient cycling. Science **155**, 424—429 (1967).

— —, D. W. FISHER, and R. S. PIERCE: Nutrient loss accelerated by clear-cutting of a forest ecosystem. Science **159**, 882—884 (1968).

BRICKER, P. O., A. E. GODFREY, and E. T. CLEAVES: Mineral-water interactions during the chemical weathering of silicates. pp. 128—142. In: Trace inorganics in water, (BAKER, R. A. Chairman): Washington, D. C.: Amer. Chem. Soc. 1968.

CLARKE, F. W.: The data of geochemistry. U. S. Geol. Surv. Bull. 770, 841 p., 1924.

CRISP, D. T.: Input and output of minerals for an area of Pennine Moorland: The importance of precipitation, drainage, peat erosion and animals. J. Appl. Ecol. **3**, 327—348 (1966).

CURLIN, J. W., and D. J. NELSON: Walker Branch watershed project objectives, facilities, and ecological characteristics. U. S., AEC Doc. ORNL-TM-2271 (1968).

DURUM, W. H., S. G. HEIDEL, and L. J. TISON: World-wide runoff of dissolved solids. Int. Assoc. Sci. Hydrol. Pub. **51**, 618—628 (1960).

GIBBS, R. J.: Amazon River: Environmental factors that control its dissolved and suspended load. Science **156**, 1734—1737 (1967).

GOLTERMAN, H. L., and R. S. CLYMO: Chemical environment in the aquatic habitat. Amsterdam: N. V. Noord-Hollandsche Vitgevers Maatschappij 1967.

HARTMAN, R. T.: Studies of plankton centrifuge efficiency. Ecology **39**, 374—376 (1958)

Interagency Committee on Water Resources. Report 1 (1940) and following at irregular intervals. St. Anthony Falls Hydraulic Laboratory. Minneapolis, Minn. (A series of reports on sampling and analysis methods for bedload and suspended sediments.)

JUANG, F. H. T., and N. M. JOHNSON: Cycling of chlorine through a forested watershed in New England. J. Geophys. Res. **72**, 5641—5647 (1967).

JUDAY, C.: A third report on limnological apparatus. Trans. Wisc. Acad. Sci. Arts, and Lit. **22**, 229 (1926).

KENNEDY, V. C.: Sediment transported by Georgia streams. U. S. Geol. Surv. Water-Supply Paper. 1668. (1964).

KOBAYASHII, J.: Chemical investigation on river waters of southeastern Asiatic countries (Report I). The quality of waters of Thailand. Berichte d. Ohara Institute **11**, 167—233 (1959).

LAMMERS, W. T.: Separation of suspended and colloidal particles from natural water. Envir. Sci. Tech. **1**, 52—57 (1967).

LEOPOLD, L. B., and T. MADDOCK JR.: The hydraulic geometry of stream channels and some physiographic implications. U. S. Geol. Surv. Prof. Paper 252, 1953.

—, M. G. WOLMAN, and J. P. MILLER: Fluvial processes in geomorphology. San Francisco: Freeman 1964.

LIKENS, G. E., F. H. BORMANN, N. M. JOHNSON, and R. S. PIERCE: The calcium, magnesium, potassium, and sodium budgets for a small forested ecosystem. Ecology **48**, 772—785 (1967).

LIVINGSTONE, D. A.: Chemical composition of rivers and lakes. U. S. Geol. Surv. Prof. Paper. 440-G., 1963.

MILLER, R. B.: Plant nutrients in hard beech. III. The cycle of nutrients. New Zealand J. Sci. **6**, 388—413 (1963).

NELSON, D. J., and D. C. SCOTT: Role of detritus in the productivity of a rock-outcrop community in a Piedmont stream. Limnol. Oceanogr. **7**, 396—413 (1962).

— Microchemical constitutents in contemporary and pre-Columbian clamshell, pp. 185—204. In: Quaternary paleoecology, (CUSHING, E. J., and H. E. WRIGHT JR. Eds.). New Haven: Yale Univ. Press. 1967.

OLSEN, S.: Recent trends in the determination of orthophosphate in water, pp. 63—105. In: Chemical environment in the aquatic habitat (GOLTERMAN, H. L., and R. S. CLYMO Eds.). Amsterdam: N. V. Noord-Hollandsche Vitgevers Maatschappij 1967.

PACKER, P. E.: Forest treatment effects on water quality, pp. 687—699. In: International Symposium on Forest Hydrology (SOPPER, W. E., and H. W. LULL Eds.). New York: Pergamon Press 1967.

REINHART, K. G., and R. S. PIERCE: Stream gaging stations for research on small watersheds. U. S. Dept. Agr. Handbook 268, 1964.

ROTHACHER, J., and N. MINER: Accuracy of measurement of runoff from experimental watersheds, pp. 705—713. In: International Symposium on Forest Hydrolygy (SOPPER, W. E., and H. W. LULL Eds.). New York: Pergamon Press 1967.

—, C. T. DYRNESS, and RICHARD L. FREDRICKSON: Hydrologic and related characteristics of three small watersheds in the Oregon Cascades. U. S. Dept. Agr. Pac. NW Forest and Range Expt. Sta. 1967.

SLACK, K. V.: Effect of leaves on water quality in the Cacapon River, West Virginia. U. S. Geol. Surv. Prof. Paper. 475-D, 181—185 (1964).

—, and H. R. FELTZ: Tree leaf control on low flow water quality in a small Virginia stream. Environ. Sci. Tech. 2, 126—131 (1968).

STRUXNESS, E. G., P. H. CARRIGAN JR., M. A. CHURCHILL, K. E. COWSER, R. J. MORTON, D. J. NELSON, and F. L. PARKER: Comprehensive report of the Clinch River study. U. S. AEC Doc. ORNL-4035 (1967).

VIRO, P. J.: Loss of nutrients and the natural nutrient balance of the soil in Finland. Pub. Forest. Res. Inst. in Finland 42, 1—45 (1953).

WEBER, C. I., and D. R. MOORE :Phytoplankton, seston and dissolved organic carbon in the Little Miami River at Cincinnati, Ohio. Limnol. Oceanogr. 12, 311—318 (1967).

Models of the Hydrologic Cycle

J. W. Curlin

Properties of Hydrologic Phenomena

The study of hydrologic processes or systems uses historical data collected from natural phenomena which are observed once, and only once, and probably will not be observed again under exactly the same conditions. All hydrologic processes are

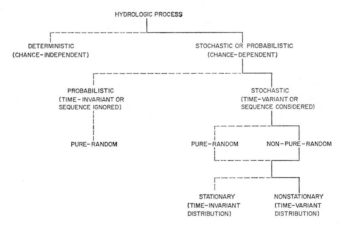

Fig. 1. Classification of hydrologic phenomena with respect to their mathematical properties

more or less stochastic, and as such are subject to the vagaries of probability. Furthermore, these stochastic processes are generally nonpure-random and also nonstationary; i.e., the probability distribution changes during the course of the process making them time-dependent. Chow (1964a) outlined a dichotomous scheme for classifying hydrologic phenomena by mathematical definition (Fig. 1). This rational classification emphasizes the mathematical complexity of analyzing and modeling hydrologic data.

Mathematically, a stochastic hydrologic system is represented by an array of nonpure-random, nonstationary variables $\chi_{(t)}$ which are a function of time t, and whose variates x_t extend in time within a period of hydrologic record, time T. Precipitation, evapotranspiration, and streamflow data are generally measured or converted to discrete observations at uniform intervals of time $t_1, t_2, t_3 \ldots t_n$; thus, the sequence of sample record forms a discrete time series of observations $x_1, x_2, x_3 \ldots x_n$, for a period of record time T at interval Δt.

The output from a catchment system depends on the nature of the input, the physical laws involved, and the nature of the system itself. As an input-output or "black box" system, the catchment responds to precipitation input in a manner determined by the complex interaction of physical and biological factors, not the least being antecedent moisture conditions.

Dooge (1967) conceived the catchment as a system consisting of a set of 3 simplified subsystems (Fig. 2). He emphasized that the hydrologic system may be considered closed only if the hydrologic cycle is described exclusively in terms of moisture movement, as opposed to considering the surplus incoming radiation and energy changes which drive the system. The system is sequential in that it consists of inputs, throughputs, and outputs.

Classical hydrologists once concerned themselves only with runoff phenomena, leaving the measurement of evapotranspiration to the climatologist and relegating the study of moisture movement in the regolith and lithosphere to the geologist and soil

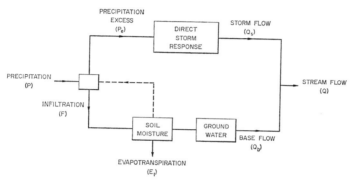

Fig. 2. A simplified box diagram of the components and pathways in a hydrologic system

scientist. The investigation of a closed system, however, requires that each major flow component be evaluated so that continuity of the system is assured. Prior to the evolution of stochastic hydrology and systems hydrology, the major emphasis was on analysis of the direct storm response subsystem. The unsaturated flow of moisture has not been studied in depth as a system and still remains one of the most difficult parts of the hydrologic cycle to quantify.

An additional problem facing the eco-hydrologist is that of interfacing the terrestrial mineral cycle with that of the water cycle. Since water is the carrier for the mineral throughput, the two subsystems are intimately coupled in a manner yet unknown.

A latent knowledge for building conceptual hydrologic models has existed for some time. But only during the past 15 years has the technology of meteorology, probability-statistics, hydrology, and computer science provided the devices for assembling, testing, and manipulaitng these models. Now we can bring to bear on the problem such powerful tools as linear and nonlinear programming, queueing theory, decision theory, power spectral analysis, Monte Carlo techniques, and multivariate analysis. The high speed, digital computer, however, must rank as the single most powerful influence.

Analysis of Direct Storm Response
Basic Runoff Models

The Instantaneous Unit-Hydrograph

By definition, *surface runoff* is the streamflow that results from precipitation which exceeds the infiltration capacity of the catchment basin. The causal relationship between precipitation and runoff distribution has received most of the attention of quantitative hydrologists. The information needed for analysis of the catchment

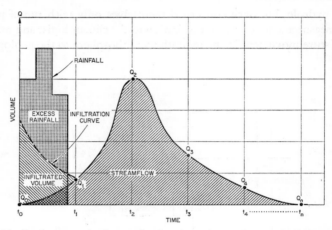

Fig. 3. An idealized storm hydrograph showing the surface runoff Q_t resulting from rainfall excess produced when precipitation rate exceeds infiltration capacity

system in the simple input-output context is precipitation records and corresponding streamflow records (the hydrograph); these data are preferably collected at the same time interval (Fig. 3).

Sherman (1932) proposed the *unit hydrograph* theory for evaluating the input-output system. It is defined as the hydrograph of direct runoff resulting from one inch of effective rainfall generated uniformly over the basin at a uniform rate during a specified period of time. The application of unit-hydrograph theory has become the focus for derivation of more sophisticated stream-response models. Implicit in unit hydrograph theory are the assumptions of linearity and time-invariance — these assumptions are seldom satisfied.

An abstraction of the unit-hydrograph was devised by Clark (1945) who assumed that the effective precipitation is applied to the catchment in zero time. This fictitous situation has become a highly useful concept called the *instantaneous unit hydrograph* (IUH). One advantage of the IUH is that the assumption of uniform distribution in time becomes irrelevant, and by decomposition of the inputs into characteristic signals, the direct runoff hydrograph can be reconstructed from the resulting model through superposition.

$$Q(t) = \int_0^t u(t-\tau) I(\tau) d\tau \qquad (1)$$

where $Q(t)$ is stream discharge at time t, $I(\tau)$ is the input function of effective precipitation, and the kernal function $u(t-\tau)$ is the instantaneous unit hydrograph at time $t-\tau$. The convolution of $I(\tau)$ and the IUH is shown in Fig. 4. Several methods have been developed for evaluating the convolution integral; these have been reviewed by DOOGE (1967) and CHOW (1964b).

Transform solutions and correlation methods both have been used successfully for deriving the IUH. One of the more common transform methods, explored by AMOROCHO and ORLOB (1961), employs the Laplace transform; others use Fourier,

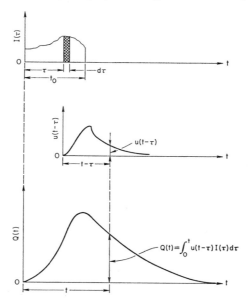

Fig. 4. A graphic explanation of the development of the Instantaneous Unit Hydrograph (IUH)

Laguerre coefficients, and harmonic coefficients. The Laplace transform applied to the convolution integral becomes

$$\overline{Q}(s) = \overline{U}(s) \cdot \overline{I}(s) \tag{2}$$

where $\overline{Q}(s)$ is the Laplace transform of the output (response transform)

$$\overline{Q}(s) = LQ(t)\int_0^\infty e^{-st}Q(t)\,dt \tag{3}$$

and likewise $\overline{U}(s)$ is the transfer function of the system, and $\overline{I}(s)$ is the driving transform. Thus,

$$\overline{U}(s) = \frac{\overline{Q}(s)}{\overline{I}(s)} \tag{4}$$

and

$$u(t) = L^{-1}\frac{\overline{Q}(s)}{\overline{I}(s)}. \tag{5}$$

The IUH, therefore, is the inverse Laplace transform of the ratio of the driving transform to the response transform. For numerical evaluation of the Laplace transform, the integral form of the function may be replaced with a summation equation by operating on small time intervals of the direct-runoff hydrograph and the effective-rainfall hydrograph.

Correlation techniques for IUH solution include the least squares method (SNYDER, 1955) and time series analysis (EAGLESON et al., 1966). The least squares solution is well adapted to computer computation through matrix operations. In matrix notation the convolution equation may be evaluated as

$$[Q]_{P,1} = [I]_{P,N} [u]_{N,1} . \tag{6}$$

The $[I]$ matrix is not square and cannot be inverted, but, if the product of the original matrix and its transpose is used, a square matrix results which can be inverted.

$$[I]^T[Q] = [I]^T[I][u] . \tag{7}$$

The resulting set of equations can be solved for values of the impulse response. Hydrologic analysis of the Walker Branch Watershed ecosystem study at Oak Ridge National Laboratory uses the least-square solution for solving the discrete form of the convolution integral (SNYDER and CURLIN, 1969).

While both the least-squares and time series methods seek to optimize the kernal function, $u_{opt}(t)$, the latter uses the Wiener-Hopf equations (WEINER, 1949) to operate on the summation form of the convolution integral Eq. (1), derived from the discrete time series of precipitation and runoff.

$$Q(i) = \sum_{t=1}^{\infty} u(t) I(i - t + 1) \Delta t \tag{8}$$

where i denotes the number representing a discrete block of precipitation or runoff from the effective-rainfall hydrograph and direct-runoff hydrograph, respectively, and Δt is the time-width of each block. The discrete time form of the Wiener-Hopf equation then reduces solution of the convolution integral to

$$\phi_{QI}(j - t) = \sum_{t=1}^{\infty} u_{opt} \phi_{II}(j - t) \tag{9}$$

where ϕ_{QI} is the cross-correlation between effective precipitation and runoff at lag $j - t$, ϕ_{II} is the autocorrelation between precipitation at lag $j - t$. The advantage of this method, according to the authors (EAGELSON et al., 1966), is the stability of the solution which reduces the oscillations often observed with the least-squares method when data are encountered which do not satisfy the assumption of linearity.

All three approaches to the evaluation of the IUH should give nearly identical results if the assumption of linearity is approximated and if system noise is not excessive.

Conceptual Models of Streamflow

In recent years unit-hydrograph theory has been extended to the development of conceptual models which describe the stream response system. These models are based on the concept of *linear reservoirs* connected by *linear channels* which simulate the storage and delay effects of a catchment basin on its runoff. A discussion of these models is found in CHOW (1964c).

A *linear reservoir* is a fictitious reservoir in which the storage S is directly proportional to the outflow Q, so that

$$S = KQ \tag{10}$$

where K is a reservoir constant. The rate of change of storage S is the difference between inflow to the reservoir I, and outflow Q, so that the continuity equation is

$$I - Q = K \frac{dS}{dt} \tag{11}$$

when $t = 0$, then

$$Q = I(1 - e^{-t/k}) . \tag{12}$$

If inflow ends at $t = 0$ (instantaneous input), then outflow at $Q_0 = S/K$, thus

$$Q = Q_0 e^{-(t-t_0)/K} \tag{13}$$

where $t - t_0$ represents the time since inflow ended. For an instantaneous unit input ($S = 1$), the outflow is the IUH of the first linear reservoir

$$u(t) = Q = Q_0 e^{-t/K} = \frac{S}{K} e^{-t/K} = \frac{1}{K} e^{-t/K} . \tag{14}$$

Similarly, a *linear channel* is a fictitious channel in which the time T required to transport a discharge Q of any magnitude through a specific reach of channel length x is constant. This implies that at any point the velocity is constant for all discharges, but may vary from point to point along the reach (DOOGE, 1959). An inflow function $I = f(t)$ has an identical outflow function except for a shift in time equal to the time of translation between inflow and outflow, such that $Q = f(t - T)$. The velocity may vary from stream section to stream section along the channel, and at any given section the relationship between discharge Q and water area A is linear.

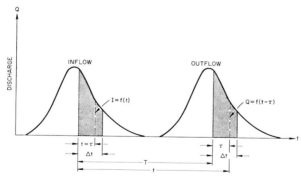

Fig. 5. Routing of a segment of the inflow-outflow hydrograph through a linear channel

If inflow volume S delivered during $\varDelta t$ is routed through a linear channel (Fig. 5), the outflow is

$$Q = S\delta(t, \varDelta t) \tag{15}$$

where

$$\delta(t, \varDelta t) = \frac{1}{\varDelta t} \tag{16}$$

is a *pulse function* (Dirac-delta function) and $t = \tau + T$. When $\varDelta t \to 0$ it is called an *impulse function* $\delta(t)$ and is analogous to the IUH for the linear channel.

Application of this concept to a catchment basin considers a linear channel analogous to spatially varied flow (Fig. 6). The total area a is divided in n subareas of size Δa_j by isochrones which are separated from the outlet by the same translation time. The flow from subarea j is equal to $i_j \Delta a_j$ where i_j is the rate of rainfall on j. Outflow from the catchment is then

$$Q = i_j \Delta a_j \delta(t - T, t) \tag{17}$$

where $T = (j - 1)\Delta t$. If discharge Q is divided by total area a and plotted against t, an area — time distribution of runoff is produced (Fig. 6) which is described by

$$\omega(t) = \sum_{j=1}^{n} \frac{i_j \Delta a_j}{a} \delta(t - T, \Delta t) . \tag{18}$$

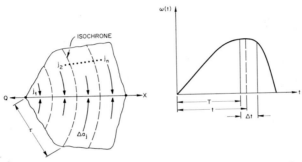

Fig. 6. Development of a time-area diagram for a linear channel. Outflow is in direction of Q

An interesting property of the function $\omega(t)$ is that as $\Delta t \to 0$ and i_j is constant, the ordinates of the time-area curve are related to the shape of the drainage basin and the diagram then depicts time-area-concentration.

NASH (1957), and in several publications thereafter, described a conceptual runoff model based on a series of linear storage reservoirs (Fig. 7). In routing the flow through this series of linear reservoirs, the outflow from the previous reservoir is considered to be the inflow to the second reservoir. Applying Eq. (14) as the input function and substituting τ as a variable for t, and similarly substituting $t - \tau$ in the kernal function, the convolution integral of Eq. (1) produces the outflow from the second reservoir q_2 such that

$$q_2 = \int_0^t \frac{1}{K} e^{-\tau/K} \frac{1}{K} e^{-(t-\tau)/K} d\tau = \frac{t}{K^2} e^{-t/K} . \tag{19}$$

The outflow q_2 then becomes the inflow for the third reservoir, etc. Finally, this procedure applied in turn gives the outflow q_n from the nth reservoir as

$$u(t) = \frac{1}{K(n-1)!} \frac{t^{n-1}}{K} e^{t/K} . \tag{20}$$

This yields the IUH of the simulated drainage basin and mathematically represents a form of the gamma distribution.

DOOGE (1959) extended the linear storage reservoir concept to include the translation of flows inherent in linear channel theory. DOOGE's model represents the catchment basin as a series of alternating storage reservoirs and channel reaches (Fig. 8). Conceptually, the drainage basin is divided into n subareas by isochrones, and each subarea is considered a linear channel connected in series by linear reservoirs.

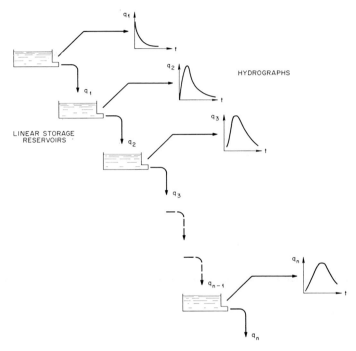

Fig. 7. A conceptual model of a series of linear storage reservoirs, each one feeding the next in sequence

The cumulative outflow from each linear channel is represented by a segment of the time-area diagram and is considered to be inflow to the next linear reservoir. If all of the linear reservoirs are assumed to be of equal size, the IUH, $u(t)$, of the simulated drainage basin can be described as

$$u(t) = \frac{S}{T} \int_0^{t \leq T} P(m, n-1)\, \omega\left(\frac{\tau}{T}\right) dm \tag{21}$$

where $m = \dfrac{t-\tau}{K}$ and S is the input volume, T is the maximum translation time of the catchment, $P(m, n-1)$ is the Poisson probability function, $\omega\left(\dfrac{\tau}{T}\right)$ is the ordinate of the time-area-concentration curve (Fig. 6), n is the number of linear reservoirs, and m is a dimensionless time variable described above.

DOOGE's model accounts for translation effects on streamflow resulting from stream channel characteristics, and is therefore conceptually superior to the Nash

18*

model (Eq. (19)). Others have devised similar models in an attempt to overcome the complexity of applying the Dooge model to practical problems. Several of these were reviewed by CHOW (1964b) including the treatment of nonlinear effects by DISKIN

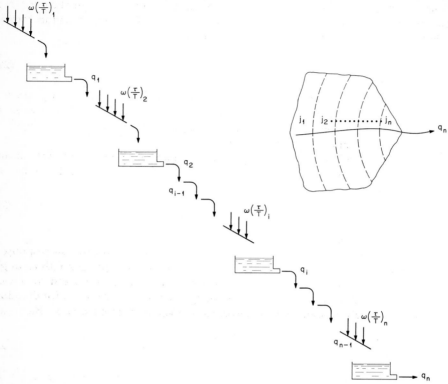

Fig. 8. A conceptual model of routing through alternating linear channels and reservoirs. An idealized catchment showing isocrones and flow segments

(1964) and KULANDAISWAMY (1964). The latter author applied system analysis to the solution of the IUH in a way analogous to the treatment of dynamic mechanical and electrical systems.

Stochastic Properties of Storm Events

Catchment models intended for simulation and synthesis must be sensitive to the cyclical and seasonal patterns of storm occurrence. The recursive properties of precipitation data contain both a short-term, high-frequency, annual periodic component and a long-term, low-frequency, component sometimes spanning a number of years. Both the periodicity and variance of precipitation events at the catchment must be included in a plausible synthesis of precipitation — runoff phenomena.

Most data generation techniques presuppose knowledge of the statistical distribution and statistical moments of the hydrologic phenomena, all of which can be gained from analysis of historical hydrologic records.

Simple Frequency Analysis

Frequency estimates may be derived from an ordered set of rainfall or streamflow data over a period of record for any uniform time base, e.g., a day, month, or year. Simple frequency formulae have been used with the justification that their accuracy matches that of most hydrologic data (BUTLER, 1957). The basic equation for calculating frequency from an ordered data set is

$$F = \left(\frac{m}{n}\right) 100 \tag{22}$$

where m is the order number, n is the total number of entries in the record, and F is the percentage of occurrences with precipitation $\geq m$. The resulting frequencies can be plotted on probability coordinates or a curve can be derived on the computer by least squares or maximum likelihood techniques.

The recurrence interval R is the average number of base-time intervals during which an event of given magnitude may be expected to occur once, and is computed by the equation

$$R = \left(\frac{1}{F}\right) 100 \tag{23}$$

where F is the frequency from Eq. (22) in percent of base intervals.

Another characteristic of precipitation data that influences streamflow response is rainfall intensity. It is defined as the average rate at which precipitation falls throughout a given period of time. When combined with units of duration and frequency, one can derive curves which estimate the average rate of precipitation i, for durations of t minutes, reached or exceeded once on the average of F base intervals (BERNARD, 1942)

$$i = \frac{KF^x}{(t+b)^n} \tag{24}$$

where K, b, x, and n are empirical coefficients which describe a family of intensity-duration curves. The exponent n is the slope of the curves, x determines the vertical spacing of the curves for various recurrence intervals T, and K determines the vertical position of the set of curves. The constants of the equation can be derived empirically in logarithmic form from sufficiently long precipitation records which include information on short period intensities such as records collected by recording raingages.

Power Spectral Analysis

The distribution of variance for a time series in the frequency domain is called a *power spectrum*. Electrical and communications engineers have found spectral analysis useful for identifying cyclical patterns, phase shifts, and lag times from random frequency data (BENDAT and PIERSOL, 1966). The technique has recently been applied to hydrologic problems for identification and isolation of the periodic component from the stochastic component of rainfall and runoff data (ROESNER and YEVDJEVICH, 1966; RODRIQUEZ-ITURBE, 1967). WASTLER (1963) also used spectral analysis for investigating the movement of pollutants in streams and estuaries.

The nonstationary properties of the hydrologic data are overcome by standardization and appropriate transformation. Once transformed, the time series are subjected to autocorrelation analysis. The autocorrelation function for N discrete data values $\{x_n\}$, $n = 1, 2, \ldots, N$, which are transformed to be stationary with $\bar{\bar{x}} = 0$, and a displacement in record time of rh, is defined by the formula

$$\hat{R}_r = R_x(rh) = \frac{1}{N-r} \sum_{n=1}^{N-r} x_n x_{n+r} \quad r + 0, 1, 2, \ldots, m \tag{25}$$

where r is the lag number in time intervals ($\varDelta t$), m is the maximum lag number, and $\hat{R}_r$ is the estimated value of the true autocorrelation coefficient at lag r, corresponding to a displacement in time rh (BENDAT and PIERSOL, 1966). Autocorrelation values, like sample correlation coefficients, may assume the limits $-1 \leq R_r \leq 1$. If the data are not standardized so that $\bar{\bar{x}} = 0$, then the autocorrelation function is termed *autocovariance* $\hat{C}_r$, and the limits of ± 1 do not apply.

Power spectral density is synonymous to variance spectral density. With the same transformed record $x(t)$ and the autocorrelation function $\hat{R}_r$ calculated by Eq. (25), a "raw" estimate of $\tilde{G}_x(f)$ of a true power spectral density function $G_x(f)$ for a frequency f in the range $0 \leq f \leq f_c$ may be derivied with the equation

$$\tilde{G}_x(f) = 2h \left[\hat{R}_0 + 2 \sum_{r=1}^{m-1} \hat{R}_r \cos\left(\frac{\pi r f}{f_c}\right) + \hat{R}_m \cos\left(\frac{\pi m f}{f_c}\right) \right] \tag{26}$$

where h is the time interval between samples, $\hat{R}_r$ is the estimated autocorrelation function at lag r, m is the maximum lag number, and $f_c = 1/2h$ or the cutoff frequency.

Eq. (26) is analogous to making a Fourier cosine transformation of the autocorrelation computed by Eq. (25). The graph of the function $\tilde{G}_x(f)$ is often termed the "periodogram" and that of $\hat{R}_r$ is called the "correlogram". The true spectral density is derived from Eq. (26) by frequency-smoothing the periodogram $\tilde{G}_x(f)$ using the Hanning method (BENDAT and PIERSOL, 1966). Specifically, this amounts to weighting the autocorrelation function nonuniformily to adjust for the fallacious lack of variability inherent in Eq. (25) when dealing with records of different length.

Power spectral analysis can be extended to treat the relationship of precipitation to runoff by substituting cross-correlations rather than autocorrelations. This results in derivation of cross-spectral density estimates which are calculated in a similar manner.

Typical correlograms and periodograms from analysis of monthly precipitation and streamflow data are shown in Fig. 9. The original data were reported by the Tennessee Valley Authority (1961) from 23 years of historical records collected monthly from White Hollow Watershed in Tennessee. Note the diminishing covariance as the time-lag separation between correlated observations becomes longer [(a) and (c)]. Another outstanding characteristic of the data is the strength of the 12-month cycle for streamflow. A similar cycle is apparent for precipitation but is not as well defined. Crosscovariance incorporates the characteristics of both precipitation and streamflow into a time-lag response curve (e).

The corresponding power spectral density curves (b), (d), and (f) transform the scale from time to frequency which emphasizes the harmonics and subharmonics resulting from the repeated occurrence of similar rainfall-runoff patterns throughout the historical records. The annual cycle and its subharmonics recurring at 6-month intervals become more distinct with power spectral analysis.

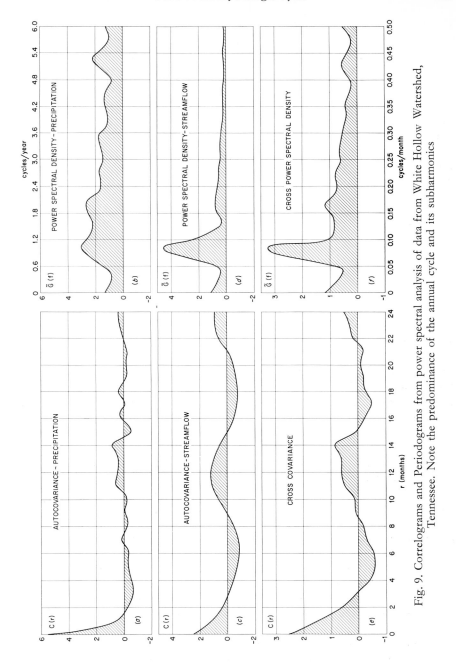

Fig. 9. Correlograms and Periodograms from power spectral analysis of data from White Hollow Watershed, Tennessee. Note the predominance of the annual cycle and its subharmonics

Hydrologic Synthesis

Although the sequential generation of hydrologic data is not new, the recent development of Monte Carlo techniques has stimulated the evolution of a number of new synthetic models. Many of these use autoregressive adaptations of Markov-chains, queueing theory and random walk theory.

Hydrologic time series can be described by various analytical functions of the general form

$$q_t = f(t) + \varepsilon_t \tag{27}$$

where q_t is the magnitude of the hydrologic event at time t, $f(t)$ is a deterministic component, and ε_t is a random component. Depending on the temporal character-istics of the time series, the functional form might be a polynomial of low degree (Eq. (28)), a Fourier sine-cosine cyclical series (Eq. (29)), or a linear autoregressive scheme with a random component (Eq. (30)).

$$q_t = \xi_0 + \xi_1 q_t + \xi_2 q_t^2 + \ldots + \xi_m q_t^m , \tag{28}$$

$$q_t = \alpha_0 + \sum_{i=1}^{m} (\alpha_i \mathrm{Cos}\lambda_i q_t + \beta_i \mathrm{Sin}\lambda_i q_t) , \tag{29}$$

$$q_t = \beta_0 + \beta_1 q_{t-1} + \beta_2 q_{t-2} + \ldots + \beta_m q_{t-m} + \varepsilon_t . \tag{30}$$

Markovian Models

The central assumption of the Markov chain for describing stochastic processes is that the present state of a system can be predicted with probability from a knowledge of the state of the system at the time immediately preceding. The first-order Markov process is described by the linear autoregressive model

$$q_t = r q_{t-1} + \varepsilon(I)_t \tag{31}$$

where q_t might be streamflow at time t, q_{t-1} is the streamflow at the preceding time interval, r is the serial correlation coefficient for streamflow, or a Markov-chain coefficient; and $\varepsilon(I)_t$ is a random uncorrelated component due to rainfall input during time t.

FIERING (1967) described the application of Markovian flow models and extended them to decision theory for use in engineering design studies. ROESNER and YEVD-JEVICH (1966) tested the efficiency of Markov models including log forms for predicting annual and monthly precipitation and runoff and found that a first-order model adequately described the time dependence of the stochastic component. CHOW (1964c) reviewed the principles of hydrologic simulation using Markov chains and cited several successful applications of the technique. Other references to this approach may be found in the proceedings of a recent international hydrology symposium sponsored by Colorado State University (1967).

The Stanford Watershed Model

One of the most interesting and comprehensive approaches to modeling the total hydrologic system is that of CRAWFORD and LINSLEY (1966). While much of the work on hydrologic models has dealt directly with stream response, the Stanford Model attempts to simulate the behavior and processes of the internal subsystems, and not only predict stream output, but also system losses from evapotranspiration and groundwater storage. A flowchart for the watershed model is shown in Fig. 10.

As in the "real world", precipitation and potential evapotranspiration are the causal inputs. By operating on these inputs with mathematical predictors derived from either the literature or experimentation, the throughput is moved through the system in a rational way. The effect of any component on the total system can, therefore, be evaluated by altering the mathematical model at a crucial junction in the flowchart.

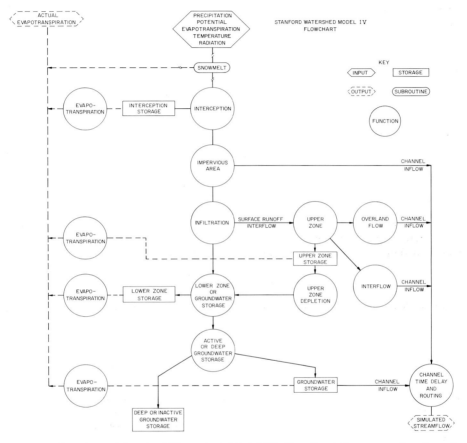

Fig. 10. Flowchart of Stanford Watershed Model IV. Details of the snowmelt subroutine are not shown

For instance, the effect of changes in soil infiltration capacity on streamflow, which could result from unwise land use, might be surmised by assuming different values for the infiltration function in the watershed model. Similarly, disturbance of the vegetation could be simulated by altering the functions for interception and actual evapotranspiration, then observing the net effect on streamflow.

It is beyond the scope of this review to treat the mathematical functions for each process in detail, however, there are several points worthy of note. All precipitation "entering" the system during a finite interval is routed through interception storage until it exceeds a preassigned volume, while evaporation from interception storage is assumed to occur at a rate equal to the potential evapotranspiration. The complex processes of infiltration are evaluated by interlocking calculations which attempt to separate the available soil moisture in the upper soil profile from that in the lower soil zones which feeds the ground-water reservoir. Evapotranspiration is introduced as an input variable but must be systematically reduced, since actual evapotranspiration seldom equals its potential rate. Evaporation from a free water surface is entered as input and is adjusted by a routine which attempts to satisfy the potential evapotranspiration from the interception storage. If the full requirement is not satisfied by interception storage, it is sought from the available soil moisture using a cumulative linear frequency distribution of evapotranspiration.

The channel subsystem recognizes three sources of input: 1) continuous overland flow, 2) interflow, and 3) groundwater flow into the stream. Time delay in translation of channel flow is accomplished by a modified adaption of the method of Clark (1945). The Stanford model treats the channel subsystem separately from the land surface; thus, the time-area curve used by Clark is redefined to represent the time of flow in the channel and is referred to as a time-delay histogram. It is found by determining the area of contributing portions of the catchment, estimating streamflows at successive points in the channel system, and calculating the time of flow to the outlet. Such a curve represents an instantaneous channel inflow analogous to a unit hydrograph, and must be adjusted to simulate inflow of finite duration. As in Clark's method the watershed model employs a reservoir routing procedure to represent the attenuation due to storage in the stream channels.

Comparison between simulated catchment response and actual response of several widely different drainage basins has shown the Stanford Watershed Model to be a reasonably good predictor. It exemplifies the potential that simulation techniques hold for watershed ecosystem research.

Ecologic-Hydrologic Models

Up to now the discussion has dealt with the state of the art of modeling hydrologic systems. As ecologists, this subject in its purest form is but of secondary interest to us. Our main objective is to relate the intricacies of the hydrologic system to that of the adjacent terrestrial system in such a manner that a total ecosystem model evolves.

The hydrologic cycle and the chemical cycles operating in a watershed ecosystem are inseparable (Fig. 11). Chemicals deposited on the catchment by rainfall and atmospheric fallout are the only sources of nutrient income (ignoring animal ingress) unless man adds chemical amendments. Chemicals released by organic decomposition and by dissolution of rocks and soil material are transported through the soil-water system, and finally emerge as streamflow to be lost from that specific drainage basin forever. To gain a useful knowledge of the fluence of chemicals through an ecosystem, one concomitantly must consider the movement of water which serves as the carrier.

Spofford (1965) approached this problem in dealing with the movement of radioactive nuclides in streams. By using applications of queueing theory, probability theory, and finite Markov chains, he attempted to queue two independent stochastic variates simultaneously: namely, water and the contaminant. This application of queueing theory in combination with a Markovian model is an example of the type of imaginative solutions we must seek to simulate and explore ecosystem processes. SPOFFORD's model was only concerned with movement after the pollutant reached the stream channel. Imagine, if you will, the complexity of including the terrestrial

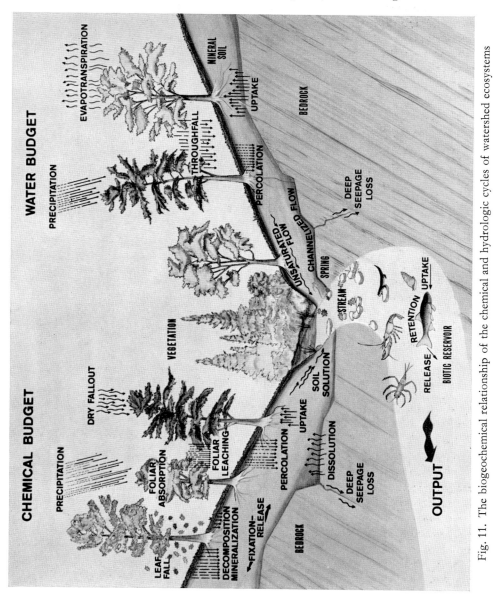

Fig. 11. The biogeochemical relationship of the chemical and hydrologic cycles of watershed ecosystems

nutrient cycles in applying his approach to movement of stable decomposition products through the catchment ecosystem where these materials enter all along the channel rather than at a discrete point. This is just one example of the problems we face in developing systems models for ecosystem analysis.

By analogy, some of the concepts used by quantitative hydrologists seem to have application to other phases of ecosystem modeling. For instance, the concept of linear storage reservoirs seemingly might be applied to the phenomenon of elemental transfer through the terrestrial system. A modification of unit hydrograph theory might apply equally well to other impulse phenomena which can be described by such transfer functions. Power spectral analysis seems to hold promise for exploring the frequency of cyclical phenomena such as seasonal loss and accretion of organic materials on the forest floor, or population dynamics of insects or mammals. The list of potential applications need only be limited by one's imagination.

Acknowledgements

Research sponsored by the US Atomic Energy Commission under contract with the Union Carbide Corporation.

References

AMOROCHO, J., and G. T. ORLOB: Nonlinear analysis of hydrologic systems. Univ. Calif. Water Resources Cent. Contr. No. 40 (1961).

BENDAT, J. S., and A. G. PIERSOL: Measurement and analysis of random data. New York: John Wiley and Sons 1966.

BERNARD, M.: Precipitation, Chap. II. In: Hydrology. (MEINZER, O. E. Ed.), New York: Dover 1942.

BUTLER, S. S.: Engineering hydrology. Englewood Cliffs, N. J.: Prentice-Hall 1957.

CHOW, V. T.: Statistical and probability analysis of hydrologic data. Sec. 8, Part I. Frequency analysis. In: Handbook of applied hydrology. (CHOW, V. T. Ed.) New York: McGraw-Hill 1964a.

— Runoff, Sec. 14. In: Handbook of applied hydrology (CHOW, V. T. Ed.), New York: McGraw-Hill 1964b.

— Statistical and probability analysis of hydrologic data, Sec. 8, Part IV. Statistical and probability analysis of hydrologic data. In: Handbook of applied Hydrology. (CHOW, V. T. Ed.) New York: McGraw-Hill 1964c.

CLARK, C. O.: Storage and the unit hydrograph. Trans. Amer. Soc. Civil Engr. 110, 1419—1446 (1945).

Colorado State University: Proceedings of the International Hydrology Symposium. Fort Collins, Col. Sept. 6—8, 1967.

CRAWFORD, N. H., and R. K. LINSLEY: Digital simulation in hydrology: Stanford watershed model IV. Stanford Univ. Tech. Rep. No. 39, 1966.

DISKIN, M. H.: A basin study of the linearity of the rainfall-runoff process in watersheds. Ph. D. Thesis, University of Illinois, Urbana, Ill. 1964.

DOOGE, J. C. I.: A general theory of the unit hydrograph. J. Geophys. Res. 64, 241—256 (1959).

— The hydrologic system as a closed system, pp. 98—113. In: Proceedings of the International Hydrology Symposium, Vol. 2, Sept. 6-8, 1967. Fort Collins, Colorado 1967.

EAGELSON, P. S., M. RICARDO, and M. FREDERIC: Computation of optimum realizable unit hydrographs. Water Res. 2, 755—764 (1966).

FIERING, M. B.: Streamflow synthesis. Cambridge, Mass.: Harvard Univ. Press 1967.

KULANDAISWAMY, V. C.: A basic study of the rainfall excess-surface runoff relationship in a basin system. Ph. D. Thesis, University of Illinois, Urbana, Ill., 1964.

NASH, J. E.: The form of the instantaneous unit hydrograph. Intern. Assoc. Sci. Hydrology. Pub. 45, 114—121 (1957).

RODRIGUEZ-ITURBE, I.: The application of cross-spectral analysis to hydrologic time series. Hydrologic Papers No. 24, Colo. State Univ., Fort Collins, Colo. 1967.

ROESNER, L. A., and V. M. YEVDJEVICH: Mathematical models for time series of monthly precipitation and monthly runoff. Hydrology Papers No. 15, Colo. State Univ., Fort Collins, Colo., 1966.

SHERMAN, L. K.: Stream flow from rainfall by the unit-graph method. Eng. News-Rec. **108**, 501—505 (1932).

SNYDER, W. M.: Hydrograph analysis by the method of least squares. Proc. Amer. Soc. Civil Engr., Hydraulics Div. **81**, 1—25 (1955).

—, and J. W. CURLIN: Walker Branch watershed project: hydrologic analysis and data processing. Oak Ridge National Laboratory ORNL-TM-4392, 1969.

SPOFFORD, W. O., JR.: Queueing model for transport of radioactive nuclides in streams, Part IV. In: Operations research in disposal of liquid radioactive wastes in streams. (THOMAS, H. A. Ed.) Harvard Water Resources Group, Harvard Univ. NYO 10447, 1965.

Tennessee Valley Authority: Forest cover improvement influences upon hydrologic characteristics of White Hollow Watershed 1936-1958. TVA RPT. No. 0-5163 A, 1961.

WASTLER, T. A.: Application of spectral analysis to stream and estuary field surveys, I. Individual power spectra. U. S. Public Health Serv. Pub. No. 999-WP-7. 1963.

WEINER, N.: The interpolation, extrapolation and smoothing of stationary time series. New York: John Wiley and Sons 1949.

Subject Index

Page numbers in bold type indicate major sections, italicized numbers refer to table and figure citations, and standard roman type numbers refer to text discussion

Geographic Index of World Ecosystems

Volume 1 of *Ecological Studies* explains problems and principles applying to the analysis of any ecosystem. It is a synthesis of methods and results mostly from the zones where dense population and advanced technology have combined to modify, drastically, the once lofty and vast deciduous and mixed forests. Most data are from the COLD-DECIDUOUS (or summergreen) temperate forest areas which are shown in Eastern and Western Eurasia and North America, by "dot in diamond" patterns on the map in the back of this book (especially zone 10, with its sections a, b and c). Evergreens occur there too, but are even more important in the COOL MIXED forests along with broadleaved deciduous and larch species at higher latitudes or altitudes (zones 7a—10a) and in the SOUTHERN MIXED forests (11—11b). In both of these complexes especially, forestry becomes an ever greater, and more scientifically demanding, commercial enterprise. This must be based on more fundamental understanding of not only the trees but of the whole forest: plant layers, animals, microbes, plus the interactions and environmental controls which make a "system".

Upper case zone groupings mentioned in the proceding paragraph are by no means the only ones on the rear map which have important deciduous forests. By definition of the UNESCO Committee on world vegetation classification, the "cold-deciduous" (i.e. winter-leafless) forests (Formation Subclasses IB 2 and IB 3 in the UNESCO geographic index given below) stand in contrast with "drought-deciduous" forests (IB 1) and open woodlands (IIB 1) which drop more or less of their foliage between seasonal rains, such as the tropical and subtropical monsoons (in zones 24, 25 and part of 26 in the TROPICAL complex). Various DRY WOODLANDS (12—12b) are transitional respectively to GRASSLANDS (zones 13—15, 26; but even these may have alluvial or canyon forests), MONTANE COMPLEXES (21—22a) and DESERTS (16—20, 27, 29).

Typical genera are mentioned throughout the book, but a few including aspens and other poplars and several birches penetrate up through zones 7, 6 and 5 of the BOREAL or Taiga belt where conifers girdle the world. As "miniature forests" (Scrub = UNESCO Class III; Dwarf Scrub = Class IV) these genera infiltrate even the woody tundra (4), mountain tundra (3) and parts of Arctic lowland tundra (2): all but the polar deserts (1).

Only a few of the numerous extremely productive alluvial forest-herbage complexes (30) were of a size mappable by BAZILEVICH (1967, 1969) on the map which was modified for the back of the present book. Additional complexes (31—34) were added to her list here because they exemplify interesting, though localized, conditions where biomass (hence carbon inventory) and/or the income rate to that inventory might be almost as high as in some tropical rainforests (in zone 23).

To avoid confusion and a possible error in seeming to suggest that we know far more about these inventories and rates than is true, it should be understood clearly that the map in the back cover is *not* based on a large number of data points, well distributed over the world. RODIN and BAZILEVICH (1967) include a map (last chapter) showing very scattered sources (up to 1964) and very uneven coverage (which persists in new studies into the 1970's). Data tables confirm such unequal coverage that those authors emphasized perhaps more strongly than anyone else the need for an essentially more unified approach in the International Biological Program. Yet their heroic compilation, with rough but reasonable adjustments for incomplete data, seems nevertheless to show the broad ranges of biomass between the contrasting ecosystems or biogeocoenoses.

Carbon adjustments have been made according to criteria discussed in Chapter 15, with some changes from new sources or assumptions. A conservative approach here allows more overlapping between adjacent classes on the ascending scale. (The scale has been differentiated to twelve Roman numerals, but each column represents six gradations, staggered with the series in the other column).

Additional uncertainties are acknowedged in interpreting the back map as a historic *reconstruction* of conditions before the drastic landscape changes of the iron age and industrial revolution. If anything, these last doubts are less troublesome than those inherent in previous summaries of present census data on organic materials, when extrapolated over the biosphere and its major zones. Yet there are grounds for hoping that such generalization for the biosphere's modern carbon inventory can be improved by better use of available data or potential sources.

Not to be confused with an inventory (in grams per square meter, or metric tons per square kilometer) are estimates of *rate* of input (or loss) per year, or fractions thereof. LIETH's map inside the front cover is based on sources essentially independent of those for the back map. It too involves many assumptions (e.g. relating cumulative yields to growing season, as discussed in Chapter 4). Later data suggest that such estimates (actually predictions of what *could be* produced after "taxes" of greenplants' own respiration are deducted) are more likely to be increased than to be decreased, as more data and interpretation are available.

Unfortunately, misuse of land and water, related to overpopulation and misdirected technology, is commonly decreasing the long-range productivity of many environments at the same time it infringes on less tangible values of environmental quality. Both aspects of environmental quality are poorly known, but both are knowable, and subject to man's collective control.

Acknowledgements

Main contributors to the UNESCO world vegetation classification system included: H. GAUSSEN, G. BUDOWSKI, H. ELLENBERG, S. EVTEEV, O. FRÄNZLE, R. GERMAIN, A. KÜCHLER, J. LEBRUN, D. MUELLER-DOMBOIS, D. POORE, V. SOCHAVA and C. TROLL. Besides SOCHAVA, I. GERASIMOV, YE. LAVRENKO and others on the Editorial Board and staff of the *Physical-Geographic Atlas of the World* contributed concepts and information which will aid the reader who is interested in biogeographic details. Particularly LEONID RODIN (and others of the Botanical Institute, Leningrad) and N. I. BAZILEVICH (and the Dokuchaev Soil Institute, Moscow) contributed to mapping and other data used here. Besides many colleagues, IBP participants, chapter authors and investigators whom they cite, I should like to thank T. AHTI (University of Helsinki) and V. Krajina (University of British Columbia, Vancouver) for discussion and reprints pertinent to modifying the map. For distortions, oversimplifications (or intentionally provocative elaborations) of the rear map, I accept responsibility.

For financial aid during preparation for the authors' conference in Gatlinburg, including partial aid for Madgwick, Satoo, and Stanhill, the Volume and Series Editors appreciate a Ford Foundation grant to Oak Ridge Associated Universities. The Terrestrial Productivity section of IBP, the U.S. Analysis of Ecosystems Program of the National Science Foundation and the U.S. Atomic Energy Commission shared the main conference costs in Gatlinburg and Oak Ridge.

J. S. Olson, Series Editor

Sources

Ahti, Tuevo, Leena Hamet-Ahti, Jaakko Jalas: Vegetation zones and their sections in northwestern Europe. Ann. Botan. Fenn. 5, 169—211 (1968).

Bazilevich, N. I.: Cartoschemes of the productivity and biological cycle of the chief types of terrestrial vegetation of the earth. Izv. Vses. Geograf. Obs. 99 (3), 190—194 (1967).

—, L. E. Rodin: Reserves of organic matter in underground sphere of terrestrial phytocoenoses. P. 4—8. In: Methods of Productivity Studies in Root Systems and Rhizosphere Organisms. (M. S. Ghilarov et al., editors.) Leningrad: Publishing House "Nauka", U.S.S.R. Acad. Sci., Soviet National Committee for International Biological Programme. 1968.

—, T. K. Gordeeva, O. V. Zalensky, L. E. Rodin, J. K. Ross: Obshchie teoreticheskie problemi biologicheskoi produktivnosti. (Basic Theoretical Problems of Biological Productivity.) Leningrad: Publishing House "Nauka", U.S.S.R. Acad. Sci. 1969.

Ellenberg, H., D. Müller-Dombois: Tentative physiognomic-ecological classification of plant formations of the earth. Ber. Geobotan. Inst. ETH, Stiftg. Rubel, Zürich 37, 21—55 (1965—1966).

Hueck, Kurt: Die Wälder Südamerikas: Ökologie, Zusammensetzung und wirtschaftliche Bedeutung. Stuttgart: G. Fischer 1966.

Krajina, Vladimir J.: Biogeoclimatic zones and classification of British Columbia. Ecology of Western North America 1, 1—112 (1965). University of British Columbia: Vancouver.

Küchler, A. W.: Manual to accompany the map: Potential Natural Vegetation of the Conterminous United States. Am. Geogr. Soc. Spec. Publ. 26 (1964).

Rodin, L. E., N. I. Bazilevich: Production and mineral cycling in terrestrial vegetation. London: Oliver and Boyd (translated from Russian). Leningrad: Publishing House Nauka 1965 (1967).

Senderova, G. M. et al. (Ed.): Fiziko-geograficheskii Atlas (Physical-geographic Atlas of the World). Moscow: USSR Acad. Sci. and Main Administration of Geodesy and Cartography. 1964. (Legend translated in Soviet Geography, 1965).

UNESCO: Study of Tropical Vegetation: Proceedings of the Kandy Symposium. Unesco, Place de Fontenoy, Paris. 1958.

The following ecosystem ordination is after the UNESCO Committee on Classification and Mapping (mainly for scales 1 : 1,000,000 and smaller; larger scale is required for many of the finer subdivisions, however). It is only slightly changed from that published provisionally by ELLENBERG and MUELLER-DOMBOIS (1965/66). It is being altered (1970) for much more complete publication by UNESCO.

IBP Index Code*		Zonal Map Legend (inside back cover)**

I. Formation Class

    A. FORMATION SUBCLASS

        1. *Formation Group*

          a. Formation

          1) Subformation

F   I. **Closed Forests** (> 5 meter tall, crowns touching in wind- except that immature, cutover and grazed forest types may be shorter or more open without being called scrub or woodland, respectively)

      A. MAINLY EVERGREEN FOREST: canopy never without foliage, although individual trees may shed leaves

Fr   1. *Tropical Rainforest* (= ombrophilous):   **23**
little or no bud protection, nor cold or drought resistance; "drip-tip" leaves common
a, b Lowland, Submontane
c–e Montane, "Subalpine" and Cloud forest
f–h Alluvial, Swamp and Bog forest

Frd   2. *Tropical and Subtropical Evergreen Seasonal Forest:*   **23, 24**
some bud protection and noticeable dry-season shedding
a–d Lowland, Submontane, Montane, dry "Subalpine"

Fdr   3. *Tropical and Subtropical Semi-Deciduous Forest:* upper **24**
canopy mostly drought-deciduous; evergreen trees in canopy layers or understory
a,b Lowland, Montane or Cloud forest

Fr   4. *Subtropical Seasonal Rainforest:*
a–h (as in IAl.)   **32**

Fr   5. *Mangrove Forest*   **28**

Fr   6. *Temperate and Subpolar Rainforest*   **33**

Fd   7. *Temperate Evergreen Seasonal Broadleaved Forest*   **31**

Fs   8. *Winter-rain Hard-Broadleaved* (Sclerophyll) *Evergreen Forest:* **11a**
a,b Well-drained, Alluvial or Swamp

Fn   9. *Coniferous Evergreen Forests:*
a. Giant evergreen conifers (> 50 m tall)   **34**
b. Conifers rounded or flattened (at maturity)   **5–11a, 20–22, 31, 34**
1) with hard-broadleaved understory (e.g. pine heath)
2) without hard-broadleaved understory
c. Conifers mostly conical   **6–10, 20–22, 31—34**
d. Conifers cylindro-conical, with short branches   **5–10a, 22**

---

   *
** see page 304.

B. MAINLY DECIDUOUS FORESTS

Fd   1. *Drought-deciduous* (Monsoon) *Forest* (tropical, subtropical):   24
           a,b Lowland-submontane, Montane and Cloud forest
        2. *Cold-deciduous Forest with Evergreens:*
Fc   a. with Evergreen broadleaved trees and climbers   10–11
Fc   b. with Hard-broadleaved evergreen shrubs   10–11b
Fcn   c. with Evergreen needle-leaved trees (cool)   6–10a
Fcn   d. with Evergreen needle-leaved trees (warm)   10c–11a
Fcn   e. with Conifers and/or broadleaved evergreens   30–34

Fc   3. *Cold-deciduous (Summergreen) Forests:*
           evergreens (if any) mostly shrubs, or scattered
           a. Temperate lowland and submontane ("nemoral")   10a–13
           b. Montane, boreal and humid-site:
              1) mainly broadleaved   7–10, 33
              2) mainly deciduous conifer (e.g. larch)   6–10
              3) mixed broadleaf and deciduous conifer   7–11, 22
           c. Subalpine or subpolar (< 20 m; commonly gnarled):
              1) herbaceous undercover (scattered shrubs)   6–11, 22
              2) dwarf shrubs dominate undercover   5–11a, 22
              3) tall scrub undercover
           d. Alluvial, flooded: 5–16, 30–34
              1) occasionally or never
              2) regularly
           e. Swamp or bog forest:
              1) mainly broadleaved   8–16, 30
              2) deciduous conifer (larch, baldcypress)   5–10; 11, 30
              3) mixed broadleaved-conifer   5–11; 30–34

        C. DRY FORESTS (commonly grading to open woodlands)

Fd   1. *Hardleaved Forests:*   25–27
           some with swollen underground bases (xylopods)
Fd   2. *Thorn Forests:*
           a,b Mixed deciduous-evergreen, deciduous   12—17; 24–27
Ds   3. *Mainly succulent Forests* (trees and /or shrubs)   12, 15–17

II. **Open Woodlands** (< 5 m tall, crowns projecting over 30% of
      surface; may be grassy, grading to savanna)

    A. MAINLY EVERGREEN WOODS

Fs   1. Evergreen broadleaved woodlands   10c–12
Fn   2. Evergreen needle- or scale-leaved woodlands:
           a. Conifers rounded, flattened or irregular (e.g. pine)
              1) with hard-broadleaved understory   11a
              2) without hard-broadleaved understory   5–11, 12a
           b. Conifers mostly conical or dense (e.g. juniper)   6–10, 12a
           c. Conifers cylindro-conical, or sheared   5–9, 20–22

    B. MAINLY DECIDUOUS WOODLANDS

Fd   1. *Drought-deciduous Woodlands*   24–27
Fcn   2. *Cold-deciduous Woodland with Evergreens* (see I B 2)   4–12
Fc   3. *Cold-deciduous Woodland* (summergreen):
           a. Broadleaved deciduous   5–15
           b. Needleleaved deciduous woodland   4–10, 22
           c. Mixed broadleaved-deciduous conifer   5–11, 22

D   C. Dry Woodlands (divided as for IC, but sparser)

    III. **Scrub** (mainly 0.5–5m; thicket or shrubland, with grass)

        A. MAINLY EVERGREEN SCRUB

          1. Evergreen Broadleaved Shrubland or Thickets: 22a–25
            a. Bamboo Thicket (or shrubland)
            b. Tuft-tree scrub (dwarf palm, tree fern)
            c. Evergreen soft-broadleaved scrub *(Hibiscus)*
            d. Evergreen Hard-broadleaved scrub (chaparral)   11a–12
            e. Evergreen semi-lignified scrub (*Cistus* heath)   10, 31–34
          2. Evergreen needle-leaved and microphyll scrub:
            a. Needle-leaved thicket, shrubs (mugo pine, Krumholz)   21–22
            b. Microphyll scrub (e.g. *Leiophyllum*, tropical ericoid) 20–22a, 31–34

        B. MAINLY DECIDUOUS SCRUB

Fd, D   1. Drought-deciduous scrub with evergreens   11a–12a, 24–27
Fd, D   2. Drought-deciduous scrub without evergreens   11b–20, 25–27
Fc   3. Cold-deciduous scrub:

            a. Temperate deciduous thicket or shrubland   6–22, 31–34
            b. Subalpine or subpolar deciduous scrub   3–9, 21–22
            c. Deciduous alluvial scrub   4–20, 29–30
            d. Deciduous peat scrub   2–11, 21–24, 28–34

        C. DESERT SHRUBLANDS

Dx   1. Mainly evergreen subdesert   16–20, 25–27
Dx   2. Deciduous subdesert: 12b–20, 25–27
            a. without succulents
Ds   b. with succulents

    IV. **Dwarf Scrub and Related Ecosystems**

        A. MAINLY EVERGREEN DWARF-SCRUB

Th   *1. Evergreen dwarf scrub thickets* (heath)   3–12a
T   *2. Evergreen dwarf shrubland*   2–22a, 31–34
Tg, Ga   *3. Evergreen dwarf scrub-herb mixture*   4–17, 20–22a

T   B. MAINLY DECIDUOUS DWARF SCRUB   5–22, 24–26

Dx   C. SEMIDESERT DWARF SHRUB   16–21, 25–27, 29

Tt   D. TUNDRAS
          1. *Moss tundra*   2–5
          2. *Lichen tundra*   1–6, 21

Tp, Ts   E. MOSSY BOGS
          1. *Raised Bogs* (oceanic, montane, subcontinental)   1–10c, 22
          2. Non-raised Bogs:
            a. Blanket bog (oceanic, submontane, montane)   2–8, 10c
            b. String bogs ("aapa")   4–10a
            c. Lake bogs   5–11, 22

* The simple letter combination code for major groups of biomes was proposed by the International Biological Program section on Terrestrial Productivity (PT) in January 1969, for indexing IBP projects, sites and ecosystems included. Additional information can be obtained from the IBP, 7 Marylebone Road, London N.W. 1, England, or c/o UNESCO, Place de Fontenoy, 75 Paris 7.

** Zone numbers are retained in the Series 1—30 as used by BAZILEVICH et al. (1969), with minor changes, subdivisions (by additional letters) and additions (31—34). This basemap and some zones have been altered from the schematic map of the *Teacher's Atlas* of the U.S.S.R. which in turn is simplified from plates 66—67, 75 and continental maps of the *Physical-geographic Atlas of the World* (SENDEROVA et al., 1964). Estimated below-ground "reserves" of carbon in roots and rhizomes (BAZILEVICH and RODIN, 1968) are subject to uncertainties of sampling and of recognizing live and dead materials consistently. Yet including even the most preliminary approximations serves well to emphasize their quantitative importance, expecially by comparison with scanty or evanescent top parts of plants in many arid and cold communities, and grasslands of dry and moist regions.

Typesetting, printing, and bookbinding: Brühlsche Universitätsdruckerei, Gießen